Temporal Data Mining

Chapman & Hall/CRC
Data Mining and Knowledge Discovery Series

SERIES EDITOR
Vipin Kumar
University of Minnesota
Department of Computer Science and Engineering
Minneapolis, Minnesota, U.S.A.

AIMS AND SCOPE

This series aims to capture new developments and applications in data mining and knowledge discovery, while summarizing the computational tools and techniques useful in data analysis. This series encourages the integration of mathematical, statistical, and computational methods and techniques through the publication of a broad range of textbooks, reference works, and hand-books. The inclusion of concrete examples and applications is highly encouraged. The scope of the series includes, but is not limited to, titles in the areas of data mining and knowledge discovery methods and applications, modeling, algorithms, theory and foundations, data and knowledge visualization, data mining systems and tools, and privacy and security issues.

PUBLISHED TITLES

UNDERSTANDING COMPLEX DATASETS: Data Mining with Matrix Decompositions
David Skillicorn

COMPUTATIONAL METHODS OF FEATURE SELECTION
Huan Liu and Hiroshi Motoda

CONSTRAINED CLUSTERING: Advances in Algorithms, Theory, and Applications
Sugato Basu, Ian Davidson, and Kiri L. Wagstaff

KNOWLEDGE DISCOVERY FOR COUNTERTERRORISM AND LAW ENFORCEMENT
David Skillicorn

MULTIMEDIA DATA MINING: A Systematic Introduction to Concepts and Theory
Zhongfei Zhang and Ruofei Zhang

NEXT GENERATION OF DATA MINING
Hillol Kargupta, Jiawei Han, Philip S. Yu, Rajeev Motwani, and Vipin Kumar

DATA MINING FOR DESIGN AND MARKETING
Yukio Ohsawa and Katsutoshi Yada

THE TOP TEN ALGORITHMS IN DATA MINING
Xindong Wu and Vipin Kumar

GEOGRAPHIC DATA MINING AND KNOWLEDGE DISCOVERY, Second Edition
Harvey J. Miller and Jiawei Han

TEXT MINING: CLASSIFICATION, CLUSTERING, AND APPLICATIONS
Ashok N. Srivastava and Mehran Sahami

BIOLOGICAL DATA MINING
Jake Y. Chen and Stefano Lonardi

INFORMATION DISCOVERY ON ELECTRONIC HEALTH RECORDS
Vagelis Hristidis

TEMPORAL DATA MINING
Theophano Mitsa

Chapman & Hall/CRC
Data Mining and Knowledge Discovery Series

Temporal
Data Mining

Theophano Mitsa

CRC Press
Taylor & Francis Group
Boca Raton   London   New York

CRC Press is an imprint of the
Taylor & Francis Group, an **informa** business

A CHAPMAN & HALL BOOK

MATLAB® is a trademark of The MathWorks, Inc. and is used with permission. The MathWorks does not warrant the accuracy of the text or exercises in this book. This book's use or discussion of MATLAB® software or related products does not constitute endorsement or sponsorship by The MathWorks of a particular pedagogical approach or particular use of the MATLAB® software.

Chapman & Hall/CRC
Taylor & Francis Group
6000 Broken Sound Parkway NW, Suite 300
Boca Raton, FL 33487-2742

© 2010 by Taylor and Francis Group, LLC
Chapman & Hall/CRC is an imprint of Taylor & Francis Group, an Informa business

No claim to original U.S. Government works

Printed in the United States of America on acid-free paper
10 9 8 7 6 5 4 3 2 1

International Standard Book Number: 978-1-4200-8976-9 (Hardback)

Library of Congress Cataloging-in-Publication Data

Mitsa, Theophano.
  Temporal data mining / Theophano Mitsa.
     p. cm. -- (Chapman & Hall/CRC data mining and knowledge discovery series)
  Includes bibliographical references and index.
  ISBN 978-1-4200-8976-9 (hardcover : alk. paper)
  1. Data mining. 2. Temporal databases. I. Title. II. Series.

QA76.9.D343M593 2010
005.75'3--dc22   2009048856

**Visit the Taylor & Francis Web site at
http://www.taylorandfrancis.com**

**and the CRC Press Web site at
http://www.crcpress.com**

To my parents, who taught me to spend every moment wisely, and to the Eternal One, who taught me that every moment is infinitely important.

Table of Contents

Preface

IMPORTANCE OF TEMPORAL DATA MINING TODAY

Temporal data are of increasing importance in a variety of fields, such as biomedicine, geographical data processing, financial data forecasting, and Internet site usage monitoring. Temporal data mining deals with the harvesting of useful information from temporal data, where the definition of *useful* depends on the application. The most common type of temporal data is time series data, which consist of real values sampled at regular time intervals. Let us examine how new initiatives in health care and business organizations increase the importance of temporal information in data today.

First, in health care, the government mandate for universal electronic medical record (EMR) adoption by 2014 will enable computer access to all chronological information about a patient's history, such as dates of lab tests and hospital admissions, and enable the automatic production of temporally initiated alerts, such as the date for a vaccination renewal. Another initiative in health care is becoming increasingly adopted: *connected health*, which really means *patient-centered* health care. In this type of health care, regular physiological monitoring, such as blood-glucose and cholesterol level monitoring, combined with data-adaptive mentoring of the patient becomes a key component and improves the patient's quality of life, while reducing hospital overload by cutting down on the number of hospital admissions.

By encouraging regular physiological monitoring, connected health hospitals and practices will increase the importance of watching trends and general temporal changes in the patient's data, which in turn will lead to the increased need for temporal data mining of health care data. The combination of electronic medical record adoption and connected health leads to a new model of health care often referred to as *Health 2.0.*

Additionally, in a recent study [Ama09], it was shown that incorporation of health care technology, such as clinical decision support, and automated notes and records led to reductions in mortality rates, costs, and complications in multiple hospitals.

Similarly, in business organizations, agility and client-centricity are principles of ever-increasing importance in today's highly competitive business world because incorporation of these two principles allows a business organization to respond quickly and efficiently to changes in clients' needs and changes in the business environment. This is achieved by having efficient and seamlessly integrated business processes throughout the value chain, starting from the supply chain and ending in customer feedback incorporation in business processes. This type of agility requires significant business reorganization, such as IT–finance integration, and incorporation of business intelligence, such as careful monitoring of trends and changes in customer purchasing patterns, as well as increased awareness of the competitive environment in which the business operates. This again translates into increased importance of temporal data patterns and temporal data mining.

Overall, the increased need nowadays for temporality incorporation in data, whether health care or business data, can be described as need for *integration of business object provenance and analysis*, where the business object can be a product or a patient's medical profile. *Provenance* refers to having a documented history of ownership of an object and is a term frequently used for fine art objects. The authors in [Mor08] use the term *electronic data provenance* to describe the need for maintaining the history of electronic data, such as design documents. An example of *integrated provenance and analysis*, in the context of temporal data, is having timestamped information regarding which engineering/marketing/sales teams are responsible for a product at different times and, for each one of those times, having information regarding key actions of these teams as well as the number of defects and the number of sales of the product. Applying temporal data mining to these data can yield valuable insights as to how different team "ownership" can affect the quality and success of the product.

SCOPE OF THE BOOK AND INTENDED AUDIENCE

This book covers the theory of temporal data mining as well as applications in a variety of fields, and its goal is twofold:

1. To provide the basic concepts as well as the state of the art in the following:

   - Incorporation of temporality in databases

   - Temporal data representation and similarity computation

   - Temporal data classification and clustering

   - Temporal pattern discovery

   - Prediction

2. To discuss the applications and state of the art advances of temporal data mining in four areas:

   - Medicine and biomedical informatics

   - Business and industrial applications

   - Web usage mining

   - Spatiotemporal data mining

Because the book covers the theory of temporal data mining starting from basic data mining concepts and advancing to state-of-the-art methods, it is intended for data mining novices, such as graduate students, as well as experienced data mining researchers who want to learn the latest advances in the temporal data mining field.

In addition, because the book provides an extensive coverage of temporal data mining applications in a variety of fields, it is also intended for biomedical researchers, financial data analysts, business managers, geospatial data analysts, and Web developers.

BOOK STRUCTURE

The book is organized as follows: Chapter 1 covers the topic of how temporal information can be incorporated in databases. Chapters 2 and 3 cover the theory of temporal data mining, specifically temporal data representation and similarity computation (Chapter 2) and classification and clustering (Chapter 3). Chapter 4 covers prediction, also known as forecasting. Although prediction is not a temporal data mining task, it is quite often the ultimate goal of temporal data mining, and therefore it was deemed sufficiently important to devote a chapter to it. Chapter 5 discusses another theoretical data mining task, temporal pattern discovery. Chapters 6–9 discuss applications of temporal data mining in medicine and bioinformatics (Chapter 6), business (Chapter 7), Web usage mining (Chapter 8), and spatiotemporal data mining (Chapter 9).

As various state-of-the-art algorithms are described in each chapter, the corresponding reference article or book is provided. All chapters have an additional bibliography section that, in addition to the references discussed in detail in the body of each chapter, provides a short description of algorithms and techniques described in other references that are relevant to the material discussed in each chapter.

Appendix A provides a description of how data mining fits the overall goal of an organization and how these data can be interpreted for the purpose of characterizing a population. Appendix B contains programs written in the Java language that implement some of the algorithms described in Chapter 1 of the book.

MATLAB® is a registered trademark of The Math Works, Inc. For product information, please contact:

The Mathworks, Inc.
3 Apple Hull Drive
Natick, MA
Tel: 508-647-7000
Fax: 508-647-7001
E-mail: info@mathworks.com
Web: http://www.mathworks.com

I would like to thank the Taylor & Francis reviewers for their valuable comments and thorough review.

REFERENCES

[Ama09] Amarisngham, R. et al., Clinical Information Technologies and InPatient Outcomes: A Multiple Hospital Study, *Archives of Internal Medicine*, vol. 169, no. 2, pp. 108–114, 2009.

[Mor08] Moreau et al., The Provenance of Electronic Data, *Communications of the ACM*, vol. 51, no. 4, pp. 52–58, 2008.

CHAPTER 1

Temporal Databases and Mediators

1.1 TIME IN DATABASES

To correctly harvest temporal information, it is important to understand how time information is incorporated in databases and data warehouses. Therefore, although the focus of this book is temporal data mining, we will devote Section 1.1 of this chapter to a discussion of temporal databases and incorporation of time in data warehouses.

Temporal database research has seen an explosive growth in the 1980s and 1990s; however, most of this research has failed to make its way to commercial database systems. In particular, there is not a well-accepted temporal query language that will allow such tasks as the extraction of temporal information from databases at different granularities or the extraction of time interval information from time instant data. These tasks are important on their own but also as a data preprocessing step, prior to data mining. Therefore, the temporal data owner is left on her own to devise a solution to extract this kind of information from a standard database system. Another recently emerging need is the extraction of *temporally semantic information*, that is, information within the context of a temporal ontology. In Section 1.2 of this chapter, we discuss the concept of a temporal database mediator, which is a computational layer placed between the user interface and the database for the discovery of temporal relations, temporal data conversion, and the discovery of semantic relationships.

TABLE 1.1   Student Database

Student ID	First Name	Last Name	Graduation Year
345622	John	Smith	2009
112367	Mary	Thompson	2008
983455	Stewart	Allen	2010

1.1.1 Database Concepts

A database system consists of three layers: *physical, logical,* and *external.* The physical layer deals with the storage of the data, while the logical layer deals with the modeling of the data. The external layer is the layer that the database user interacts with by submitting database queries. A database model depicts the way that the database management system stores the data and manages their relations. The most prevalent models are the *relational* and the *object-oriented.* For the relational model, the basic construct at the logical layer is the *table,* while for the object-oriented model it is the *object.*

Because of its popularity, we will use the relational model in this book. Data are retrieved and manipulated in a relational database, using *SQL.* A relational database is a collection of tables, also known as *relations.* The columns of the table correspond to attributes of the relational variable, while the rows, also known as *tuples,* correspond to the different values of the relational variable. An example is shown in Table 1.1. Table 1.2 contains common database terminology related to the physical and logical layers for the relational model.

Other frequently used database terms are the following:

Constraint: A rule imposed on a table or a column.

Trigger: The specification of a condition whose occurrence in the database causes the appearance of an external event, such as the appearance of a popup.

View: A stored database query that hides rows and/or columns of a table.

TABLE 1.2   Correspondence between Logical and Physical Database Terms

Logical Term	Physical Term
Relation	Table
Unique ID	Primary key
Tuple	Row
Attribute	Column

1.1.2 Temporal Databases

Temporal databases are databases that contain time-stamping information. Time-stamping can be done as follows:

- With a *valid time*, which is the time that the element information is true in the real world. For example, "The patient was admitted to the hospital on 5:15 a.m., March 3, 2005."

- With a *transaction time*, which is the time that the element information is entered into the database.

- Bi-temporally, with both a valid time and a transaction time.

Time-stamping is usually applied to each tuple; however, it can be applied to each attribute as well. Databases that support time can be divided into four categories:

- *Snapshot databases*: They keep the most recent version of the data. Conventional databases fall into this category.

- *Rollback databases*: They support only the concept of transaction time.

- *Historical databases*: They support only valid time.

- *Temporal databases*: They support both valid and transaction times.

In this book, we differentiate between two types of temporal entities that can be stored in a database: *intervals* and *events*.

- *Interval*: A temporal entity with a beginning time and an ending time.

- *Event*: A temporal entity with an occurrence time.

Note that transaction time is always of type *event*, while valid time can be of type *interval* or *event*. In addition to interval and event, another type of a temporal entity that can be stored in a database is a *time series*. As it will also be defined in Chapter 2, a time series consists of a series of real-valued measurements at regular intervals. Other frequently used terms related to temporal data are the following:

- *Granularity*: It describes the duration of the time sample/measurement. For example, the granularity can be week or day.

TABLE 1.3   Uncoalesced Table

Patient ID	First Name	Last Name	Hospitalization Time
234779	Mary	Ferguson	(2001-03-10, 2001-03-15)
234779	Mary	Ferguson	(2001-03-15, 2001-03-20)
112788	Gary	Lindell	(2002-02-11, 2002-02-25)

- *Anchored data:* Time is represented using absolute values, such as January 20, 1999, 3:15 a.m. Anchored data can be used to describe either the time of an occurrence of an event or the beginning and ending times of an interval.

- *Unanchored data:* They are used to represent the duration of an interval, such as 2 weeks.

- *Data coalescing:* The replacement of two tuples A and B with a single tuple C, where A and B have identical nontemporal attributes and adjacent or overlapping temporal intervals. C has the same nontemporal attributes as A and B, while its temporal interval is the union of A's and B's temporal intervals. An example is shown in Tables 1.3 and 1.4.

1.1.3 Time Representation in SQL

Anchored time data are represented using the TIME, DATE, and TIMESTAMP data types. Unanchored time data are represented using the INTERVAL data type. The specific formats for each data type are as follows:

- DATE: The format is YYYY-MM-DD and it represents a date using YEAR, MONTH, and DAY.

- TIME: The format is HH:MM:SS[.sF] and it represents time using the fields HOUR, MINUTE, SECOND, where F is the fractional part of the SECOND value.

- TIMESTAMP: The format is YYYY-MM-DD HH:MM:SS[.sF] and it describes both a date and time, with seconds precision s.

- INTERVAL: The format is either YEAR-MONTH or DAY-TIME.

TABLE 1.4   Coalesced Table

Patient ID	First Name	Last Name	Hospitalization Time
234779	Mary	Ferguson	(2001-03-10, 2001-03-20)
112788	Gary	Lindell	(2002-02-11, 2002-02-25)

1.1.4 Time in Data Warehouses

A data warehouse (*DW*) is a repository of data that can be used in support of business decisions. Many data warehouses have a *Time* dimension and therefore they support the idea of valid time. Also data warehouses contain snapshots of historical data and inherently support the idea of transaction time. Therefore, a *DW* can be considered as a temporal database, because it inherently contains bi-temporal time-stamping. Time affects the structure of the warehouse also. This is done by gradually increasing the granularity coarseness as we move further back in the time dimension of the data. Data warehouses, therefore, *inherently support coalescing.*

Despite the fact that data warehouses inherently support the notion of time, they are not equipped to deal with temporal changes in *master data*. For example, let us assume that a business data warehouse has a dimension Partners. Let us assume that originally the Partners dimension consists of {BioData, NuSoftware, MetaData}. In 1998, BioData and MetaData merge under the name of BiomedData. A way to deal with this is to time-stamp the data schema and provide transformation to handle user queries. For example, if the user submits a query about the stock price of BioData in May 1999, the transformation function maps the query to the stock price of BiomedData in May 1999.

1.1.5 Temporal Constraints and Temporal Relations

[Chi04] discusses reasoning about temporal constraints, which deal with the handling of relations among temporal entities. Temporal constraints can be either *qualitative* or *quantitative*. Regarding *quantitative temporal constraints*, variables take their values over the set of temporal entities and the constraints are imposed on a variable by restricting its set of *possible values*. In *qualitative temporal constraints*, variables take their value from a set of temporal relations. For example, in Allen's seminal work [All83], variables take their values from a set of 13 temporal relations, which are shown in Table 1.5 for two time intervals X and Y.

The *after* relationship is denoted as $bi(X,Y)$ to indicate that it is the inverse of the *before* relationship. Specifically, $bi(X,Y) = b(Y,X)$. The same explanation can be applied to the other operators, whose notation ends with the letter *i*. For example, $di(X,Y) = d(Y,X)$.

In later work, Allen and Hayes [All90] expand on the previous time interval-based theory and add points as entities of interest. There are two kinds of point entities: *points* and *moments*. *Points* are defined as meeting

TABLE 1.5   Notation for Allen's Temporal Relationships

Name	Notation
Before	$b(X,Y)$
Overlaps	$o(X,Y)$
During	$d(X,Y)$
Meets	$m(X,Y)$
Starts	$s(X,Y)$
Finishes	$f(X,Y)$
Equals	$e(X,Y)$
After	$bi(X,Y)$
Overlapped-by	$oi(X,Y)$
Contains	$di(X,Y)$
Met by	$mi(X,Y)$
Started by	$si(X,Y)$
Finished by	$fi(X,Y)$

places of *periods*, while *moments* are non-decomposable, very small time periods. Also, *meets* is defined as the one primitive relationship and all other are derived from it. In [Cam07], Campos et al. discuss qualitative temporal constraints to extract more complete and representative patterns. A fuzzy temporal constraint network formalism is used and temporal constraints are used not just for representation but also for reasoning.

1.1.6 Requirements for a Temporal Knowledge-Based Management System

Koubarakis [Kou90] provides a list of requirements that must be fulfilled by a temporal knowledge-based management system:

- To be able to answer real-world queries, it must be able to handle large temporal data amounts.

- It must be able to represent and answer queries about both quantitative and qualitative temporal relationships. An example of quantitative temporal relationship is "Patient Jones must be administered drug X two hours before his operation." An example of qualitative relationship is "Patient Norton is to be operated on after Patient Jones."

- It must be able to represent causality between temporal events. For example, patient Norton's post-traumatic stress disorder is the result of a burglary in his house last year.

- It must be able to distinguish between the history of an event and the time the system learns about the event. In other words, it must be able to distinguish between valid and transaction times.

- It must offer an expressive query language that also allows updates.

- It must be able to express persistence in a parsimonious way. In other words, when an event happens, it should change only the parts of the system that are affected by the event.

1.1.7 Using XML for Temporal Data

Using XML to model temporal data and perform temporal queries is an idea that is gaining momentum. This is discussed in [Bun04], [Ger04], [Gra05], [Ama00], [Zha02], [Gao03b], [Riz08], and [Wan08]. The authors in [Bun04] introduce the concept of using hierarchical time-stamping in XML documents. In [Ger04], a multidimensional XML model is proposed whose dimensions are applied to the elements and attributes of the XML document as a way to represent temporal information. In [Ama00], a temporal data model is introduced that utilizes *XPath*™, while in [Gao03b] the authors propose a generalization of *XQuery*. *XPath* is a language that allows the selection of nodes from an XML document, while *XQuery* is a query language that queries collections of XML data.

[Riz08] proposes a data model for modeling historical information in an XML document. The data model can be used as a schema against which the consistency of incoming documents can be checked. Temporal queries are performed using *TXPath*, which is a temporal extension of *XPath 2.0* and returns sequences of (node, interval) pairs.

[Wan08] discusses a novel architecture called *ArchIS*, which achieves the following:

- It uses XML to model the evolution history of a database.

- *XQuery* is used to perform temporal queries. *XQuery* is easily extensible and can be used to perform powerful and complex temporal queries. As the authors note, the important advantage of using *XQuery* for temporal queries is that there is no need to introduce new constructs in the language to perform powerful queries, such as temporal projection, temporal slicing, temporal snapshot, and temporal aggregate.

- Temporal clustering and indexing techniques are used to manage the actual historical data.

1.1.8 Temporal Entity Relationship Models

The *entity relationship model* (*ER*) represents the world as a set of entities and the relations among those entities. The *ER* model can be used in the database design process to model the database needs of an organization using a diagram and then the *ER* diagram is mapped to a relational schema. Temporal *ER* diagrams are either mapped directly to relational schemas or first mapped to a regular diagram, which is then mapped to a relational schema.

The *ER* design model is very popular today, both in the research community and in industry. For this reason, there have been a number of *ER* extensions to model temporal aspects of a database. A thorough survey of temporal *ER* models can be found in [Gre99]. Specifically, the following models are surveyed in the article:

- *EER:* Enhanced Entity Relationship model

- *RAKE:* Relationships, Attributes, Keys, and Entities model

- *TERM:* Temporal Entity Relationship Model

- *MOTAR:* Model for Objects with Temporal Attributes and Relationships

- *TEER:* Temporal EER model

- *STEER:* Semantic Temporal EER model

- *ERT:* Entity-Relation-Time model

- *TER:* Temporal ER model

- *TempEER:* Temporal EER model

- *TempRT:* Kraft's model

References for each model are provided in [Gre99]. The models are evaluated according to 19 design criteria chosen by Gregersen and Jensen, such as temporal functionality, provision of a query language, graphical notation provision, graphical editor provision, and mapping algorithm availability.

In [Gre06], the author addresses the problem that the semantics of most of the aforementioned models are not clearly defined. For this reason, the author focuses on the *TIMEER* model and develops formal semantics for it. The *TIMEER* model extends the *EER* model, mentioned above, by providing temporality for entities, relationships, super-classes, sub-classes, and attributes.

1.2 DATABASE MEDIATORS

This section discusses the use of a temporal database mediator to discover temporal relations, implement temporal granularity conversion, and also discover semantic relationships. This mediator is a computational layer placed between the user interface and the database. Figure 1.1 shows the different layers of processing of the user query: "Find all patients who were admitted to the hospital this February." The query is submitted in natural language in the user interface, and then, in the Temporal Mediator (TM) layer, it is converted to an SQL query. It is also the job of the Temporal Mediator to perform temporal reasoning to find the correct beginning and end dates of the SQL query.

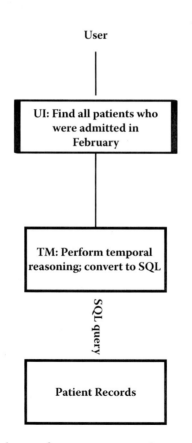

FIGURE 1.1   Different layers of user query processing.

1.2.1 Temporal Relation Discovery

The discovery of temporal relations has applications in temporal queries, constraint, and trigger implementation. The following are some temporal relations:

- A column constraint that implements an *after* relation between events. For example, the surgery date has to be after the hospital admission date of the patient.

- A query about a *before* relation between events: Was Ed Jones released from the hospital *before* John Smith?

- A query about an *equal* relationship between intervals: Did Jones spend an *equal number of days* in the hospital as Smith?

- A database trigger about a *meets* relationship between an interval and an event or between two intervals: For example, if the patient is released from the hospital the same day she has a certain type of operation, implement a database trigger.

Let us see now how we can implement the query about the *before* relationship using a mediator. It would be desirable to have a user interface that allows the user to express her query at a higher level than *SQL*, utilizing natural language concepts. A possible realization of the user interface is shown in Table 1.6, where the highlighted items show the implementation of the aforementioned query about the *before* temporal relationship.

The user interface could be implemented as an *applet* or *servlet*. The mediator is implemented as a *Java*™ program (see Appendix B) that utilizes *JDBC*™ *(Java Database Connectivity)* to access the database and submits an *SQL* query that extracts the release dates of the two patients. *JDBC* is an *API* that allows *Java* programs to execute *SQL* statements and retrieve data from databases. In addition to the *JDBC API*, a *Java*

TABLE 1.6   User Interface for the Implementation of a Temporal Query

Patient	Event	Temporal Relation	Patient	Event
Ed Jones	Admission	**Before**	Ed Jones	Admission
Paul Lorenzo	Surgery	After	Paul Lorenzo	Surgery
Tom Frier	Post-op	Meets	Tom Frier	Post-op
John Smith	**Release**	Overlaps	**John Smith**	**Release**

program that needs to access a specific database management system has to import and register the appropriate driver. In the program, we import the *Oracle*™ *driver* and the corresponding importation and registration statements are

```
import oracle.jdbc.driver.*;
```

and

```
DriverManager.registerDriver (new oracle.jdbc.driver.
OracleDriver());
```

Java offers some useful features for temporal queries. Note that as new releases of *Java* become available, some of the classes/methods mentioned below might change or become deprecated. The programs in Appendix B were compiled using JDK 1.6.0.

- A class *Date*, which represents an instance in time using year, month, day, hour, minute, second, and millisecond information. This class also offers methods for temporal relation discovery. These are the functions *before(), after(), equals(),* and *compareTo().*

- A class *Time*, which represents a time using hour, minute, and second information.

- A class *Timestamp* that contains date and time information.

- A class *Calendar* that also offers temporal data representation.

- The class *ResultSet,* which models the data retrieved from an *SQL* query, has methods *getDate(), getTime(),* and *getTimeStamp()* that return *Date, Time,* and *TimeStamp* information.

The programs in Appendix B show some of the functionality of Java in regards to extracting temporal information. Program 1 utilizes the *getDate()* function of the *ResultSet* class to get the release date as a *Date* object and the *before()* function of the *Date* class to compute the temporal relation between the release dates.

Program 2 converts anchored data to unanchored data. Similarly to Program 1, *getDate()* of *ResultSet* is used to get the release and admission dates of two patients. Then the *Date* objects are converted to *Calendar* objects. Finally, the method *get(Calendar.DAY_OF_YEAR)* of class *Calendar* is used to extract the number of days that each date corresponds to, and eventually find the number of days that each patient stayed in the hospital.

1.2.2 Semantic Queries on Temporal Data

Traditional temporal data mining focuses heavily on the harvesting of one specific type of temporal information: *cause/effect* relationships through the discovery of association rules, classification rules, and so on. However, in this book we take a broader view of temporal data mining, one that encompasses the discovery of structural and *semantic* relationships, where the latter is done within the context of *temporal ontologies*. While the discovery of structural relationships will be discussed in the next chapter, the discovery of semantic relationships is discussed in this section, because it is very closely intertwined with the representation of time inside the database. Regarding terminology, an *ontology* is a model of real-world entities and their relationships, while the term *semantic relationship* denotes a relationship that has a meaning *within* an ontology.

Ontologies are being developed today for a large variety of fields ranging from the medical field to geographical systems to business process management. As a result, there is a growing need to extract information from database systems that can be classified according to the ontology of the field. It is also desirable that this information extraction is done using natural language processing (*NLP*).

In this section, we will discuss the discovery of a semantic relationship of the hierarchical type. A hierarchical relationship constitutes an "is a member of" relationship within a temporal ontology. For example, in the ontology that describes the stages of human life, *childhood* has an "is a member of" relationship with *lifetime*.

Let us consider the temporal ontology shown in Figure 1.2. It shows the geologic eras *Cenozoic* and *Mesozoic* and their corresponding periods. These periods are *Quarternary, Neogene,* and *Paleogene* for the *Cenozoic* era and *Cretaceous, Jurassic,* and *Triassic* for the *Mesozoic* era. Figure 1.2 also shows the duration of these periods. For example, the duration of the Neogene period is 24–1.8 million years ago. Because the *Quarternary, Neogene,* and *Paleogene* periods have an "is member of" relationship with the *Cenozoic* era, we say that they are subclasses of the ontology class *Cenozoic.*

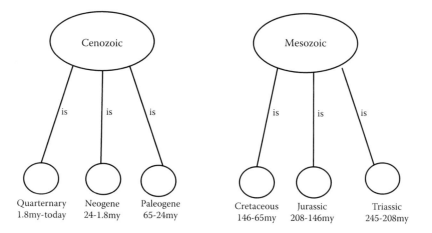

FIGURE 1.2   Geologic era ontology.

A number of tools exist today for the creation of ontologies. The most widely used one is *Protégé*™ [Pro09], which offers a graphical user interface for the specification of classes, subclasses, and their relationships. *Protégé* uses the semantic language *OWL* to specify the classes and relationships of an ontology. Once the ontology is created, it can be checked for consistency with a reasoner. For example, if we specified that *Neogene* is a subclass of *Cenozoic* and *Cenozoic* is a subclass of *Neogene*, this inconsistency would be caught by the reasoner. The geologic era ontology can be expressed in an XML file, as shown in the Appendix B.

Because of the simplicity of the geologic era ontology, there was not a need for reasoner validation and, therefore, use of a sophisticated ontology language, such as *OWL*, was not deemed necessary. XML offers an attractive alternative, because of the prevalence of XML technology and its flexibility in expressing various types of ontological information. In the XML file shown in Appendix B, the root element is the *Genealogy* element. The concept of the ontological relationship class–subclass can be mapped to the XML relationship parent–child element.

The eras are represented by children elements of the *Genealogy* element, while the periods are represented by children elements of the eras elements. The properties of the period element, *name, beginDate, endDate*, are expressed as child elements of period, while the era that each period belongs to is expressed as an XML attribute called *parent*. Let us assume now that we have a database where each tuple has the following attributes: location, name of fossil and date of fossil (in million years ago). An example is shown in Table 1.7.

TABLE 1.7 Fossils Database

CITY	COUNTRY	FOSSIL ID	DATE (in million years)
Tucson	US	Fossil 1	150
Adwa	Ethiopia	Fossil 2	250
Santorini	Greece	Fossil 3	170
Palermo	Italy	Fossil 4	50
Ghanzi	Botswana	Fossil 5	63
Cairo	Egypt	Fossil 6	210

Let us assume now that a user wants to extract the following information from the database:

1. How many fossils do we have from the Jurassic period?

2. How many fossils do we have from the Mesozoic era?

These are all *semantic* queries, because they have meaning inside the geologic temporal ontology. The problem we are faced with is how to extract this semantic information from a database that has no semantic information, just a date for each fossil. Program 3 in Appendix B shows the first step in solving this problem, which is the most difficult as well. This step consists of parsing the XML file to extract the date range for the *Jurassic* period and the date range for the *Mesozoic* era. Knowing the date ranges, the second step, which is not shown, is to write a program similar to Programs 1 and 2 that utilizes the *JDBC API* to submit a *SELECT SQL* query to the database.

Let us examine how Program 3 in Appendix B works. The program utilizes the *DOM* (Document Object Model) application interface to retrieve and examine the values of elements in the XML document. The *DOM API* is a language-independent interface that has been developed by the W3C™ *DOM* working group, which models a document as a hierarchy of nodes and whose purpose is to extract information and manipulate the content and style of documents. The adopted version of *DOM* in *Java* offers useful methods, such as *getElementsByTagName(String s),* which allows the extraction of elements with a specific tag name, and *getNodeValue(),* which returns the value of a node. For the extraction of the date range for the *Mesozoic* range, we find the minimum end date and maximum beginning date of the periods that belong to the *Mesozoic* era.

1.3 ADDITIONAL BIBLIOGRAPHY

A review of database concepts can be found in [Opp04], while a review of data warehouse systems can be found in [Han05]. A significant amount of literature exists on the topic of temporal databases. Surveys of this work can be found in [Cel99], [Dat03], [Etz98], and [Sno06]. [Jen98] contains a glossary of temporal database concepts. In [Gol09], one can find a very recent review on temporal data warehousing issues, such as data/schema in the data warehouse and data mart. Another reference that discusses changes of master data regarding temporality is [Ede02].

1.3.1 Additional Bibliography on Temporal Primitives

In this chapter we have discussed the representation of temporal phenomena in terms of temporal intervals, based on Allen's temporal interval-based time theory. [Sch08] discusses fuzzification of Allen's temporal interval relations. Besides time intervals, researchers have approached temporal phenomena representation in other ways. One of them is change-based and it is based on the intuitive notion that time changes constantly. Change indicators are the primitive entities in one of these theories [Sho88]. In two other types of change-based representations, the situation calculus [McC69] and the event calculus [Kow86], actions and events are the primitives, respectively.

Other researchers have focused on points as the primitives in the representation of temporal phenomena, particularly for phenomena that represent continuous change [McD82], [Gal90], [Sho87]. In other work, intervals are represented as ordered pairs of points [Lad87].

1.3.2 Additional Bibliography on Temporal Constraints and Logic

The reader interested in learning more about temporal logic can find an excellent review in [Gab00]. Also, more information about temporal logic and temporal mining frameworks can be found in [Dea87], [Aln94], [Fre92], [Vil82], [Rai99], and [Sar95]. In [Kur94], the author discusses the incorporation of fuzzy logic in temporal databases, to deal with vague events such as "Sales increased by 150% in the last days of the month." In [Bit04], Bittner addresses the issue of approximate qualitative temporal reasoning. Goralwalla et al. in [Gor04] discuss granularity as an integral feature of both anchored and unanchored data. In their work, the authors model granularity as a unit unanchored temporal primitive. In [Vie02], the authors discuss the syntax and semantics of a fuzzy temporal constraint logic. This way the authors are able to express interrelated events using fuzzy temporal constraints.

1.3.3 Additional Bibliography on Temporal Languages and Frameworks

Although a widely used temporal query language does not exist today, *TSQL2* represents the most serious effort in this arena and it has integrated more than fifteen years in temporal database research. The interested reader can find more about *TSQL2* in [Sno06], including a tutorial on the language. Also an excellent review of temporal knowledge base systems and temporal logic theories can be found in [Kou90]. The authors in [Elm93] and [Pis93] discuss the incorporation of temporal concepts in object-oriented databases. A thorough survey of join operators in temporal databases is performed in [Gao03a].

In [The94], the authors discuss the *ORES* temporal *DBMS* that supports temporal data classification, grouping, and aggregation according to the *Entity Relationship Time* data model. Another work that discusses grouping and aggregation of temporal data is [Dum98]. A temporal query language and the temporal *DBMS* system called *TEMPOS*, which has some basic temporal *OLAP* capabilities, is introduced in [Fau99]. In [Are02], a framework for answering queries about the hypothetical evolution of a database is presented. The framework can help answer queries of the form: "Have the data in the database always satisfied condition A?" In [Mor01], the architecture of a system that combines temporal planning, plan execution, and temporal reasoning is described. The temporal reasoning layer allows the maintenance of temporal constraints and the better tracing of plan execution.

In [Kag08], the authors discuss how one can design and optimize constructs for complex pattern search, using *SQL-TS*, which is an extension of *SQL* that can perform temporal queries. Specifically, they propose a search algorithm, called *RSPS*, which can speed up queries up to 100 times by minimizing repeated passes over the same data. In [Sta06], the authors describe how to handle current time in native XML databases. In [Unn09], the authors describe how to implement temporal coalescing in temporal databases implemented on top of relational database systems. Their results show that the performance of temporal coalescing using *SQL 2003* is better than temporal coalescing performed using *SQL 1992*.

For more information about ontologies, one can learn about *OWL* in [OWL04] and about *Protégé* in [Pro09]. A reference that specifically discusses the use of XML for ontological descriptions is [Phi04]. In [Owl06], the incorporation of time in *OWL* is discussed, to meet the temporal needs of Web services. There are several resources for a review of the *Java* language and the *JDBC API* used in the programs of this chapter. The author specifically utilized [Dei05] and [All00]. Finally, the reader interested in learning more about the *DOM API* in *Java* is referred to [All00].

REFERENCES

[All83] Allen, J.F., Maintaining Knowledge about Temporal Intervals, *Communications of the ACM,* vol. 26, no. 11, pp. 832–843, 1983.

[All90] Allen, J.F. and P. J. Hayes, Moments and Points in an Interval-Based Temporal Logic, *Computational Intelligence,* vol. 5, no. 4, pp. 225–238, November 1990.

[Aln94] Al-Naemi, S., A Theoretical Framework for Temporal Knowledge Discovery, *Proceedings of the International Workshop Spatio-Temporal Databases,* pp. 23–33, 1994.

[All00] Allamaraju, S. et al., *Professional Java Server Programming J2EE Edition,* Wrox Press, 2000.

[Ama00] Amagasa, T., M. Yoshikawa, and S. Uemura, A Data Model for Temporal XML Documents, *DEXA,* 2000.

[Are02] Arenas, M.O. and L. Bertossi, Hypothetical Temporal Reasoning in Databases, *Journal of Intelligent Information Systems,* vol. 19, no. 2, pp. 231–259, 2002.

[Bit04] Bittner, T., Approximate Qualitative Temporal Reasoning, *Annals of Mathematics and Artificial Intelligence,* Springer, vol. 36, pp. 39–80, 2004.

[Bun04] Buneman, P. et al., Archiving Scientific Data, TODS, vol. 29, no. 1, pp. 2–42, 2004.

[Cam07] Campos, M., J. Palma, and R. Marin, Temporal Data Mining with Temporal Constraints, *Artificial Intelligence in Medicine, Lecture Notes in Computer Science,* Springer, vol. 4594, pp. 67–76, 2007.

[Cel99], Celko, J., *Joe Celko's Data and Databases: Concepts in Practice*, Morgan Kaufmann, 1999.

[Chi04] Chittaro, L. and A. Montanari, Temporal Representation and Reasoning in Artificial Intelligence: Issues and Approaches, *Annals of Mathematics and Artificial Intelligence,* vol. 28, no. 1–4, 2004.

[Dat03] Date, C.J., H. Darwin, and N.A. Lorentzos, *Temporal Data and the Relational Model*, Morgan Kaufmann, 2003.

[Dea87] Dean, T.I. and D.V. McDermott, Temporal Database Management, *Artificial Intelligence,* vol. 32, no. 1, pp. 1–55, 1987.

[Dei05] Deitel, H.M. and P.J. Deitel, *Java: How to Program*, Pearson Education, 2005.

[Dum98] Dumas, M., M.C. Fauvet, and P.C. Scholl, Handling Temporal Grouping and Pattern-Matching Queries in a Temporal Object Model, *Proc. 7th International Conference Information and Knowledge Management,* 1998.

[Ede02] Eder, J., C. Koncilia, and H. Kogler, Temporal Data Warehousing: Business Cases and Solutions, *Proc. Of the International Conference on Enterprise Information Systems,* pp. 81–88, 2002.

[Elm93] Elmasri, R., V. Kouramajian, and S. Fernando, Temporal Database Modeling: An Object-Oriented Approach, *CIKM'93*, ACM, pp. 574–585, 1993.

[Etz98] Etzion, O., S. Jajodia, and S. Sripada, *Temporal Databases: Research and Practice (Lecture Notes in Computer Science)*, Springer, 1998.

[Fau99] Fauvet, M.C. et al., Analyse de Donnees Geographiques: Application des Bases de Donnees Temporelles, Revue Internationale de Geomatique, 1999.

[Fre92] Freksa, C., Temporal Reasoning Based on Semi-Intervals, *Artificial Intelligence*, vol. 54, pp. 199–227, 1992.

[Gab00] Gabbay, D.M., M. Finger, and M.A. Reynolds, *Temporal Logic: Mathematical Foundations and Computational Aspects*, Oxford University Press, 2000.

[Gal90] Galton, A., A Critical Examination of Allen's Theory of Action and Time, *Artificial Intelligence*, vol. 42, pp. 159–188, 1990.

[Gao03a] Gao, D. et al., Join Operations in Temporal Databases, *VLDB Journal*, vol. 14, pp. 2–29, 2003.

[Gao03b] Gao, D. and R.T. Snodgrass, Temporal Slicing in the Evaluation of XML Queries, *VLDB Journal*, vol. 35, pp. 632–643, 2003.

[Ger04] Gergatsoulis, M. et al., Representing and Querying Histories of Semistructured Databases Using Multidimensional OEM, *Inf. Syst.*, vol. 29, no. 6, pp. 461–482, 2004.

[Gol09] Golfarelli, M. and S. Rizzi, A Survey on Temporal Data Warehousing, *International Journal of Data Warehousing and Mining*, vol. 5, no. 1, pp. 1–17, 2009.

[Gor04] Goralwalla, I.A. et al., Temporal Granularity: Completing the Puzzle, *Journal of Intelligent Information Systems*, Springer, vol. 16, no. 1, pp. 41–63, January 2001.

[Gra05] Grandi, G., F. Mandreoli, and P. Tiperio, Temporal Modeling and Management of Normative Documents in XML Format, *Data Knowledge Engineering*, vol. 54, no. 3, pp. 327–354, 2005.

[Gre99] Gregersen, H. and C.S. Jensen, Temporal Entity-Relationship Models—A Survey, *IEEE Transactions on Knowledge and Data Engineering*, vol. 11, no. 3, pp. 464–497, 1999.

[Gre06] Gregersen, H., The Formal Semantics of the Time ER Model, *Proceedings of the 3rd Asia-Pacific Conference on Conceptual Modeling*, pp. 35–44, 2006.

[Han05] Han, J. and M. Kamber, *Data Mining: Concepts and Techniques*, 2nd edition, Morgan Kaufmann, 2005.

[Jen98] Jensen, C.S. and C.E. Dyreson (eds.), A Consensus Glossary of Temporal Database Concepts, Feb 1998 version, *Temporal Databases*, pp. 367–405, 1998.

[Kag08] Kaghazian, L., D. McLeod, and R. Sadri, Scalable Complex Pattern in Sequential Data, *Proceedings of the 17th ACM Conference on Information and Knowledge Management*, pp. 1467–1468, 2008.

[Kou90] Koubarakis, M., Reasoning about Time and Change: A Knowledge Base Management Perspective, Citeseer, 1990.

[Kow86] Kowalski, R.A. and M.J. Sergot, A Logic-Based Calculus of Events, *New Generation Computing*, vol. 1, no. 4, pp. 67–95, 1986.

[Kur94] Kurutach, W., FITMod: A Fuzzy Interval-Based Temporal Model for Temporal Databases, *Proceedings of the 1994 Second Australian and New Zealand Conference on Intelligent Information Systems,* 1994.

[Lad87] Ladkin, P., Models of Axioms for Time Intervals, *Proceedings of AAI-87,* pp. 234–239, 1987.

[Mar07] Martin, C. and A. Abello, The Data Warehouse: Temporal Database, Citeseer.

[McC69] McCarthy, J. and P.J. Hayes, Some Philosophical Problems from the Standpoint of Artificial Intelligence, in B. Meltzer and D. Mitchie, eds, *Machine Intelligence,* pp. 463–502, Edinburg University Press, 1969.

[McD82] McDermott, D., A Temporal Logic for Reasoning about Processes and Plans, *Cognitive Science,* vol. 6, pp. 101–155, 1982.

[Mor01] Morris, R. and L. Khatib, General Temporal Knowledge for Planning and Data Mining, *Annals of Mathematics and Artificial Intelligence,* vol. 33, pp. 1–19, 2001.

[Opp04] Oppel, A., *Databases Demystified,* McGraw-Hill, 2004.

[Owl04] *OWL Web Ontology Language.* http://www.w3.org/TR/owl-features/, 2004.

[Owl06] *Time Ontology in OWL.* http://www.w3.org/TR/owl-time/, 2006.

[Phi04] Philippi, S. and J.K. Koehler, Using XML Technology for the Ontology-Based Semantic Integration of Life Science Databases, *IEEE Transactions on Information Technology in Biomedicine,* vol. 8, no. 2, pp. 154–159, June 2004.

[Pis93] Pissinou, N., K. Makki, and Y. Yesha, On Temporal Modeling in the Context of Object Databases, *Proceedings of SIGMOD,* vol. 22, no. 3, pp. 8–15, September 1993.

[Pro09] Protégé. http://protege.stanford.edu/, 2009.

[Rai99] Rainsford, C.P., Accomodating Temporal Semantics in Data Mining and Knowledge Discovery, Ph.D. thesis, University of South Australia, *1999.*

[Riz08] Rizzolo, F. and A.A. Vaisman, Temporal XML: Modeling, Indexing, and Query Processing, *The VLDB Journal,* vol. 17, no. 5, pp. 1179–1212, 2008.

[Sar95] Saraee, M.H. and B. Theodoulidis, Knowledge Discovery in Temporal Databases: The Initial Step, *Proceedings of the Conference on Deductive Object Oriented Databases, Post Conference Workshop Knowledge Discovery in Database and DOOD,* K. Ong, S. Conrad, and T.W. Ling, eds, pp. 17–22, 1995.

[Sch08] Schockaert, S., M.D. Cock, and E.E. Kerre, Fuzzifying Allen's Temporal Interval Relations, *IEEE Transactions on Fuzzy Systems,* vol. 16, no. 2, pp. 517–533, 2008.

[Sho87] Shoham, Y., Temporal Logics in AI: Semantic and Ontological Considerations, *Artificial Intelligence,* vol. 33, pp. 89–104, 1987.

[Sho88] Shoham, Y and N. Goyal, Temporal Reasoning in Artificial Intelligence, in H. Strob, and AAAI eds, *Exploring Artificial Intelligence,* pp. 419–439, Morgan Kaufmann, 1988.

[Sno06] Snodgrass, R., and C. Jensen, *Temporal Databases,* Morgan Kaufmann, 2006.

[Sta06] Stantic, B., G. Governatori, and A. Sattar, Handling of Current Time in Native XML Databases, *Proceedings of 17th Australian Database Conference,* 2006.

[The94] Theodoulidis, B. et al., The ORES Temporal Database Management System, *Proceedings of ACM SIGMOD Int. Conference on Management of Data,* R.T. Snodgrass and M. Winslett, eds., p. 511, 1994.

[Unn09] Unnikrishnan K. and K.V. Pramod, On Implementing Temporal Coalescing in Temporal Databases Implemented on Top of Relational Database Systems, *Proceedings of International Conference on Advances in Computing, Communication and Control,* pp. 153–156, 2009.

[Vie02] Viedma. M.A.C. and R.M. Morales, Syntax and Semantics for a Fuzzy Temporal Constraint Logic, *Annals of Mathematics and Artificial Intelligence,* vol. 36, no. 4, pp. 357–380, 2002.

[Vil82] Vilain, M.B., A System for Reasoning about Time, *Proceedings of National Conference Artificial Intelligence,* pp. 197–201, 1982.

[Wan08] Wang, F., C. Zaniolo, and X. Zhou, ArchIS: An XML-based Approach to Transaction-Time Temporal Database Systems, *The VLDB Journal,* vol. 17, no. 6, pp. 1445–1463, 2008.

[Zha02] Zhang, S. and C. Dyreson, Adding Value to XPath, *DNIS,* 2002.

CHAPTER 2

Temporal Data Similarity Computation, Representation, and Summarization

An important task in temporal data mining, in fields ranging from medicine to finance, is the computation of similarity between time series data or a series of events. Similarity computation is an important part of time series indexing, classification, and clustering. For example, in medicine, comparison of patient physiological signals with "normal" signals can be useful in disease detection. In another example, identification of patients with similar series of events can be useful in choosing treatment. In finance, similarity of stock value fluctuations with older stock data can be useful in prediction. The results of the similarity computation depend on the temporal data representation scheme.

This chapter is organized as follows: Section 2.1 discusses temporal data types and ways to preprocess them to achieve more meaningful results. Section 2.2 discusses similarity metrics for time series. Section 2.3 presents time series data representation schemes, while Section 2.4 discusses time series summarization methods, that is, ways to represent time series in a very compact way. Section 2.5 discusses similarity between series of events, Section 2.6 discusses similarity computation for semantic temporal objects, and Section 2.7 discusses temporal knowledge representation in case-based reasoning systems. Finally, additional bibliography is presented in Section 2.8.

2.1 TEMPORAL DATA TYPES AND PREPROCESSING

2.1.1 Temporal Data Types

Temporal data can be of three types:

- *Time series.* They represent ordered real-valued measurements at regular temporal intervals. A time series $X = \{x_1, x_2, ..., x_n\}$ for $t = t_1, t_2, ..., t_n$ is a discrete function with value x_1 for time t_1, value x_2 for time t_2, and so on. Time series data consist of varying frequencies where the presence of noise is also a common phenomenon. A time series can be *multivariate* or *univariate*. A multivariate time series is created by more than one variable, while in a univariate time series there is one underlying variable. Another differentiation of time series is *stationary* and *nonstationary*. A stationary time series has a mean and a variance that does not change over time, while a nonstationary one has no salient mean and can decrease or increase over time.

- *Temporal sequences.* These can be timestamped at regular or irregular time intervals. An example of a temporal sequence is the timestamped sequence of purchases of a customer on a Web site.

- *Semantic temporal data.* These are defined within the context of an ontology. For example, "senior" and "middle-aged" are defined within the context of a human lifetime ontology.

Finally, an *event* can be considered as a special case of a temporal sequence with one time-stamped element. Similarly, a series of events is another way to denote a temporal sequence, where the elements of the sequence are of the same types semantically, such as earthquakes, alarms, etc.

2.1.2 Temporal Data Preprocessing

2.1.2.1 Data Cleaning

In order for the data mining process to yield meaningful results, the data must be appropriately preprocessed to make it "clean," which can have different meanings depending on the application. Here we will deal with two aspects of data "cleanliness": handling missing data and noise removal.

Other aspects, such as integration of data from different databases or data reduction using data cube aggregation, are outside the scope of this book.

2.1.2.1.1 Missing Data   A problem that quite often complicates time series analysis is missing data. There are several reasons for this, such as malfunctioning equipment, human error, and environmental conditions. The specific handling of missing data depends on the specific application and the amount and type of missing data. There are two approaches to deal with missing data:

1. *Not filling in the missing values.* For example, in the similarity computation section of this chapter we discuss a method that computes the similarities of two time series by comparing their local slopes. If a segment of data is missing in one of the two time series, we simply ignore that piece in our similarity computation.

2. *Filling in the missing value with an estimate.* For example, in the case of a time series and for small numbers of contiguous missing values, we can use *data interpolation* to create an estimate of the missing values, using adjacent values as a form of imputation. The greater the distance between adjacent values used to estimate the intermediate missing values, the greater the interpolation error. The allowable interpolation error and, therefore, the interpolation distance vary from application to application. The simplest type of interpolation that gives reasonable results is linear interpolation. Assume we have points x_1 and x_3, with values $f(x_1)$ and $f(x_3)$, respectively. The value for point x_2 is missing, where x_2 is somewhere in between x_1 and x_3. We can use linear interpolation to estimate $f(x_2)$ as follows:

$$f(x_2) = f(x_1) + \frac{(x_2 - x_1)(f(x_3) - f(x_1))}{x_3 - x_1} \qquad (2.1)$$

An example of using linear interpolation in temporal data mining is discussed in [Keo98]. Another type of interpolation that gives superior results to linear interpolation without being too computationally expensive is *spline interpolation*. It uses low order polynomials to approximate the missing intervals. *Cubic interpolation* is used in [Per00].

2.1.2.1.2 Noise Removal   *Noise* is defined as *random error* that occurs in the data mining process. It can be due to several factors, such as faulty measurement equipment and environmental factors. We will discuss two methods of dealing with noise in data mining: binning and moving-average smoothing.

In *binning*, the data are divided into buckets or bins of equal size. Then the data are smoothed by using either the mean, the median, or the boundaries of the bin. An example is given below.

> **Example:** Assume that we have the following time series data: 10, 11, 11, 15, 16, 19, 23, 24, 24, 26, 27, 28, 30, 32, 35, 36, 36, 39, 41, 45.
>
> Dividing into 4 bins of size 5 and using mean smoothing, where the mean is defined as the average of the values in the bin, we get the following:
>
> Bin 1: 12.6, 12.6, 12.6, 12.6, 12.6   Bin 2: 23.2, 23.2, 23.2, 23.2, 23.2
> Bin 3: 30.4, 30.4, 30.4, 30.4, 30.4   Bin 4: 39.4, 39.4, 39.4, 39.4, 39.4
>
> Using median smoothing, which is the middle value in the bin, we get the following:
>
> Bin 1: 11, 11, 11, 11, 11   Bin 2: 24, 24, 24, 24, 24
> Bin 3: 30, 30, 30, 30, 30   Bin 4: 39, 39, 39, 39, 39

Let us now examine *moving-average smoothing*. Moving-average (or rolling-average) smoothing is used frequently in finance to smooth out short term fluctuations in stock prices. Let us assume we have a time series with 5 points, as shown in Figure 2.1, and we want to compute the 3-point central moving average (*CMA*), where we use both past and future values around the point of interest. For example, the *CMA* at time 2 (denoted with a star on the graph) is the average of the values for time points 1 and 3. The *CMA* at time 4 is the average of times 3 and 5, and so on.

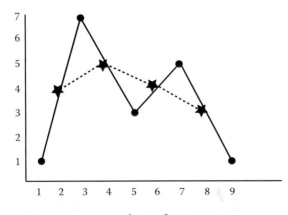

FIGURE 2.1   Moving-average smoothing of a time series.

2.1.2.2 Data Normalization

In *data normalization*, the data are scaled so that they fall within a pre-specified range, such as [0–1]. Normalization allows data to be transformed to the same "scale" and, therefore, allows direct comparisons among their values. For example, assume we want to compare the number of "above average" temperatures in two cities during the month of February, where in one city temperatures are measured in Celsius and in the other they are measured in Fahrenheit. This can be easily done with normalization. The importance of normalization is studied extensively in the literature [Keo02a]. We will examine two ways to do normalization:

Min-max normalization: To do this type of normalization, we need to know the minimum (x_{min}) and the maximum (x_{max}) of the data:

$$x_{norm} = \frac{x - x_{min}}{x_{max} - x_{min}} \tag{2.2}$$

Z-score normalization: Here, the mean and the standard deviation of the data are used to normalize them:

$$x_{norm} = \frac{x - \mu}{\sigma} \tag{2.3}$$

As noted in [Han05], z-score normalization is useful, in cases of outliers in the data, that is, data points with extremely low or high values that are not representative of the data and could be due to measurement error. As we can see, both types of normalization preserve the shape of the original time series, but z-score normalization follows the shape more closely.

Example: Let us assume that we are given time series data that represent the loss/gain of weight of a person on a diet. The data are collected each week for 8 weeks:

Data: –2, –3, –1, 0, 0, –2, +1, –1

Normalization using min-max normalization:

Min = –3, Max = 1
Normalized Data: 0.25, 0, 0.5, 0.75, 0.75, 0.25, 1, 0.5

Normalization using z-score normalization:

Mean = –1, standard deviation = 1.3
Normalized Data: –0.77, –1.53, 0, 0.77, 0.77, –0.77, 1.54, 0

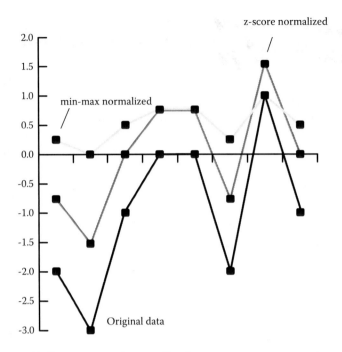

FIGURE 2.2   Different types of normalization.

The original and normalized data are also shown in Figure 2.2. As we can see, both types of normalization preserve the shape of the original time series.

2.2   TIME SERIES SIMILARITY MEASURES

If we have N time series signals, the similarity problem, using a distance measure, can be expressed in two different ways:

1. Find all pairs of time series that have a distance that is less than $\in$.

2. *Indexing* or *query by content.* There are two approaches to this kind of search: (1) By range: Find all time series that have distance less than $\in$ from a specific time series. (2) Find the *m* closest neighbors for a specific time series.

The above cases fall in the category of *whole series matching.* Another type of similarity computation is referred to as *subsequence matching,* where a shorter sequence is matched against a longer sequence, by sliding the former along the latter. As described in [Fal94] and in [Keo00b], the following are included in the desirable properties of an *indexing* scheme:

- Its speed should be significantly greater than performing a sequential scan of the database and it should need little space overhead.

- It should allow no false dismissals. Note that false alarms are not typically that dangerous, because they can always be detected in postprocessing.

- It should be able to handle queries of various lengths and allow insertions and deletions without requiring the index to be rebuilt.

- The index should be able to be built in a reasonably timely fashion and handle different distance measures.

Several techniques have been proposed for the indexing of time series and most of them involve tree structures, such as *R-trees* [Cha99] and *kd-trees* [Den98]. A semantic hierarchy for indexing is proposed in [Li98].

The success of a query to find a similar time series in a database can be measured using the precision and recall measures. *Precision* is defined as the proportion of returned time series from the query that is indeed similar to the input time series. *Recall* is the proportion of time series similar to the input time series, which is retrieved from the database.

We will examine three widely used types of time series similarity measures: (1) *distance-based similarity*, (2) *dynamic time warping*, and (3) *longest common subsequence*. In addition we will briefly look at other similarity measures, less widely adopted.

2.2.1 Distance-Based Similarity

2.2.1.1 Euclidean Distance

The *Euclidean distance* between two time series $X = \{x_1, x_2, ..., x_n\}$ and $Y = \{y_1, y_2, ..., y_n\}$ is defined as

$$EU_{DIS} = \sqrt{(x_1 - y_1)^2 + (x_2 - y_2)^2 + \cdots + (x_n - y_n)^2} \qquad (2.4)$$

The utility of the *Euclidean distance* metric is limited to time series that have the same baseline, scale, and length. Also the two time series should not have any gaps. In addition, the Euclidean distance is sensitive in the presence of noise and variations in the time axis.

The Euclidean distance has been used by a large number of researchers for indexing purposes, such as in [Fal94] and [Agr95b].

2.2.1.2 Absolute Difference

This metric is defined as

$$AD = \sum_{i=0}^{N-1} |x_i - y_i| \tag{2.5}$$

Given that the absolute difference metric does not square the differences, its advantage over Euclidean distance is that it does not emphasize dissimilarities as much.

2.2.1.3 Maximum Distance Metric

This metric is defined as

$$D_{max} = \max |x_i - y_i| \; where \; i = 0, N-1 \tag{2.6}$$

As can be seen from its definition, this metric emphasizes the most dissimilar features.

2.2.2 Dynamic Time Warping

In most of the aforementioned similarity metrics, an assumption is that the time series are aligned on the X-axis. However, this is not always a valid assumption. An example of similar, yet misaligned time series, is shown in Figure 2.3.

The solution to this problem is dynamic time warping (DTW). This is a nonlinear distance measure that originated in the speech community and maps one time series onto another. It has found wide applicability in many

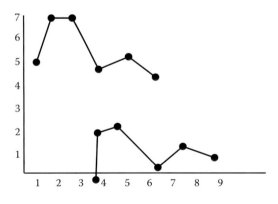

FIGURE 2.3   An example of similar, yet misaligned, time series.

data mining tasks such as clustering, classification, and indexing and a wide area of applications, varying from medicine to robotics [Keo05].

After it was shown that requiring the two time series to have the same length and constrained amount of warping [Keo05] results in *no false* dismissals and *DTW exact* indexing, *DTW* became a very powerful tool in the data mining literature [Rat04]. It is interesting to note that, in classification problems dealing with very large data sets, the classification accuracy of the *DTW* and the *Euclidean distance* were found to be very similar [Shi08]. However, in small data set classification problems, *DTW* produces better classification accuracy.

Despite its wide use, there are still some prevalent misconceptions about the *DTW*. An excellent reference for these misconceptions and proof of their dismissal is [Rat04]. One of the problems of the *DTW* distance is that it is computationally intensive and one of the misconceptions is that the speed of *DTW* can be improved. [Rat04] shows that the state of the art is currently at the *asymptotic limit* of improving *DTW*'s speed. Another misconception is that the 10% constraint on warping inherited from the speech community is necessary. This is *actually much higher* than the constraint needed for data mining applications.

The warp is computed by finding the point-to-point alignment that minimizes the error between the two time series. As an example, let us assume that we have a time series $X = \{x_1, x_2,..., x_m\}$ with length m and another time series with length k, $Y = \{y_1, y_2,..., y_k\}$. The warping path $W = \{w_1, w_2,..., w_N\}$, where $max(k, m) \leq N < k + m - 1$, is found using dynamic programming [Keo01] to compute the cumulative distance $CUM\_DIST(i, j)$:

$$CUM\_DIST(i, j) = dist(i, j) + min\{CUM\_DIST(i - 1, j - 1),$$
$$CUM\_DIST(i - 1, j), CUM\_DIST(i, j - 1)\} \qquad (2.7)$$

where the $k \times m$ matrix *dist.* is defined such that the (i, j) element contains the distance of Xi and Yj. The warping path satisfies the following conditions:

- It starts and finishes in diagonally opposite corner sides of the matrix.

- The cells in the warping path have to be adjacent (including diagonally).

- The points in W have to be placed monotonically in space.

Computation of Equation 2.7 leads to the derivation of an optimum warping path, which minimizes the following warping cost:

$$COST = \min\left(\sqrt{\sum_{n=1}^{N} w_n}\right)\bigg/ N \qquad (2.8)$$

A warping window can be defined as two lines parallel to the diagonal. As demonstrated in [Keo02a], restricting the size of the warping window not only speeds up computation, because only a part of the matrix needs to be computed, but also tightens the lower bounding property. This work by Keogh resulted into *DTW* being widely adopted by the research community and also in a system for indexing historical handwritten documents [Man03].

A further improvement to the *DTW* is discussed in [Chu02]. It is called *Iterative Deepening Dynamic Warping*. In this algorithm, a probabilistic model describing the distribution of the distance approximation errors for any level of approximation error is used. Also, the user only needs to submit a tolerance for false dismissals in addition to her query. The experimental results show that the lower the specified tolerance, the smaller the number of the resulting false dismissals, which, however, impacts somewhat the computational speed.

In [Fu08] Fu et al. propose the combination of *DTW* with *Uniform Scaling* in many real world problems in order to achieve meaningful results. This combination is of particular importance in areas where human variation exists. Examples are biometrics, motion-capture, handwritten recognition, and *query by humming* where users search music by providing some notes. The authors present a characteristic example: gait recognition algorithms. In these algorithms, *DTW* is frequently used; however, there is also need for uniform scaling because of the variation of the gait length in different gait cycles.

Finally, in [Sak05], the authors propose a fast algorithm for the computation of similarity using *DTW*. *DTW* is usually computationally expensive because distances for all possible paths have to be computed to find the exact warping distance. This new algorithm guarantees that there will be no false dismissals and it excludes the paths that will not yield successful search results using a novel dynamic programming method. This is achieved by using a new lower bounding measure that approximates the time warping distance and a search algorithm that searches in multiple granularities and continuously improves the distance approximations. Experiments on real and synthetic data have shown that the algorithm is at least one order of magnitude faster than the best existing method and sometimes up to 222 times faster.

2.2.3 The Longest Common Subsequence

In contrast to the Euclidean distance, the longest common subsequence (*LCSS*) is a measure that is tolerant to gaps in the two time series. A discussion of *LCSS* can be found in [Das97]. The *LCSS* assumes the same baseline and scale for the two time series. However, it is superior to *DTW* in the following areas [Vla05], [Vla03]:

- *LCSS* handles noisy data and outliers better.

- The *DTW* can distort the actual distance between points in the time series by overfitting.

- The computational complexity of *DTW* is significant and therefore its scalability is not very good.

Here is an example of computing the longest common subsequence of two time series:

$X = \{1, 34, 9, 12, 22, 100, 45, 65\}$

$Y = \{2, 34, 10, 12, 22, 55, 96, 23\}$

$LCSS = \{34, 12, 22\}$

An example of using *LCSS* can be found in [Agr92], where enough non-overlapping pairs of similar subsequences determine that the overall time series are similar.

2.2.4 Other Time Series Similarity Metrics

In [Alt06], a new metric for the time series similarity computation of clinical data, the *slope–base similarity computation,* is proposed. The slope wise comparison metric measures and quantizes the difference in slope between two time series. There are five possible differences that can be assigned: 0, 0.25, 0.5, 0.75, and 1. Given points a_1, a_2 on one time series and points b_1, b_2 on another, we compute the artificial slopes, *SA* and *SB*, as follows:

$$SA = \frac{a_2 - a_1}{|a_2| + |a_1|} \tag{2.9}$$

$$SB = \frac{b_2 - b_1}{|b_2| + |b_1|} \tag{2.10}$$

Then the artificial slope is compared to a positive and negative threshold to determine the quantized difference to be assigned.

Let *pt* denote the positive threshold, *nt* the negative threshold, and *diff* the quantized threshold. Assuming *SA* < *nt*, the possible values for the quantized difference are as follows, depending of the value of SB: (a) if SB < nt, then diff = 0 (b) if SB >= nt, diff = 0.25 (c) if SB < pt diff = 0.75 (d) if SB >= pt diff = 1.

Another time series similarity metric divides the time series into segments, known as *envelopes,* and the individual similarity of the envelopes is computed by determining whether the difference of individual points is above a certain threshold. Then, a *yes/no vote* is computed for each envelope and the overall similarity is computed by determining how many envelopes are similar [Agr96], [Bol97].

Finally, we briefly mention these other distance measures:

- The *DISSIM* distance measure [Fre07] computes the similarity of two time series at different sampling rates.

- *EDR* [Che05]: This distance measure uses the *edit distance* between two time series, and it uses a threshold pattern to quantify the edit distance between two points to 0 or 1.

- *ERP* [Che04b]: Here a constant reference point is used to measure the distance between the gaps of two time series.

- *TQuEST* [Aβf06]: Here a time series is converted to a sequence of threshold-crossing intervals. The points within each interval have a value greater than a threshold ε.

- *SpADe* [Che07]: In this algorithm, the similarity is computed by finding matching patterns between two time series.

- *Probabilistic models* [Ge00], [Keo01d]. Here, an underlying probability distribution of the time series is used to guide the similarity computation. For example, in [Ke01d], piecewise linear segmentation is first used to represent local time series features, such as peaks, as deformations from a template. Then, these are integrated with a global model about the expected locations of these local features.

- *Multivariate time series similarity measures* [Yan04]: The measure here is based on *Principal Component Analysis*, where the *eigenvalues* of the time series are extracted.

- *Similarity of multiattribute motions* [Li04]: *Singular Value Decomposition (SVD)* that takes into account the different rates and durations of each motion is used for the similarity computation.

2.3 TIME SERIES REPRESENTATION

Time series representation schemes will be divided into the following categories [Rat05]: (1) *nonadaptive*, (2) *data adaptive*, (3) *model based*, and (4) *data dictated*. There are primarily two desirable properties in a representation scheme: (1) It must not allow any false dismissals. In other words, if two series are to be found similar in the original space, they should also be found similar in the transformation space. This is also known as the *lower bounding* property, which means that we can define a distance measure on the representation-transformed data that is guaranteed to be less than or equal to the distance in the original data space [Shi08]. (2) It must reduce the dimensionality of the similarity search problem.

It is interesting to note here a framework, called *GEMINI*, that played a significant role in temporal data mining research [Fal94]. This has framework can exploit any representation scheme that performs dimensionality reduction to perform efficient indexing. However, one very important property is that the distance measure in the index space should satisfy the lower bounding property.

A work with surprising results regarding time series representation accuracy is that of Palpanas et al. [Pal04]. In this work, a wide variety of representation schemes is applied to a large number of time series with no great difference in approximation accuracy. Therefore, the authors note, when one needs to choose a representation scheme for a time series, other criteria such as the visual appearance of the representation should be considered.

2.3.1 Nonadaptive Representation Methods

The most commonly used methods in this category are (1) the Discrete Fourier Transform (*DFT*), (2) the Discrete Wavelet Transform (*DWT*), and (3) Piecewise Aggregate Approximation (*PAA*). All these methods satisfy the lower bounding property and therefore can be used in the aforementioned *GEMINI* framework.

2.3.1.1 Discrete Fourier Transform

The *DFT* represents a time series in the frequency domain. The *DFT* coefficients F_k of a time series $X = \{x_0, x_2, ..., x_{n-1}\}$ are complex numbers given by

$$F_k = \sum_{i=0}^{N-1} x_i e^{-j2\pi i k/N} \qquad (2.11)$$

where $k = 0, 1, ..., N - 1$.

The advantage of using the *DFT* in signal processing is that a fast algorithm exists for its computation, known as the fast fourier transform (*FFT*). The computational complexity of the *FFT* is $O(n \log n)$. Regarding time series representation, the advantage of using the *DFT* is that only the first few *DFT* coefficients, which correspond to the lowest frequencies in the time series, need to be kept for an adequate representation of most time series. For example, using the *DFT* on random walk data and keeping only 10% percent of the coefficients results in retaining of 90% of the energy [Vla05].

The *DFT* does not allow any false dismissals because it preserves the Euclidean distance between two time series objects. This last statement is a direct result of *Parseval's* theorem, which states that the energy of a signal in the space domain (or time domain in our case) is equal to the energy of the signal in the frequency domain.

Following the computation of the *DFTs*, the similarity of the two time series can be determined by computing the difference between the real and imaginary parts of the transform coefficients. Another way is to compute the difference between the squares of the magnitudes of their *DFTs*. The square of the magnitude of the *DFT*, $|X(f)|^2$, approximates the signal's *power spectrum*, which describes the frequency energy of the signal. Finally, although the *DFT* guarantees no false dismissals, it might generate false alarms if the number of coefficients kept is too small, such as three.

The *DFT* achieves significant dimensionality reduction and, when plugged into the *GEMINI* framework, it achieves a speed increase over sequential scanning in the range of 3 to 100. As mentioned in [Vla05], however, the *DFT* is problematic for bursty signals and mixed signals with flat and busy areas.

2.3.1.2 Discrete Wavelet Transform

The wavelet transform utilizes basis functions known as wavelets, which are functions that allow localization of a time series in *both frequency and space*. Wavelets allow the analysis of a time series at different scales, also known as *resolutions*. This is a significant advantage over the Fourier transform, whose basis functions (sines and cosines) do not allow localization in space. A class of wavelet functions can be created from a function, ψ, known as the *mother wavelet*. The children wavelets are created by translation and contraction (or dilation) of the mother wavelet as shown below:

$$\psi(t)^{s,\tau} = \frac{1}{\sqrt{s}} \psi\left(\frac{t-\tau}{s}\right) \tag{2.12}$$

where s is the scaling constant used for contraction or dilation and τ is the translation constant used to slide a window over the time series. The result of using a scaling factor to generate the children wavelets is that time series can be analyzed in a multiresolution way. Specifically, low frequencies of the signal are examined at low-resolution views that could span the entire time series, while high frequencies are examined at high-resolution views.

One of the simplest wavelets is the *Haar* wavelet. The *Haar* wavelet transform is the most widely used transform in data mining because of its simplicity [Vla05]. The *Haar* scaling function $\varphi(t)$ is a step function defined as

$$\varphi(t) = 1 \quad t \in (0,1)$$
$$\varphi(t) = 0 \text{ elsewhere} \tag{2.13}$$

The *Haar* mother wavelet is defined as

$$\psi(t) = 1 \quad t \in [0, 1/2)$$
$$\psi(t) = -1 \quad t \in [1/2, 1) \tag{2.14}$$
$$\text{Elsewhere } \psi(t) = 0$$

The computational complexity of the *Haar* wavelet transform is $O(n)$. Similar to the *Fourier* transform, the wavelet transform also preserves the *Euclidean* distance after proper normalization [Cha03] and, therefore, guarantees no false dismissals.

The *DWT* is a discretized version of a wavelet. We will examine a technique for the fast computation of the *Haar* wavelet transform, known as the *ordered fast Haar* wavelet transform. In this technique, a series of averages and differences is computed on the data. The averages represent views of the data at a coarse resolution, while the differences show changes in the data and represent views of the data at a fine resolution. The averages and the differences are computed over a window of size 2. Note that the data have to be a power of 2. The averages and differences are computed on progressively averaged views of the data. The computations continue until we are left with one value in the averaged data set. The final average value and the computed differences become the wavelet transform coefficients.

Let us now work through an example. The time series X has the following values: $X = \{1, 5, 3, 9, 10, 4, 2, 8\}$.

2.3.1.2.1 First Pass   Compute the averages of consecutive values:

$$Av = \left\{ \frac{1+5}{2}, \frac{3+9}{2}, \frac{10+4}{2}, \frac{2+8}{2} \right\} = \{3, 6, 7, 5\} \tag{2.15}$$

Compute the differences of consecutive values:

$$Co = \left\{ \frac{1-5}{2}, \frac{3-9}{2}, \frac{10-4}{2}, \frac{2-8}{2} \right\} = \{-2, -3, 3, -3\} \tag{2.16}$$

This is stored as

$$St = \{3, 6, 7, 5; -2, -3, 3, -3\} \tag{2.17}$$

2.3.1.2.2 Second Pass   Compute the average values of the stored average values above:

$$Av = \left\{ \frac{3+6}{2}, \frac{7+5}{2} \right\} = \{4.5, 6\} \tag{2.18}$$

Compute the differences of the stored average values:

$$Co = \left\{ \frac{3-6}{2}, \frac{7-5}{2} \right\} = \{-1.5, 1\} \tag{2.19}$$

Store the new averages along with the computed differences:

$$St = \{4.5, 6; -1.5, 1; -2, -3, 3, -3\} \tag{2.20}$$

2.3.1.2.3 Third Pass   Compute the final average value, which is also the overall average of all values in the time series:

$$Av = \left\{ \frac{4.5+6}{2} \right\} = \{5.25\} \tag{2.21}$$

Compute the final difference:

$$Co = \left\{ \frac{4.5-6}{2} \right\} = \{-0.75\} \tag{2.22}$$

The set below, which consists of the overall average and all the computed coefficients, represents the Haar wavelet coefficients:

$$St = \{5.25; -0.75; -1.5, 1; -2, -3, 3, -3\} \tag{2.23}$$

Chan and Fu, [Cha99] and [Cha03], have done a significant amount of work in using wavelets to perform similarity search. Specifically, in [Cha03], the authors propose to use wavelets to reduce the cost of computing the warping distance for *DTW*. This is done by using a low resolution warping distance, which is based on *Haar* wavelet coefficients and is easy to compute, yet is a close match to the warping distance.

In [Pop02], the authors conducted a study of the effect of different transformations on the performance of similarity search over time series data. Comparing the *DFT*, the *Haar wavelets*, the *Daubechies wavelets*, and *PAA* (a nontransformation method to be discussed next), it was found that the *Daubechies wavelet* achieved the highest precision.

2.3.1.3 Piecewise Aggregate Approximation (PAA)

[Keo00b] and [Lin03] discuss a representation method that achieves significant dimensionality reduction of the time series. In this method, the time series is divided into k segments of equal length and then each segment is replaced with a constant value, which is the average value of the segment. Then these average values are grouped in a vector, which represents the signature of the time series. An example is shown in Figure 2.4. The signature of the time series is $\left(3, \frac{8}{3}, \frac{10}{3}, 5 \right)$.

Despite the fact that *PAA* is significantly simpler than the *DFT* and the wavelet transform, its dimensionality reduction ability rivals the ability

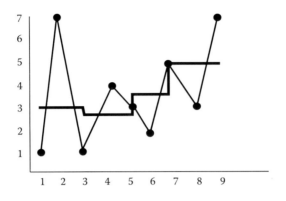

FIGURE 2.4   *PAA* representation of a time series.

of these techniques. In addition, the index can be built in linear time and it allows the performance of queries, whose time length is shorter than the one for which the index was built. For example, a doctor might be interested in finding the daily temperature variations of a patient while an index has been built only for weekly variations. This is impossible with the *DFT* and *DWT*.

Also [Keo00b] shows that the speed of a query when *PAA* is combined with a *weighted Euclidean distance* is significantly higher than the speed of *PAA* combined with simple *Euclidean distance*. In addition, as [Keo05] discusses, combining *PAA* with *Dynamic Time Warping* degenerates in a special case to Euclidean indexing with *PAA*, which is considered to be state of the art in terms of flexibility and efficiency.

A modification of the *DTW* that applies *DTW* to a *PAA* approximation is discussed in [Keo00a]. The algorithm is called *PDTW* and it produces very similar results to the classical *DTW* algorithm; however, it achieves a speed up of one to three orders of magnitude. *PDTW* produces accurate results for classification and clustering problems; however, it has the problem that the user has to choose a compression rate for the dimensionality reduction. Too high a compression rate means increased probability of false dismissals; too low a compression rate means slower computational speed.

2.3.2   Data-Adaptive Representation Methods

2.3.2.1   Singular Value Decomposition of Time Sequences

In [Kor97], a method for supporting efficient query in large data sets of time sequences is proposed. As is discussed in the article, there is always a tradeoff when we need to perform ad hoc queries on very large data sets. Although we might want to keep the entire data set on one disk, if we decide to compress the data, we might have significant difficulty in index-ing or accessing the data. Singular value decomposition *(SVD)* is offered as a solution to the aforementioned problem: By keeping only a few most important coefficients, we can achieve good compression while still being able to reconstruct an arbitrary value with a small reconstruction error. The purpose of *SVD* is to find the principal components, that is, the fea-tures that have high discriminatory power. The formal definition of *SVD*, given an $N \times M$ real matrix W is as follows [Kor97]:

$$W = U \times F \times V^T \tag{2.24}$$

where U is a column orthonormal $N \times r$ matrix, r is the rank of the matrix W, F is a diagonal $r \times r$ matrix, and V is a column-orthonormal $M \times r$ matrix. The *SVD* is a global transformation in the sense that it is applied to the entire data set and the goal is to find the axes that best describe the data variance. The data set is rotated such that the first axes have the maximum possible variance, the second axes have the maxi-mum possible variance orthogonal to the first, and so on. Its most sig-nificant drawbacks are computational complexity and that the addition of a single time series requires recomputation. It has the advantages that it satisfies the lower bounding property and has optimal dimensionality reduction in a Euclidean sense [Vla05].

2.3.2.2 Shape Definition Language, CAPSUL, and SOM

When only coarse information about the overall shape of a time series is important for its representation, a time series can be represented using a limited vocabulary that describes the different gradients in it. An exam-ple of such a vocabulary is the *Shape Definition Language* (SDL), which is introduced in [Agr95]. The *SDL* vocabulary is {*Up, up, stable, zero, down, Down*}. The meaning of each word in the vocabulary is as follows:

- *Up*: Strong upward gradient

- *up*: Moderately strong upward gradient

- *stable*: small upward or downward gradient

- *zero*: no change

- *down*: Moderately strong downward gradient

- *Down*: Strong downward gradient

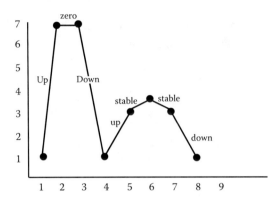

FIGURE 2.5   *SDL* representation of a time series.

An example of using *SDL* to describe a shape is shown in Figure 2.5. The similarity of two curves represented using *SDL* words can be computed by using a variant of the longest common sequence metric. In other words, the similarity of two time series is proportional to the length of the longest common sequence of words in their *SDL* representation.

Another example of a symbol-based language is *CAPSUL* (Constraint-Based-Pattern-Specification-Language) [Rod02]. *CAPSUL* allows the expression of more complex pattern relationships, such as periodic patterns. Another category of symbol-based representation is based on *SOMs* (self-organizing maps). A *self-organizing map* is a type of neural network that is trained using unsupervised learning to produce a discretized representation of the input space. This representation constitutes the map. Conversion of time series into *SOMs* consists of three steps [Ant01]:

- A new series composed of the differences between consecutive values is created.

- Delay embedding is used to input windows of size *d* to the *SOM* and output the winner node.

- Finally each node is associated with a symbol.

2.3.2.3 Landmark-Based Representation
In the landmark model proposed in [Per00], perceptually important points of a time series are used to represent it. The perceptual importance depends on the specific type of the time series. For example, Figure 2.6 shows a coarse approximation of a normal heartbeat and its perceptually important points.

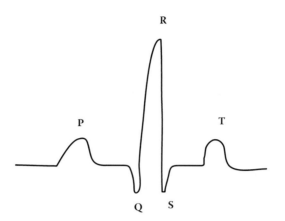

FIGURE 2.6 Perceptual points of a heartbeat.

However, not all time series have a well-known expected shape. In general, good choices for landmarks are local maxima and minima as well as inflection points. The advantage of using the landmark-based method is that this time representation is invariant to amplitude scaling and time warping. As above, a variant of the longest common sequence metric can be used to compute the similarity of time series represented using landmarks.

Another work that uses important points for the dimensionality reduction/compression and indexing of a time series is [Pra02]. Important points considered are minima and maxima, while other points are discarded. The authors define a similarity measure for two numbers a, b and then they extend this to time series. Their similarity is defined as

$$sim(a, b) = 1 - 2\frac{|a - b|}{\max(|a|, |b|)} \tag{2.25}$$

The similarity between two time series $X = \{x_1, x_2, ..., x_N\}$ and $Y = \{y_1, y_2, ..., y_N\}$ is defined as

$$\frac{1}{N}\sum_{i=1}^{n} sim(x_i, y_i) \tag{2.26}$$

The work of [Pra02] is an extension of the work described in [Gur99] and [Las01] that focuses on the detection of change points. [Gur99] focuses on finding changes in trends of a data stream, while [Las01] represents a time series with features such as signal to noise ratio and slope and then identifies change points of these features.

The use of a small alphabet for time series dimensionality reduction and then pattern matching in similarity computation is also proposed in [Hua99], [Jon97], [Lam98], [Par99], and [Qu98].

2.3.2.4 Symbolic Aggregate Approximation (SAX) and iSAX

As shown in [Lin03], following *PAA*, we can further discretize the resulting series using a string alphabet. The technique is known as *SAX* and it produces symbols with *equiprobability*. Given that a normalized time series has a Gaussian distribution, we can divide it into equal size areas. Points at the two ends of these equal areas are called *breakpoints*. Then we can convert the *PAA* series to a series of strings by the following procedure: All *PAA* coefficients that are less than the smallest breakpoint are assigned the string *a,* the coefficients greater than the smallest breakpoint and smaller than the second breakpoint are assigned the string *b,* and so on.

One unique feature of the *SAX* method is that the distance measure in the symbolic space lower bounds the distance in the original time series space. A distance function that satisfies the lower bound criterion and returns the minimum distance between the original time series of two words can be defined as [Lin03]:

$$MINDIST(Q, C) = \sqrt{\frac{N}{W}} \sqrt{\sum_{i=1}^{W} dist(q_i, c_i)} \tag{2.27}$$

where *N* is the size of the time series, *W* is the size of the codeword, and *dist* is the distance between the *PAA* symbols q_i and c_i, which can be obtained from a lookup table. In [Lin03], the *SAX* representation technique has successfully been used in clustering, classification, query by content, and anomaly detection. The reason for its success is that it has the very attractive features of lower bounding the Euclidean distance and the *DTW* distance, while achieving dimensionality reduction.

An extension to *SAX*, known as *iSAX*, which is a multiresolution indexing method suitable for data sets that are several orders of magnitude bigger than most data sets considered in the temporal mining literature is described in [Shi08]. This is a problem of increasing importance, especially in biology with the explosive growth of gene sequence time series expressions, in medicine with the increasing adoption of *Electronic Health*

Records, and in astronomy in projects such as *MACHCO* [Keo00b]. *iSAX* allows for *very fast exact* queries and also for *fast approximate* queries and satisfies the lower bounding property. The multiresolution approach has the same foundations as wavelets and allows for representations of various granularities. [Shi08] also describes an algorithm that combines approximate and exact searches to find the ultimate correct answer to a search query.

2.3.2.5 Adaptive Piecewise Constant Approximation (APCA)

This representation scheme was introduced in [Keo01c], and it satisfies the lower bounding property. This algorithm uses constant value segments of varying lengths to approximate a time series, such that the approximation errors are minimal. *APCA* is indexed using a multidimensional structure and then the index can be used for efficient range *k–Nearest Neighbors* queries. *APCA* uses segments of *varying lengths*, while *PAA*, a technique developed by the same team as *APCA*, uses segments of constant lengths. The authors in [Keo01c] mention that whether *PAA* or *APCA* results in a smaller error depends on the type of the time series data. In experimental results using ECG and a "mixed bag" signal, the CPU cost of *APCA* was compared against a linear scan of the database, a DFT-based method, and a wavelet-based method.

APCA has the advantage that it can *adapt* the number of segments to local detail, such that a single segment can be used to approximate an area of low detail, while many segments can be used to approximate an area of high detail. However, for *PAA* only one number is required, while for *APCA* two numbers are required (*length, value*). Also as noted in [Vla05], *APCA* is good for bursty signals.

2.3.2.6 Piecewise Linear Representation (PLA)

In this category of time series representation, we will discuss three schemes: *Sliding Window*, *Bottom-Up*, and *Top-Down*. These schemes share the following characteristics: (1) the end result is the approximation of a time series with line segments, and (2) a cost function is applied to the creation of each line segment. In the case of the *Sliding Window* algorithm, we start at the leftmost point of the time series and then grow the segment point-by-point, calculating a cost function at each growing step. This cost function is the error between the line segment and the original corresponding piece of the time series. For the computation of the error, we could use the *Euclidean distance,* for example.

In the case of the *Bottom-Up* algorithm, the time series is segmented into the smallest size possible line segments. So, if the time series contains *n* points, then it is originally segmented into line segments of size *n/2*. Then these segments are continuously merged. At each merging, a cost function is checked. If the cost does not exceed a certain limit, then the merging is completed, otherwise it is stopped. Finally, the *Top-Down* algorithm is the reverse of the bottom up, where we start with the coarsest line segments and split them in different ways, while we choose the splitting configuration that minimizes a cost function.

PLA schemes are discussed in detail in [Las05]. Also significant work on *PLA* has been done by Keogh [Keo99], [Keo97]. As mentioned in [Keo99], the *PLA* scheme results in data compression that is 3 orders of magnitude greater than that of the *DFT*. *PLA* is widely used in the medical and financial domains.

A comparison of time series segmentation schemes is presented in [Keo01b], where it is shown that the *Sliding Window* approach generally does not produce satisfactory results, the *Top-Down* approach produces reasonable results but does not scale well, while the *Bottom-Up* approach produces the best results and good scalability.

An application of *PLA* to *amnesic* representations of time series is described in [Pal04], which allows the online approximation of streaming time series, with the ability to let the user specify approximation quality according to time. Amnesic approximations more accurately answer queries about the recent past because of their higher importance in many applications, such as environmental monitoring and denial of service attacks. The authors propose an algorithm, *grAp-R*, which merges the segments that result in the least approximation error. The technique was tested successfully on 40 real and synthetic data sets.

2.3.3 Model-Based Representation Methods

2.3.3.1 Markov Models for Representation and Analysis of Time Series

Markov models represent a time series in a stochastic way. *Hidden Markov models (HMM)* are *Markov models* whose parameters are unknown. A detailed description of *HMMs* can be found in [Rab89]. A first order *HMM* is completely described with the following parameters:

- Number of states

- The state transition probability distribution, that is, the probability that the system will go from one state to the next

- The density of the observations

- The initial state probability distribution

In order for a Markov modeled system to be fully modeled probabilistically, we need to specify the current state and previous states. A special case is a Markov chain, where we need to know only the current and predecessor state. The problem of estimating the parameters of an *HMM*, given the observations, can be viewed as a *maximum likelihood estimation problem*. The problem, however, does not have a global solution. Iterative algorithms, such as the *Baum–Welch* algorithm, guarantee only convergence to a local maximum. The parameters that are estimated by this algorithm are the initial state probability, the state transition parameters, the mean, and the covariance of the Gaussian of each state.

In the time series literature, [Dim92] proposes an approach to model an auto-regressive time series using Markov models. In [Pap98], the state space matrix is considered the Cartesian product of smaller state spaces which yields a computationally efficient way for the estimation of the parameters of the Markov model.

2.3.4 Data Dictated Representation Methods

2.3.4.1 Clipping

In [Bag06], the benefits of time series clipping are explored. *Clipping* transforms a time series into a *series of bits*, based on whether each point is less or more than the average. This representation method satisfies the lower bounding property.

The authors show that the discriminatory power of clipped time series, besides being a very effective compression method, can be asymptotically equivalent to that of the raw data. The results show that clipping is a representation method that achieves consistently better results for query and clustering than the *DFT* and the *discrete wavelet transform*.

2.3.5 Comparison of Representation Schemes and Distance Measures

In [Din08], an extensive comparison of representation and distance measures of time series is performed. The conclusions of the article can be summarized as follows:

- The tightness of the lower bounding property has very little effect on the various data sets.

- In time series classification applications, *elastic distance measures,* such as the *DTW* and *LCSS,* produce similar results with the *Euclidean* distance when the data size is large. For small data sets, elastic measures can be significantly better than the Euclidean distance.

- Getting more data whenever possible is always a good idea, especially when the current accuracy is not satisfactory. If getting more data is not possible, then using other schemes might help.

- The *DTW* measure performs very well in comparison with *edit-based* techniques such as *LCSS, EDR,* and *ERP.*

- The constraint of the warping size window for elastic measures can reduce the computational cost without a tradeoff for accuracy.

2.3.6 Need for Time Series Data Mining Benchmarks

In [Keo02b], Keogh and Kasetty make a very interesting observation regarding temporal data mining research articles. Specifically, the authors conducted an exhaustive set of experiments on more than two dozen papers on temporal algorithms using a wide variety of data sets. Their conclusion was that there is a serious need in the research community to establish time series data mining benchmarks. A list of recommendations is provided below:

- A wide range of data sets should be used for the validation of algorithms (unless created for a specific set of data).

- Implementation bias should be avoided by careful design of the experiments.

- If possible, data and algorithms should be freely provided so that they can be duplicated by other researchers.

- New similarity measures should be compared to established, yet simple metrics, such as the *Euclidean distance* and *dynamic time warping.*

2.4 TIME SERIES SUMMARIZATION METHODS

Summarization uses *global characteristics* of a time series to represent it. Its advantage is significant dimensionality reduction. It is common that more than one summarization technique is used to represent a time series, thus leading to the creation of a *feature vector,* which consists of various global characteristics of the time series. In the next chapter, we discuss a clustering technique that uses global characteristics as the basis for time series clustering.

2.4.1 Statistics-Based Summarization

Assume we have a time series with N values $X = \{x_1, x_2,..., x_n\}$. Then the following statistical measures can be used to characterize it.

2.4.1.1 Mean

This statistic shows the average value of the time series values and is defined as

$$\mu = \frac{\sum_{i=1}^{N} x_i}{N} \tag{2.28}$$

The mean is also known as the first moment.

2.4.1.2 Median

This statistic shows the middle value in an ordered set of time series values if the number of values is odd. If the number of values is even, then the median is the average of the middle two in the ordered set.

2.4.1.3 Mode

This statistic shows the most frequent value in the time series.

2.4.1.4 Variance

This statistic shows the amount of variation of the time series values around the mean and it is defined as

$$\sigma^2 = \frac{\sum_{i=1}^{N} (x_i - \mu)^2}{N-1} \tag{2.29}$$

This statistic is also known as the second moment. Another frequently used measure is the standard deviation, which is defined as the square root of the variance.

2.4.2 Fractal Dimension–Based Summarization

Since the fractal dimension was introduced by Mandelbrot [Man82] in 1982, it has found several applications in diverse areas ranging from medicine to ecology. There is more than one way to compute the fractal dimension of a curve, with the most popular being the *box counting method*. The method we will discuss here utilizes the variance of the time series as a measure of self-similarity.

This method is preferable to the box-counting method for the computation of the fractal dimension of a time series because the box counting method utilizes a box as its basic shape, which is better suited for 2-D entities, such as images. The formula below [Led94] shows how to compute the variance of a time series, X, where $x(t_i)$ is the value of the time series at time t_i, $x(t_i + d)$ is a point on the time series with distance d from t_i, and $N(d)$ is the number of all points on the time series with distance d.

$$Var = \frac{\displaystyle\sum_{i=1}^{N(d)} (x(t_i) - x(t_i + d))^2}{2N(d)} \qquad (2.30)$$

The plot of the log of the variance versus the log of the distance gives us the *semi-variogram* of the time series. The fractal dimension, *FD*, then is estimated as

$$FD = \frac{4 - s}{2} \qquad (2.31)$$

where s is the slope of the semi-variogram.

2.4.3 Run-Length–Based Signature

A *run-length* is a sequence of consecutive values in a time series that have the same value. Figure 2.7 shows a time series with two run-lengths (for value 5 and value 7). Therefore, the run-length signature of this time series is (5, 2), (7, 2). This means that we have a run-length of size 2 at level 5 and we also have a run-length of size 2 for level 7.

In addition to the run-length signature, additional measures can be extracted from the *run-length matrix*. In this matrix, element $r(i, j)$ indicates that gray level i has a run-length of length j. The measures that can be defined on this matrix are described below [Upp02].

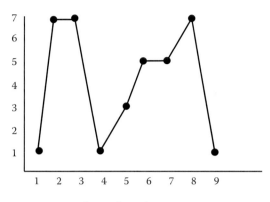

FIGURE 2.7   A time series with run lengths.

2.4.3.1 Short Run-Length Emphasis

This metric places emphasis on short run-lengths and it is defined below where NL is the number of values (levels) in the time series and NR is the number of run-lengths:

$$SRE = \frac{\displaystyle\sum_{i=1}^{NL}\sum_{j=1}^{NR}\frac{r(i,j)}{j^2}}{\displaystyle\sum_{i=1}^{NL}\sum_{j=1}^{NR}r(i,j)} \tag{2.32}$$

where $r(i, j)$ is the run-length for value i with length j [Upp02].
High values of this metric indicate the presence of short run-lengths.

2.4.3.2 Long Run-Length Emphasis

This metric places emphasis on long run-lengths and it is defined as

$$LRE = \frac{\displaystyle\sum_{i=1}^{NL}\sum_{j=1}^{NR}j^2 r(i,j)}{\displaystyle\sum_{i=1}^{NL}\sum_{j=1}^{NR}r(i,j)} \tag{2.33}$$

High values of this metric indicate the presence of long run-lengths.

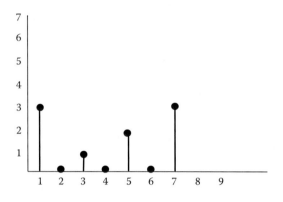

FIGURE 2.8   Histogram of the time series shown in Figure 2.7.

2.4.4 Histogram-Based Signature and Statistical Measures

A *histogram* is a graph that depicts the frequency of the different values in a time series. An example of using histogram data to summarize a time series is described in [Che04a]. The *x* axis of the graph depicts all the possible values in a time series. The frequency of each value $X(t_i)$ in the time series is depicted using a vertical bar, crossing the *x* axis at $X(t_i)$ and having a height that corresponds to the number that this value appears in the time series. An example is shown in Figure 2.8, which depicts the histogram of the time series shown in Figure 2.7. Therefore, the histogram-based signature of the time series is (1, 3), (2, 0), (3, 1), (4, 0), (5, 2), (6, 0), (7, 3).

In addition to the histogram-based signature, a time series can be characterized by the following statistical measures defined on the histogram, where we assume a histogram with *M* values, *μ* as the mean, and *f(m)* as the frequency of the *m*th level:

Skewness: Skewness measures the asymmetry of the histogram's shape and it is defined as

$$sk = \frac{\sum\limits_{m=1}^{M}(f(m)-\mu)^3}{M\sigma^3} \tag{2.34}$$

Kurtosis: Kurtosis measures the peakedness of the histogram and it is defined as

$$kur = \frac{\sum\limits_{m=1}^{M}(f(m)-\mu)^4}{M\sigma^4} \qquad (2.35)$$

Entropy: The entropy measures the amount of "information" or randomness in the histogram. It is defined as

$$Entropy = -\sum\limits_{i=1}^{M} \Pr(m)\log(\Pr(m)) \qquad (2.36)$$

where $\Pr(m)$ is the probability of level m.

Lowest-fifth percentile: This metric is equal to the time series value below which the 5% of the histogram values lie.

Upper-fifth percentile: This metric is equal to the time series value below which the 95% of the histogram values lie.

In addition to the above measures, a graphical representation known as the *box plot* is quite often used to represent *quartile* information along with statistical information regarding the mean and the median of the time series. A box plot is shown in Figure 2.9.

2.4.5 Local Trend-Based Summarization

Here, a variation of the moving average smoothing scheme is used, which utilizes a *moving* window *within which* a linear regression of the time series is performed. The result of the linear regression is a line $f_i = a_i\,t + b$, which minimizes the least squares error between the line and the points

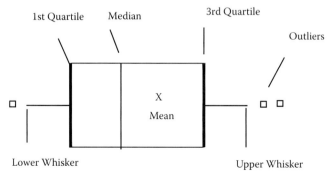

FIGURE 2.9   A box plot.

of the time series within the window. The slope of the best fit line is computed as

$$a_i = \frac{\sum (t_i - t_{mean})(f_i - f_{mean})}{\sum (t_i - t_{mean})^2} \qquad (2.37)$$

where t_{mean} and f_{mean} are the means of the time values and time series values, respectively, within the window. Then the *local-trend-based association* similarity of two time series $X = \{x_1, x_2, \ldots, x_n\}$ and $Y = \{y_1, y_2, \ldots, y_n\}$ can be computed as

$$ls = \frac{\displaystyle\sum_{i=1}^{m} a_{yi} a_{xi}}{\sqrt{\displaystyle\sum_{i=1}^{m} a_{yi}^2 \sum_{j=1}^{m} a_{xi}^2}} \qquad (2.38)$$

where $m = n - k + 1$ and k is the size of the window.

Further information on using local trends to represent and compute the similarity between two time series can be found in [Bat05].

2.5 TEMPORAL EVENT REPRESENTATION

2.5.1 Event Representation Using Markov Models

In many areas of research, a question that often arises is, what is the probability that a specific event B will happen after an event A? For example, in earthquake research, one would like to determine the probability that there will be a major earthquake, after certain types of smaller quakes. In speech recognition, one would like to know, for example, the probability that the vowel *e* will appear after the letter *h*. *Markov* models offer a way to do that.

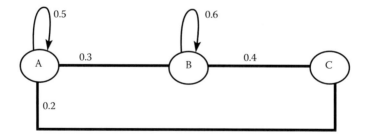

FIGURE 2.10   A Markov diagram.

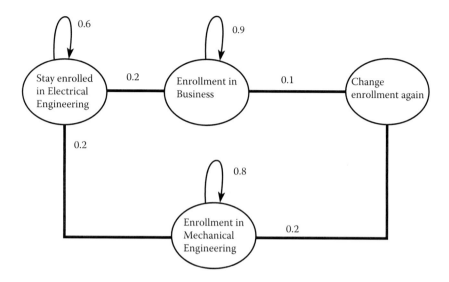

FIGURE 2.11 A Markov diagram that describes the probability of program enrollment changes.

Here we will use the fact that a *Markov* model can be represented using a graph that consists of vertices and arcs. The vertices show states, while the arcs show transitions between the states. Let us explain the *Markov* diagram shown in Figure 2.10. The numbers above the arcs are probabilities that the system will transition from one state to another. So for example, the probability of transitioning from state A to state B is 0.3. The probability of staying in state A is 0.5. Note that the sum of the probabilities of all arcs leaving a node is 1.

A more specific example is shown in Figure 2.11, which shows the probability of changing majors in college.

So we see that the probability of staying enrolled in electrical engineering is 0.6, while the probability of changing major from electrical engineering to business studies is 0.2 and the probability of switching from electrical engineering to mechanical engineering is also 0.2. As we can see, the sum of probabilities leaving the "Electrical eng. enrollment" is 1. Further information on utilizing *Markov* models in event modeling can be found in [Dun03].

2.5.2 A Formalism for Temporal Objects and Repetitions

As discussed in Chapter 1, Allen's temporal algebra is the most widely accepted way to characterize relationships between intervals. Some of these relationships are *before meets, starts, during, overlaps, finishes,* and *equals,* and they are respectively written as *.b., .m., .s., .d., .o., .f.,* and *.e.* Their inverse relationships can be expressed with a negative power of −1 (or with a letter *i* following the original symbol, as shown in Chapter 1). For example, *.e*⁻¹. denotes the inverse of the *equals* relationship, which is *not equal*.

Let us now consider the following example: How can we express in a concise way, using Allen's temporal algebra, the fact that "Following enrollment in the Circuits class, 10 students dropped out of the Electrical Engineering program?" The authors in [Cuk04] propose a formalism for the description of temporal relationships including repetition. Based on this formalism, the answer is

$$(\{dropped\ out\ .b^{-1}.\ enrollment\},\ 10) \tag{2.39}$$

In the same formalism, the expression *{dropped out .b⁻¹. enrollment}* is called a *temporal object*, while the expression *({dropped out .b⁻¹. enrollment}, 10)* is called a *temporal loop*, which is a repetition of a *temporal object*. In another example, let us see how we can express the following: "Lunch sometimes is served during the meeting and sometimes after the meeting." The answer is

$$(\{lunch\ .d.\ OR\ .b^{-1}.\ meeting\}) \tag{2.40}$$

2.6 SIMILARITY COMPUTATION OF SEMANTIC TEMPORAL OBJECTS

As discussed in Chapter 1, semantic entities are these that exist within the context of an ontology. Semantic similarity has meaning only within the context of the ontology. In this section, we will examine the semantic similarity of temporal data that belong to an ontology, which can be expressed in the form of an "is-a" taxonomy. Referring to the fossil example of the previous chapter, let us assume that we want to find the temporal similarity of a fossil that belongs to the *Neogene* period and another one that belongs to the *Paleogene* period. Both periods are members of the genealogic era ontology discussed in Chapter 1 and have an "is-a" relationship with *Cenozoic*.

A similarity measure that utilizes information theory can be used to define the similarities of semantic temporal entities as long as probabilities can be attached to the various temporal entities of an "is-a" taxonomy. Let us assume two temporal entities, t_1 and t_2, with respective probabilities $P(t_1)$ and $P(t_2)$. Let us also assume that $lca(t_1, t_2)$ represents their lowest common ancestor in the taxonomy. Then their semantic similarity can be defined as

$$ss = \frac{2\log P(lca(t_1, t_2))}{\log P(t_1) + \log P(t_2)} \tag{2.41}$$

In our fossil example, the similarity of the two fossils can be defined as

$$ss = \frac{2\log P(Cenozoic)}{\log P(Neogene) + \log P(Paleogene)} \tag{2.42}$$

2.7 TEMPORAL KNOWLEDGE REPRESENTATION IN CASE-BASED REASONING SYSTEMS

In case-based reasoning systems, a library of cases is used to classify a new case by finding the existing case that is most similar to the new case. A thorough explanation of case-based reasoning (*CBR*) can be found in [All94], while classification will be discussed in detail in the next chapter. The main idea in *CBR* systems is to operate in three phases when a new case comes in.

1. *Retrieve*. In this phase the most similar case is retrieved from the case library.

2. *Reuse*. Here, the old case is adapted within the context of the new case.

3. *Retain*. Here, a determination is made as to what to retain in the case library from the problem just solved.

Case-based reasoning is quite often combined with rule-based reasoning, which is based on a model of the problem's domain.

CBR systems typically consist of cases that are snapshots in time. An example of a *CBR* case library is a library of patient cases. Although similarity of symptoms and other characteristics, such as age and gender, can be useful in finding similarities between library cases and a new case, temporal characteristics of the old case, such as progress of disease, can be very useful in correctly classifying the new case.

The authors of [Jae02] address the problem of temporal knowledge representation in a case-based system. They apply their algorithm to *Creek* [Aam94], which is a system that combines case-based reasoning with general domain knowledge. They use Allen's temporal algebra, discussed in Chapter 1, to express temporal relationships between temporal intervals found in cases. Specifically, Allen's temporal algebra is used to create a temporal network to represent *case stories*. For example, if this is a library of patient cases, an example of a case story is the following: "Patient Jones had high fever for three hours, during which he received no medication. Then he received medication and when his temperature was measured an hour later, it was significantly lower."

In the classical *Creek* system, there are three metrics that are used to determine similarity: (1) *activation strength* based on the similarity of cases, (2) *explanation strength* based on general domain knowledge, and (3) *matching strength*, which is based on a combination of the former two to determine a similarity degree. The authors of [Jae02] added an additional temporally oriented measure of similarity, the *temporal path strength*.

2.8 ADDITIONAL BIBLIOGRAPHY

2.8.1 Similarity Measures

A thorough tutorial on time series similarity measures is presented in [Gun00]. In addition, two excellent tutorials are presented in [Vla05] and [KeoURL]. In [Keo99a], a method that exploits relevance feedback on the similarity of two time series is discussed. In this method, the user has the ability to rank the results of his original query from −3 to 3 and then the query is repeated. Because similarity can be subjective, the authors developed the idea of creating a user profile by learning the preferences (or sensitivity) of each user to distortions and fitting a beta distribution to the data. The idea of using a variant of the *LCSS* to measure similarity between word-based representations of time series is discussed in [Cub02].

In [Zuo07], a similarity computation measure, called the *General Hierarchical Model* is proposed that places the points of the time series in hierarchies. Points are compared only against points that belong in the same hierarchy. In [Jin02], the authors propose a technique to measure the similarity of time series based on partial information and not the entire time series. In [Tos05], the similarity of time series is computed using second order moments. An extensive discussion of semantic similarity can be found in [Mag06].

In [Gre06], an efficient and effective similarity measure is computed, which is called *DSA* (Derivative time series Segment Approximation). The authors describe the main steps of the technique as follows:

- Capture the significant trends of the time series by computing the first derivatives.

- Identify the segments that consist of tight derivative points.

- Obtain a lower dimensional, yet fine grained, representation of the original time series by segment approximation.

In experimental results, *DSA* was implemented as a representation and similarity computation scheme and compared against state of the art methods, such as *DTW, ERP, EDR,* and *LCSS. DSA* had a better time performance than the other techniques and overall performed as well as or better than major competing methods.

Dynamic time warping applications in data mining can be found in [Keo00a], [Ber94], and [Yi98]. In [Mor07], the authors propose *FTSE,* which is a fast method for the evaluation of the threshold values used in such techniques as *LCSS* and *EDR.* They show that using *FTSE,* these techniques can be applied an order of magnitude faster than using traditional dynamic programming (i.e., value-by-value comparisons).

Finally, in [Kah04], the authors address the problem of similarity search for arbitrary length queries and propose a novel indexing method that consists of a *multiresolution index structure (the MR index).* Experimental results show that the method is 3 to 15 times faster than other techniques, after the indexing structure has been compressed to reduce storage space. The authors also prove their method to be optimal for nearest neighbor and range queries. Specifically, their range query algorithm splits a query into nonoverlapping subqueries performed in different resolutions. The *k–nearest neighbor algorithm* works in two phases. The first phase searches the index structure to find an upper bound for the distance of the query to the kth nearest neighbor. The range query is performed in the second phase, using this upper bound as the query distance.

2.8.2 Dimensionality Reduction

In [Agr93], the authors explore the *DFT*'s utility as a means to perform efficient time series similarity searches and note its advantage of reducing the similarity search problem's dimensionality, in addition to preserving the Euclidean distance between time series objects. They save the *DFT* coefficients in an R^* tree structure, which is a tree structure for efficient indexing.

A typical dimensionality reduction technique, when using *DFT* or *DWT,* is to keep only the first few coefficients. An alternative technique was proposed by Wu et al. in [Wu00], where the coefficients with the largest energy are preserved. However, when dealing with several time series, this can be problematic, since different coefficients will have the highest energy for each time series and much storage will be needed. In [Mör03], Mörchen proposes a novel dimensionality reduction for time series where the same set of coefficients is retained for all time series. This results in considerable storage savings. The positions of the coefficients are chosen using a criterion function that is computed from all time series and is based on energy preservation. Experimental results on medical signals showed that an additional 48% of the energy was preserved with the new method. However, on data sets with significant low frequency presence, such as financial data, where the traditional method is already near optimum, only small improvements were noted.

In [Meg05] the authors propose *PVQA* (Piecewise Vector Quantization Approximation), a novel dimensionality reduction technique that uses vector quantization to separate a series into segments of equal length and then each codeword is represented by the closest codeword in the codebook.

2.8.3 Representation and Summarization Techniques

An article that surveys temporal data mining techniques, including representation is [Ant01]. An extensive discussion of the wavelet topic can be found in [Nie01]. In [Lu08], the authors propose a novel method for the computation of pair-wise time series distances. The method represents each time series with a smooth curve in a reproducing kernel *Hilbert space* and can be applied to time series with unequal lengths.

In [Gio03], the authors propose a representation scheme for multivariate time series where the time series is divided into k segments, where each one is affected by a number of different sources. An application of the variance-based fractal dimension in landscape characterization is discussed in [Led94].

In [Mör05a] and[Mör05b], Mörchen and Ultsch propose a novel time series discretization method that takes into account the temporal order of values, by optimizing the persistence of the resulting states. The method is called "Persist" and is based on the Kullback–Leibler measure between the marginal and self-transition probability distributions of the discretization symbols. This way a meaningful discretization is achieved that takes into account the internal structure of the time series.

The authors in [Bat07] examine the role of *Computing with Words and Perceptions* (CWP) models in time series data mining, with particular emphasis on economic and finance problems. They also discuss the integration of *CWP* models with expert knowledge to construct decision support systems (*DSS*). Specifically, it is noted that to be able to put *DSS* on top of time series databases, one needs to formalize human perceptions about time, time series values, and patterns. Because many of these perceptions have a degree of vagueness, such as "in the near future," "sometime last week," "increasing temperature," and so on, fuzzy sets can play an important role to model these perceptions.

2.8.4 Similarity and Query of Data Streams

Similarity computation in streaming data is a significantly more complex problem than similarity computation in time series of fixed length. A large number of applications today require the use of streaming time series, such as monitoring of stock data and computer network monitoring.

Such a method is discussed in [KonURL]. Each time series is expressed as a vector in multidimensional space, where each vector changes in time because new values are appended. The *DFT* then is used to perform dimensionality reduction. The computation of the *DFT* is done incrementally to avoid recomputation. The R^* tree access method is used to index the vectors corresponding to the time series. The Euclidean distance is used to compute the similarity between the time series. The authors implemented their algorithm on a database composed of 50,000 streams, generated via a random walk process.

In [Gop08], index trees are used to query time series streams. Specifically, the author addresses the problem of answering approximate k-NN queries. A novel mechanism is proposed for the maintenance of the tree which is a challenge because the underlying data change. The main idea of the algorithm is to use *pivots*, since quite often query results are already contained in older stream data stored in the database. The authors propose two versions of the algorithm, where one emphasizes accuracy of the query results and the other emphasizes the cost reduction of tree maintenance by deferring tree updates. Finally, [Lax05] describes a method that utilizes *Markov* models to detect frequent episodes in data streams.

REFERENCES

[Aam94] Aamodt, A., *Explanation-Driven Case-Based Reasoning: Topics in Case-Based Reasoning,* S. Wess et al., eds., Springer-Verlag, pp. 274–288, 1994.

[Aβf06] Aβfalg, J. et al., Similarity Search on Time Series Based on Threshold Queries, *Lectures Notes in Computer Science, Advances in Database Technology—EDBT,* Springer, pp. 276–294, 2006.

[Ant01] Antunes, C.M. and A. L. Oliveira, Temporal Data Mining: An Overview, *Proceedings of the Knowledge Discovery and Data Mining*, Workshop on Temporal Data Mining, pp. 1–13, 2001.

[Agr92] Agrawal, R., K.-I. Lin, H.S. Sawhney, and K. Shim, Fast Similarity Search in the Presence of Noise, Scaling, and Translation in Time-Series Databases, *Proceedings of Conference on Very Large Databases (VLDB)*, pp. 490–501, Zurich, Switzerland, September 1995.

[Agr93] Agrawal, R., C. Faloutsos, and A. Swami, Efficient Similarity Search in Sequence Databases, *Proceedings of the 4th International Conference on Foundations of Data Organization and Algorithms*, pp. 69–84, 1993.

[Agr95a] Agrawal, R. et al., Querying Shapes of Histories, *The 21st VLDB Conference*, Switzerland, pp. 502–514, 1995.

[Agr95b] Agrawal, R. et al., Fast Similarity Search in the Presence of Noise, Scaling, and Translation in Time-Series Databases, *Proceedings of the 21st International Conference on Very Large Databases*, pp. 490–501, 1995.

[Agr96] Agrawal, R. et al., The Quest Data Mining System, *Proceedings of the 2nd ACM International Conference on Knowledge Discovery and Data Mining*, pp. 244–249, 1996.

[Alb06] Albright, S.C., W. L. Winston, and C. Zappe, *Data Analysis & Decision Making*, Thomson Higher Education, 2006.

[All94] Allemang, D., Combining Case-Based Reasoning and Task-Specific Architectures, *IEEE Expert*, pp. 24–33, October 1994.

[Alt06] Altiparmak, F. et al., Information Mining over Heterogeneous and High Dimensional Time Series Data in Clinical Trials Databases, *IEEE Transactions on Information Technology in Biomedicine*, vol. 10, no. 2, pp. 254–263, April 2006.

[Ant01] Antunes, C.M. and A.L. Oliveira, Temporal Data Mining: An Overview, *Lecture Notes in Computer Science*, Workshop on Temporal Data Mining, 2001.

[Bag06] Bagnall, A. et al., A Bit Level Representation for Time Series Data Mining with Shape Based Similarity, *Data Mining and Knowledge Discovery*, vol. 13, pp. 11–40, 2006.

[Bat05] Batyrshin, I. et al., Association Networks in Time Series Data Mining, *Annual Meeting of the North American Fuzzy Information Processing Society*, pp. 754–759, 2005.

[Bat07] Batyrshin, I.Z, J. Kacprzyk, and L. Sheremetov, *Perception-Based Data Mining and Decision Making in Economics and Finance*, Springer, 2007.

[Ber94] Berndt, D. and J. Clifford, Using Dynamic Time Warping to Find Patterns in Time Series, *AAAI-94 Workshop on Knowledge Discovery in Databases*, Seattle, Washington, 1994.

[Bol97] Bollobas, B. et al., Time Series Similarity Problems and Well-Separated Geometric Sets, *Proceedings of the 13th Annual Symposium on Computational Geometry*, pp. 454–456, 1997.

[Cha99] Chan, K. and A.W. Fu, Efficient Time Series Matching by Wavelets, *Proceedings of the International Conference on Data Engineering*, pp. 126–133, 1999.

[Cha03] Chan, F.K.P. and A.W. Fu, Haar Wavelets for Efficient Similarity Search of Time Series, with and without Time Warping, *IEEE Transactions on Knowledge and Data Engineering,* vol. 15, no. 3, pp. 686–705, May/June 2003.

[Che04a] Chen, L. and M.T. Oszu, Multi-Scale Histograms for Answering Queries over Time Series Data, *Proceedings of the 20th IEEE International Conference on Data Engineering,* p. 838, 2004.

[Che04b] Chen, L. and T.R. Ng, On the Marriage of Lp-Norms and Edit Distance, *VLDB,* pp. 792–803, 2004.

[Che05] Chen, L., M. Ozsu, and V. Oria, Robust and Fast Similarity Search for Moving Object Trajectories, *Proceedings of the SIGMOD Conference,* Baltimore, Maryland, pp. 491–502, 2005.

[Che07] Chen, Q. et al., Indexable PLA for Efficient Similarity Search, *Proceedings of the 33rd International Conference on Very Large Databases,* Vienna, Austria, pp. 435–446, 2007.

[Chu02] Chu, S. et al., Iterative Deepening Dynamic Time Warping for Time Series, *2nd SIAM International Conference on Data Mining,* 2002.

[Cub02] Cuberos, F.J. et al., Qualitative Similarity Index, QR, 2002. Also in www.qrg.northwestern.edu/papers/Files/qr-workshops/QR02/Papers/QR2002%20-%20Cuberos.pdf

[Cuk04] Cukierman, D.R. and J.P. Delgrande, The SOL Time Theory: A Formalization of Structured Temporal Objects and Repetition, *Proceedings of the 11th International Symposium on Temporal Representation and Reasoning,* pp. 28–35, 2004.

[Das97] Das, G., D. Gunopulos, and H. Mannila, Finding Similar Time Series, *Proceedings of 1st European Symposium on Principles of Data Mining and Knowledge Discovery,* pp. 88–100, 1997.

[Den98] Deng, K., OMEGA: On Line Memory-Based General Purpose System Classifier, Ph.D. thesis, Carnegie Mellon University, 1998.

[Dim92] Dimitriadis, A., Modeling Time Series with Auto-Regressive Markov Models, M.A. thesis, Portland State University, 1992.

[Din08] Ding, H. et al., Querying and Mining of Time Series Data: Experimental Comparison of Representations and Distance Measures, *Proceedings of VLDB Conference,* pp. 1542–1552, August 23–28, Auckland, New Zealand, 2008.

[Dun03] Dunham, M., *Data Mining,* Pearson Education, NJ, 2003.

[Fal94] Faloutsos, C., M. Ranganathan, and Y. Manolopoulos, Fast Subsequence Matching in Time-Series Databases, *Proceedings of ACM SIGMOD Conference on Management of Data,* Minneapolis, Minnesota, 1994.

[Fre07] Frentzos, E., K. Gratsias, and Y. Theodoridis, Index-Based Most Similar Trajectory Search, *Proceedings of the International Conference on Data Engineering,* 2007.

[Fu08] Fu, A.W. et al., Scaling and Time Warping in Time Series Querying, *The VLDB Journal,* vol. 17, no. 4, pp. 899–921, 2008.

[Ge00] Ge, X. and P. Smyth, Deformable Markov Model Templates for Time Series Pattern Matching, *Proceedings of ACM SIGKDD 2000,* Boston, Massachusetts (USA), pp. 81–90, August 2000.

[Gio03] Gionis A. and H. Mannila, Finding Recurrent Sources in Sequences, *Proceedings of RECOMB 2003*, Berlin Germany, pp. 123–130, April 2003.

[Gop08] Gopalkrishnan, V., Querying Time-Series Streams, *Proceedings of the 11th International Conference on Extending Database Technology*, pp. 547–558, 2008.

[Gre06] Greco, S., M. Ruffolo, and A. Tagarelli, Effective and Efficient Similarity Search in Time Series, *Proceedings of CIKM*, Arlington, Virginia, pp. 808–809, 2006.

[Gun00] Gunopoulos, D. and G. Das, Time Series Similarity Measures, *Proceedings of Conference on Knowledge Discovery in Data*, pp. 243–307, 2000.

[Gur99] Guralnik, V. and J. Srivastava, Event Detection from Time Series Data, *Proceedings of the 5th ACM SIGKDD International Conference on Knowledge Discovery and Data Mining*, pp. 33–42, 1999.

[Han05] Han, J. and M. Kamber, *Data Mining: Concepts and Techniques*, Morgan-Kaufmann, 2005.

[Hua99] Huang, Y.W. and P.S. Yu, Adaptive Query Processing for Time-Series Data, *Proceedings of the 5th International Conference on Knowledge Discovery and Data Mining*, pp. 282–286, 1999.

[Jae02] Jaere, M.D., A. Aamodt, and P. Skalle, Representing Temporal Knowledge for Case-Based Prediction, *Proceedings of the 6th European Conference on Advances in Case-Based Reasoning, Lecture Notes in Computer Science*, vol. 2416, pp. 174–188, 2002.

[Jin02] Jin, X., Y. Lu, and C. Shi, Similarity Measure Based on Partial Information of Time Series, *Proceedings of SIGKDD*, pp. 544–549, 2002.

[Jon97] Jönsson, H.A. and D.Z. Badal, Using Signature Files for Querying Time Series Data, *Proceedings of the 1st European Symposium on Principles of Data Mining and Knowledge Discovery*, pp. 211–220, 1997.

[Kah04] Kahveci, T. and A.K. Singh, Optimizing Similarity Search for Arbitrary Length Time Series Queries, *IEEE Transactions on Knowledge and Data Engineering*, vol. 16, no. 4, pp. 418–433, 2004.

[Kar07] Karamitopoulos, L. and G. Evangelidis, Current Trends in Time Series Representation, http://era.teipir.gr/era1/b.3.information_management-session/full_papers/b.3.3.doc, 2007.

[Keo97] Keogh, E.J., A Fast and Robust Method for Pattern Matching in Time Series Databases, *Proceedings of the 9th Int. Conf. on Tools with Artificial Intelligence*, pp. 578–584, 1997.

[Keo98] Keogh, E.J. and M.J. Pazzani, An Enhanced Representation of Time Series Which Allows Fast and Accurate Classification, Clustering, and Relevance Feedback, *Proceedings of the 4th ACM International Conference on Knowledge Discovery and Data Mining*, pp. 239–243, 1998.

[Keo00a] Keogh, E. and M.Pazzani, Scaling up Dynamic Time Warping for Data Mining Applications, In *6th ACM SIGKDD Int. Conference on Knowledge Discovery and Data Mining*, pp. 285–289, Boston, 2000.

[Keo00b] Keogh, E. et al., Dimensionality Reduction for Fast Similarity Search in Large Time Series Databases, *Knowledge and Information Systems*, vol. 3, no. 3, pp. 263–286, 2000.

[Keo01a] Keogh, E. and M.J. Pazzani, Derivative Dynamic Time Warping, *Proceedings of the SIAM Conference on Data Mining*, 2001.

[Keo01b] Keogh, E. et al., An Online Algorithm for Segmenting Time Series, *Proceedings of the IEEE International Conference on Data Mining*, pp. 289–296, 2001.

[Keo01c] Keogh, E. et al., Locally Adaptive Dimensionality Reduction for Indexing Large Time Series Databases, *Proceedings of ACM SIGMOD Conference on Management of Data*, pp. 151–162, 2001.

[Keo01d] Keogh E., and P. Smyth, A Probabilistic Approach to Fast Pattern Matching in Time Series Databases, *Proceedings of KDD 1997*, Newport Beach, California, pp. 24–30, August 1997.

[Keo02a] Keogh, E., Exact Indexing of Dynamic Time Warping, *VLDB*, pp. 406–417, 2002.

[Keo02b] Keogh, E. and S. Kasetty, On the Need for Time Series Data Mining Benchmarks: A Survey and Empirical Demonstration, *Proceedings of the 8th ACM SIGKDD International Conference on Knowledge Discovery and Data Mining*, pp. 102–111, July 23–26, 2002.

[Keo05] Keogh, E. and C.A. Ratanamahatana, Exact Indexing of Dynamic Time Warping, *Knowledge and Information Systems*, Springer, vol. 17, no. 3, pp. 358–386, 2005.

[Keo99a] Keogh, E. and M.J. Pazzani, Relevance Feedback Retrieval of Time Series Data, Proceedings of 22nd ACM Conference on R&D in Information Retrieval, pp. 183–190, 1999.

[Keo99] Keogh, E.J. and M. Pazzani, Scaling Up Dynamic Time Warping to Massive Datasets, *Proceedings of 3rd European Conf. on Principles of Data Mining and Knowledge Discovery*, pp. 1–11, 1999.

[KeoURL] Keogh, E.J., SAX and Shape Tutorial, http://www.cs.ucr.edu/~eamonn/tutorials.html

[KonURL] Kontaki, M. and A.N. Papadopoulos, Similarity Range Queries in Streaming Time Series, http://delab.csd.auth.gr/papers/PRIS04kpm.pdf

[Kor97] Korn, F., H.V. Jagadish, and C. Faloutsos, Efficiently Supporting Ad Hoc Queries in Large Datasets of Time Sequences, *Proceedings of the ACM International Conference on Management of Data*, pp. 289–300, 1997.

[Lam98] Lam, S.K. and M.H. Wong, A Fast Projection Algorithm for Sequence Data Mining, *Data and Knowledge Engineering*, vol. 28, no. 3, pp. 321–339, 1998.

[Las01] Last, M., Y. Klein, and A. Kandel, Knowledge Discovery in Time Series Databases, *IEEE Transactions on Systems, Man, and Cybernetics*, Part B, vol. 31, no. 1, pp. 160–169, 2001.

[Las05] Last, M., A. Kandel, and H. Bunke, eds, *Data Mining in Time Series Databases*, World Scientific Press, 2005.

[Lax05] Laxman, S., P.S. Sastry, and K.P. Unnikrishnan, Discovering Frequent Episodes and Learning Hidden Markov Models: A Formal Connection, *IEEE Transactions on Knowledge and Data Engineering*, vol. 17, no. 11, pp. 1505–1517, 2005.

[Led94] Leduc, A., Y.T. Prairie, and Y. Bergeron, Fractal Dimension Estimates of a Fragmented Landscape: Sources of Variability, *Landscape Ecology*, vol. 9, no. 4, pp. 279–286, 1994.

[Li98] Li, C.S., P.S. Yu, and V. Castelli, MALM: A Framework for Mining Sequence Database at Multiple Abstraction Levels, *Proceedings of the 7th International Conference on Information and Knowledge Management*, pp. 267–272, 1998.

[Li04] Li C., P. Zhai, and S.Q. Zheng, Prabhakaran B., Segmentation and Recognition of Multi-Attribute Motion Sequences, *Proceedings of ACM Multimedia 2004*, New York, pp. 836–843, October 2004.

[Lin03] Lin, J. et al., A Symbolic Representation of Time Series with Implications for Streaming Algorithms, *DMKD*, pp. 2–11, San Diego, CA, 2003.

[Lu08] Lu, Z. et al., A Reproducing Kernel Hilbert Space Framework for Pairwise Time Series Distances, *Proceedings of the 25th International Conference on Machine Learning*, pp. 624–631, 2008.

[Mag06] Maguitman, A.G., et al., Algorithmic Computation and Approximation of Semantic Similarity, *World Wide Web*, vol. 9, no. 4, publisher: Kluwer Academic Publishers, December 2006.

[Man03] Manmatha, R. and T.M. Rath, Indexing Handwritten Historical Documents—Recent Progress, *Proceedings of the Symposium on Document Image Understanding*, 2003.

[Man82] Mandelbrot, B., The Fractal Geometry of Nature, W. H. Freeman, San Francisco, 1982.

[Meg05] Megalooikonomou V., G. Li, and Q. Wang, A Dimensionality Reduction Technique for Efficient Similarity Analysis of Time Series Databases, *Proceedings of CIKM 2004*, Trondheim, Norway, pp. 160–161, August–September 2005.

[Mör03] Mörchen, F., Time Series Feature Extraction for Data Mining Using DWT and DFT, Technical Report No. 33, Dept. of Mathematics and Computer Science, University of Marburg, Germany, 2003.

[Mör05a] Mörchen, F., and A. Ultsch, Finding Persisting States for Knowledge Discovery in Time Series, In *From Data and Information Analysis to Knowledge Engineering—Proceedings 29th Annual Conference of the German Classification Society (GfKl 2005)*, Magdeburg, Germany, Springer, Heidelberg, pp. 278–285, 2005.

[Mör05b] Mörchen F., and A. Ultsch, Optimizing Time Series Discretization for Knowledge Discovery, *Proceedings of ACM SIGKDD 2005*, Chicago, Illinois, pp. 660–665, August 2005.

[Mor07] Morse, M. and J. M. Patel, An Efficient and Accurate Method for Evaluating Time Series Similarity, *Proceedings of the ACM SIGMOD Conference on Management of Data*, pp. 569–580, 2007.

[Nie01] Nieveregelt, Y., *Wavelets Made Easy*, 2nd edition, Birkhäuser, 2001.

[Pal04] Palpanas, T. et al., Online Amnesic Approximation of Streaming Time Series, *ICDE*, Boston, MA, pp. 338–349, 2004.

[Pap98] Papageorgiou, C.P., Mixed Memory Markov Models for Time Series Analysis, *CIFER*, pp. 165–170, 1998.

[Par99] Park, S., D. Lee, and W.W. Chu, Fast Retrieval of Similar Subsequences in Long Sequence Databases, *3rd IEEE Conference on Knowledge and Data Engineering Exchange Workshop*, pp. 60–67, 1999.

[Per00] Perng, C.S., et al., Landmarks: A New Model for Similarity-Based Pattern Querying in Time Series Databases, *Proceedings of the 16th International Conference on Data Engineering*, pp. 33–42, 2000.

[Pop02] Popivanov, I. and R. J. Miller, Similarity Search over Time-Series Data Using Wavelets, *Proceedings of International Conference on Data Engineering*, pp. 212–221, 2002.

[Pra02] Pratt, K.B. and E. Fink, Search for Patterns in Compressed Time Series, *International Journal of Image and Graphics*, vol. 2, no. 1, pp. 89–106, 2002.

[Qu98] Qu, Y., C. Wang, and X. S. Wang, Supporting Fast Search in Time Series for Movement Patterns in Multiple Scales, *Proceedings of the 7th International Conference on Information and Knowledge Management*, pp. 251–258, 1998.

[Rab89] Rabiner, L., A Tutorial on Hidden Markov Models and Selected Applications in Speech Recognition, *Proceedings of the IEEE*, vol. 22, no. 2, pp. 257–285, 1989.

[Rat04] Ratanamahatana, C.A. and E. Keogh, Everything You Know about Dynamic Time Warping, *3rd Workshop on Mining Temporal and Sequential Data, in conjunction with the 10th ACM SIGKDD International Conference on Knowledge Discovery and Data Mining*, Seattle, WA, August 22–24, 2004.

[Rat05] Ratanamahatana, C.A. et al., A Novel Bit Level Time Series Representation with Implications for Similarity search and Clustering, PAKDD 05, 2005.

[Rod02] Roddick, J. and M. Spiliopoulou, A Survey of Temporal Knowledge Discovery Paradigms and Methods, *IEEE Transactions on Knowledge and Data Engineering*, vol. 14, no. 4, pp. 750–767, July/August 2002.

[Sak05] Sakurai, Y., M. Yoshikawa, and C. Faloutsos, FTW: Fast Similarity Search under the Time Warping Distance, *Proceedings of the 24th ACM SIGMOD-SIGACT-SIGART Symposium on Principles of Database Systems*, pp. 326–337, 2005.

[Sak05] Sakurai, Y., M. Yoshikawa, and C. Faloutsos, FTW: Fast Similarity Search under the Time Warping Distance, *Proceedings of PODS*, Baltimore, Maryland, 2005.

[Shi08] Shieh, J. and E. Keogh, iSAX: Indexing and Mining Terabyte Sized Time Series, *Proceedings of the 14th ACM SIGKDD International Conference on Knowledge Discovery and Data Mining*, pp. 623–631, 2008.

[Tos05] Toshniwal, D. and R.C. Joshi, Finding Similarity in Time Series Data by Method of Time Weighted Methods, *Australian Database Conference*, pp. 155–164, 2005.

[Upp02] Uppaluri, R., et al., Method and Apparatus for Analyzing CT Images to Determine the Presence of Pulmonary Tissue Pathology, U.S. patent #US6466687, Issue Date: 10/15/2002.

[Vla03] Vlachos, M., et al., Indexing Multi-Dimensional Time Series with Support for Multiple Distance Measures, *Proceedings of the 9th ACM International Conference on Knowledge Discovery and Data Mining*, pp. 216–225, 2003.

[Vla05] Vlachos, M., A Practical Time Series Tutorial with MATLAB, http://www.cs.ucr.edu/~mvlachos/publications.html 9th European Conference of Practices in knowledge and Data Discovery, Portugal, 2005.

[Wu00] Wu, Y., D. Agrawal, and A.E. Abbadi, A Comparison of DFT and DWT Based Similarity Search in Time Series Databases, *Proceedings of the 9th Int. Conference on Information and Knowledge Management*, pp. 488–495, 2000.

[Yan04] Yang K., and C. Shahabi, A PCA-Based Similarity Measure for Multivariate Time Series, *Proceedings of 2nd ACM International Conference on Multimedia Databases,* Arlington, Virginia, pp. 65–74, November 2004.

[Yi98] Yi, B.K., H.V. Jagadish, and C. Faloutsos, Efficient Retrieval of Similar Time Sequences under Time Warping, *Proceedings of the International Conference of Data Engineering (ICDE),* pp. 201–208, 1998.

[Zuo07] Zuo, X. and X. Zin, General Hierarchical Model to Measure Similarity of Time Series, *Proceedings of SIGMOD*, pp. 14–18, March 2007.

CHAPTER 3

Temporal Data Classification and Clustering

Classification and clustering are two major machine learning tools and two of the most important areas of data mining, because they have many practical applications such as medical diagnosis, fraud detection, and predicting financial trends. *Classification is the task of assigning a new sample to a set of previously known classes, while clustering is the task of grouping samples into clusters of similar samples.* This is the reason that classification is known as *supervised learning* and clustering is known as *unsupervised learning.*

The chapter is organized as follows: In Sections 3.1 and 3.2, we examine general classification and clustering techniques, that is, techniques that were developed for nontemporal data. The reason is twofold: first, to familiarize the reader with classification and clustering concepts. Second, after one or more of the *temporal* representation schemes that we learned in the previous chapter are applied to the time series data, the output representation can be directly used as input to a *nontemporal* classification or clustering technique. For example, a time series can be represented using a feature vector that consists of the *Fourier transform coefficients* of a time series and its *mean, skewness,* and *kurtosis.* It is this *feature vector* that can be the input to a nontemporal classification or clustering algorithm. An example of using a feature vector to characterize a time series and then feeding the feature vector as input to a clustering algorithm is described in [Wan05].

In Section 3.3, we present outlier analysis techniques and measures of cluster validity. In Section 3.4., classification and clustering techniques developed specifically for time series data are described. In Section 3.5, additional bibliography on general classification and clustering techniques as well as time series classification and clustering techniques is provided.

3.1 CLASSIFICATION TECHNIQUES

In classification, we assume we have some domain knowledge about the problem we are trying to solve. This domain knowledge can come from domain experts or sample data from the domain, which constitute the *training* data set. In this section, we will examine classification techniques that were developed for nontemporal data. However, they can be easily extended, as the examples will demonstrate, to temporal data.

3.1.1 Distance-Based Classifiers

As the name implies, the main idea in this family of classifiers is to classify a new sample to the nearest class. Each sample is represented by N features. There are different implementations of this type of classifier, based on (1) how each class is represented and (2) how the distance is computed from each class. There are two variations on how to represent each class. First, in the *K–Nearest Neighbors* approach, all samples of a class are used to represent the class. Second, in the *exemplar-based approach*, each class is represented by a representative sample, commonly referred to as the *exemplar* of this class. Regarding the distance of the unknown sample to the different classes, any of the distance measures defined in the previous chapter can be used. The most widely used metric to compute the distance of the unknown class sample to the existing classes is the Euclidean distance. However, for some classification techniques other distance measures are more suitable. An example is the *1–Nearest Neighbor classifier* (described below) for which the *Dynamic Time Warping* metric is the best [Eru07]. As an example, we assume we have a new sample x of unknown class with N features, where each ith feature is denoted as x_i.

Then the Euclidean distance from a class sample, whose the ith feature is denoted as y_i, is defined as

$$EU_{DIS} = \sqrt{\sum_{i=0}^{N-1}(x_i - y_i)^2} \qquad (3.1)$$

3.1.1.1 K–Nearest Neighbors

In this type of classifier, the domain knowledge of each class is represented by all its samples. The new sample X, whose class we do not know, is classified to the class that has the K nearest neighbors to it. K can be 1, 2, 3, and so on. Because all the training samples are stored for each class, this is a computationally expensive method. Let us look at a temporal example. We have a 2009 date, 6/22, and two classes of dates, "Sunny Day" and "Cloudy Day." We would like to know using a 1-nearest neighbor classifier, whether the date belongs to the "Sunny Day" or "Cloudy Day" class. The training data consist of randomly selected days in each of the 12 months and are shown in Table 3.1.

Let us now compute the Euclidean distance of our sample to each one of the two classes: First, the Euclidean distances to the sunny day samples are

$$EU_{DIStosunny} = \sqrt{diff\left(6/22 \text{ to } 4/23\right)^2} = 59 \tag{3.2}$$

$$EU_{DIStosunny} = \sqrt{diff\left(6/22 \text{ to } 6/12\right)^2} = 10 \tag{3.3}$$

$$EU_{DIStosunny} = \sqrt{diff\left(6/22 \text{ to } 8/6\right)^2} = 44 \tag{3.4}$$

$$EU_{DIStosunny} = \sqrt{diff\left(6/22 \text{ to } 9/12\right)^2} = 81 \tag{3.5}$$

TABLE 3.1   Training Data for Sunny–Cloudy Classes

Month	Day	Class
1	10	Cloudy
2	3	Cloudy
3	15	Cloudy
4	23	Sunny
5	18	Cloudy
6	12	Sunny
7	4	Cloudy
8	6	Sunny
9	12	Sunny
10	8	Cloudy
11	26	Cloudy
12	8	Sunny

$$EU_{DIStosunny} = \sqrt{diff(6/22 \ to \ 12/8)^2} = 167 \qquad (3.6)$$

Then the distance to the cloudy day class can be computed as

$$EU_{DIStocloudy} = \sqrt{diff(6/22 \ to \ 1/10)^2} = 162 \qquad (3.7)$$

$$EU_{DIStocloudy} = \sqrt{diff(6/22 \ to \ 2/3)^2} = 138 \qquad (3.8)$$

$$EU_{DIStocloudy} = \sqrt{diff(6/22 \ to \ 3/15)^2} = 98 \qquad (3.9)$$

$$EU_{DIStocloudy} = \sqrt{diff(6/22 \ to \ 5/18)^2} = 34 \qquad (3.10)$$

$$EU_{DIStocloudy} = \sqrt{diff(6/22 \ to \ 7/4)^2} = 11 \qquad (3.11)$$

$$EU_{DIStocloudy} = \sqrt{diff(6/22 \ to \ 10/8)^2} = 107 \qquad (3.12)$$

$$EU_{DIStocloudy} = \sqrt{diff(6/22 \ to \ 11/26)^2} = 155 \qquad (3.13)$$

We see that the sample closest to the incoming sample belongs to the sunny class. Therefore 6/22 is classified to the sunny class. This can also be seen in Figure 3.1, which depicts the sampled days of each month in terms of days from the beginning of the year.

The S above some of the samples indicates that this sample belongs to the sunny class. The X symbol on the graph (value 173) corresponds to June 22, the day we want to classify. If we draw a straight line from X, we will see that the sample that is closest is the one from June that belongs to the sunny class.

A variation of the *K-NN* technique is to not treat the different samples within a class equally, but instead assign a *weight* to each sample, based on either its importance or its distance from the unknown sample. For example, a weight that could be used is 1/(*Eudis* + 1) where *Eudis* is the distance from the sample to the sample of unknown class.

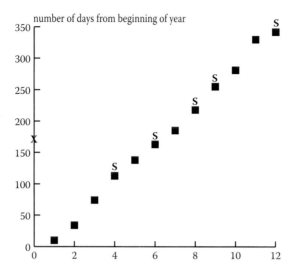

FIGURE 3.1   Diagram depicting the sunny and cloudy day classes.

Several important considerations about the nearest neighborhood algorithm are described in [Wu08]:

- The algorithm's performance is affected by the choice of K. If K is small, then the algorithm can be affected by noise points. If K is too large, then the nearest neighbors can belong to many different classes.

- The choice of the distance measure can also affect the performance of the algorithm. Some distance measures are influenced by the dimensionality of the data. For example, the *Euclidean distance*'s classifying power is *reduced* as the number of attributes increases.

- The error of the *K-NN* algorithm asymptotically approaches that of the *Bayes error*.

- *K-NN* is particularly applicable to classification problems with multimodal classes. As an example, [Wu08] mentions that for gene expression profiles *K-NN* outperformed *Support Vector Machines* (which are described below).

Additional work has been done to improve the performance of the *K-NN*. An example is the work by Hart [Har68], which removes some training objects without reducing the classification accuracy of *K-NN*. This process is known as *condensing* [Wu08]. Improvement of the *K-NN* is also presented in [Tou02], using *proximity graphs*.

3.1.1.2 Exemplar-Based Nearest Neighbor

In this type of classifier, each class is represented by a representative sample known as the *exemplar*. Typically, the features of the exemplar are computed as the mean values of the features of the training samples in each class. So, given a class Y with three features, its exemplar will be given by

$$Y_{ex} = \{\overline{y_1}, \overline{y_2}, \overline{y_3}\} \tag{3.14}$$

where the horizontal bar indicates the mean value of a feature.

Let us now examine the previous example with the sunny–cloudy day classes and apply the exemplar-based approach. The exemplar of the sunny class is *August 1* and the exemplar of the cloudy class is *May 24*. Our sample date of June 22 is closer to the exemplar of the cloudy class, and therefore, *it is now classified in the cloudy class.*

Figure 3.2 shows two classes, the samples inside them, and a sample of unknown class. In this case, the *1-NN* and the exemplar algorithms will classify correctly the sample to class 2. However, Figure 3.3 shows a case where the two algorithms will yield different results. Specifically, the exemplar algorithm will assign the unknown sample to class 2, but the *1-NN* algorithm will classify the unknown sample to class 1.

3.1.2 Bayes Classifier

The *Bayes* classifier is a statistical classifier, because classification is based on the computation of probabilities, and domain knowledge is also expressed in a probabilistic way. In this book, we will discuss the most widely used type of Bayes classifier, known as the *naïve Bayes classifier,* where we assume that the different features vary independently of each other; that is, they are "conditionally independent." Bayesian classification is discussed in [Dud73]. Violation of the feature independence in the naïve Bayes classifier and its effects on classifier accuracy are discussed in [Dom96]. In this section, we will discuss the naïve Bayes classifier in detail, because of its simplicity and effectiveness which make it very popular in such areas as text classification and spam filtering [Wu08].

Before we examine how classification using this type of classifier is performed, let us examine some basic concepts:

- *A posteriori probability:* Let us assume that we have a domain with two classes, C and D. The probability that a sample S of unknown class will belong to class C is known as *a posteriori probability* and is expressed as $P(C|S)$. As it can be easily inferred by the reader, the

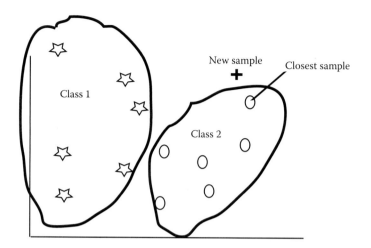

FIGURE 3.2 A case where *K-NN* and exemplar-based clustering yield the same results.

a posteriori probability is key to deciding which class to assign the new sample to. So how is this probability computed? It is done using *Bayes' theorem*, which is discussed below.

- Bayes' theorem:

$$P(C|S) = \frac{P(S|C)P(C)}{P(S)} \tag{3.15}$$

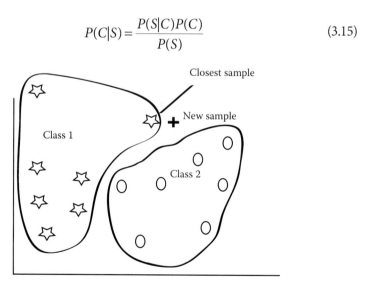

FIGURE 3.3 A case where *K-NN* and exemplar-based clustering yield different results.

where $P(S|C)$ is the probability that class C will contain sample S, $P(C)$ is the probability that a sample will belong to class C, and $P(S)$ is the prior probability of S.

Now let us assume that we are given N classes C_1, C_2,..., C_N. A new sample S with K features and values $s_1,...s_k$ is assigned to the ith class if

$$P(C_i|S) > P(C_j|S) \text{ for } j=1..., N \text{ and } i \neq j \qquad (3.16)$$

In other words, the new sample S is assigned to the class with *maximum a posteriori probability*. The *a posteriori probability* is computed using Bayes' theorem as shown above. Because the denominator, $P(S)$, is the same for all posteriori probabilities, it can be factored out, so the new sample S is assigned to the class i for which:

$$P(S|C_i)P(C_i) > P(S|C_j)P(C_j) \text{ for } j=1...,N \text{ and } i \neq j \qquad (3.17)$$

The probabilities of each class, $P(C_i)$, are computed from the training samples as

$$P(C_i) = \frac{N(C_i)}{N_{total}} \qquad (3.18)$$

where N_{total} is the total number of training samples and $N(C_i)$ is the number of training samples that belong to class C_i. The probability $P(S|C_i)$ is computed as

$$P(S|C_i) = \prod_{m=1}^{N} P(S_m|C_i) \qquad (3.19)$$

where $P(S_m|C_i)$ is the probability that for given a sample from class C, its mth feature has the value S_m. The assumption here is that the features are independent, which is the premise of the naïve Bayes classifier.

The computation of the probabilities $P(S_m|C_i)$ depends on whether the feature is a categorical or continuous variable. If the feature is a categorical variable, then

$$P(S_m|C_i) = \frac{N(S_{mi})}{N_{itotal}} \qquad (3.20)$$

where N_{itotal} is the total number of training samples that belong to class C_i, and $N(S_{mi})$ is the number of training samples that belong to class C_i and have the value S_m for the mth feature. If the mth feature is a continuous variable and we know its probability density function, we can use this function to compute $P(S_m|C_i)$. If the probability density function of the feature is not known, which is quite often the case, we typically assume that it follows a *Gaussian* distribution and therefore;

$$(S_m|C_i) = \frac{1}{\sqrt{2\pi}\sigma} e^{\frac{-(S_m-\mu)^2}{2\sigma^2}} \tag{3.21}$$

where μ is the mean of the distribution of S_m within class C_i and σ is the standard deviation of the distribution. Note that in the above equation S_m is a variable that represents the distribution of values of the mth feature inside class C_i and P is the distribution of S_m.

Let us now look at an example. Let us assume that a doctor, Dr. Clark, wants to decide whether to enroll a heart patient, Mr. Aberdeen, in a clinical trial for a fictitious drug called *Heartex* which reduces the amount of cardiac calcification dramatically. However, the clinical trial so far has shown that *Heartex* bears the risk of a *transient ischemic attack (TIA)*, which is a medical condition with stroke-like symptoms. The patient has to be on the drug for 4 months in order for it to be effective. The clinical trials showed that some of the patients had a *TIA* during the time they were using the drug. Some others suffered a *TIA* within 3 months after stopping the usage of the drug. After that, there was negligible correlation between the occurrence of a stroke and the usage of Heartex. The risk factors for developing a stroke due to *Heartex* were the *presence of hypertension, age, and presence of diabetes mellitus.*

Problem definition
Given the data shown in Table 3.2, what is the likelihood that Mr. Aberdeen will not develop a TIA from the usage of Heartex, given that he is 72, has no diabetes and no hypertension?

Let us now examine the problem from a classification point of view. There are 3 classes that are differentiated temporally, using the temporal relations *during* and *after.*

Class 1: Patients develop a stroke *during* the usage of *Heartex.*

TABLE 3.2   Training Data for Heartex

Age	Hypertension	Diabetes Mellitus	Class
$<= 70$	Yes	Yes	1
$70 < a < 80$	No	Yes	3
$>= 80$	Yes	Yes	2
$<= 70$	Yes	No	3
$70 < a < 80$	Yes	No	3
$>= 80$	No	Yes	1
$70 < a < 80$	No	Yes	2
$70 < a < 80$	No	No	3
$>= 80$	No	Yes	1
$>= 80$	No	No	3
$70 < a < 80$	Yes	No	3
$<= 70$	No	Yes	3

Class 2. Patients develop a stroke within 3 months *after* the usage of Heartex.

Class 3. Patients do not develop a stroke due to *Heartex*.

There are three categorical features:

1. Age is a categorical feature with three possible categories:

    a. age $<= 70$. We will call this category A.

    b. $70 < $ age $ < 80$. We will call this category B.

    c. age $>= 80$. We will call this category C.

2. Hypertension is a categorical variable with value Yes (category D0) and value No (category D).

3. Presence of diabetes mellitus is a categorical variable with value Yes (category E) and No (category F).

Based on this class information, to find Mr. Aberdeen's class, regarding the development of a *TIA*, we need to maximize $P(BDF|Class) P(Class)$.

Let us now see if this class is class 3. The probability of class 3 can be computed from the table as

$$P(Class\ 3) = \frac{7}{12} \tag{3.22}$$

and

$$P(BDF|Class\ 3) = P(B|Class\ 3)\ P(D|Class\ 3)\ P(F|Class\ 3) \tag{3.23}$$

assuming that the three features age, diabetes, and hypertension are independent of each other.

$$P(B|Class\ 3) = \frac{4}{7} \tag{3.24}$$

$$P(D|Class\ 3) = \frac{4}{7} \tag{3.25}$$

$$P(F|Class\ 3) = \frac{5}{7} \tag{3.26}$$

Therefore,

$$P(BDF|Class\ 3)\ P(Class\ 3) = P(B|Class\ 3)\ P(D|Class\ 3)$$
$$P(F|Class\ 3)\ P(Class\ 3) = 0.136 \tag{3.27}$$

For class 2,

$$P(BDF|Class\ 2)\ P(Class\ 2) = 0 \tag{3.28}$$

because

$$P(F|Class\ 2) = 0 \tag{3.29}$$

And for class 1,

$$P(BDF\ |Class\ 1)\ P(Class\ 1) = 0 \tag{3.30}$$

because

$$P(B|Class\ 1) = 0 \tag{3.31}$$

Therefore, because $P(Class\ 3|BDF)$ is the highest probability, Mr. Aberdeen is most likely not to suffer a minor stroke because of *Heartex*. However, it is important to stress here that this conclusion is as valid as permitted from the training set. The training set here is very small, and as the training set grows it is possible that the conclusion will be different. Also note that although *P(Class 3|BDF)* is the highest, it is still too small a number (0.136) to inspire real confidence that Mr. Aberdeen will not suffer a *TIA*. Again, this is a result of the small training set.

3.1.3 Decision Trees

Decision trees are widely used in classification because they are easy to construct and use. The first step in decision tree classification is to build the tree. A popular tree construction algorithm is *ID3*, which uses information theory as its premise and was first proposed in [Qui86]. An improvement to the *ID3* algorithm that can handle noncategorical data is *C4.5* and it is discussed in [Qui93]. A critical step in the construction of the decision tree is how to order the splitting of the features in the tree. In the *ID3* algorithm, the guide for the ordering of the features is *entropy*. Given a data set, where each value has probabilities $p_1, p_2,..., p_n$, its *entropy* is defined as

$$H(p_1, p_2,..., p_n) = \sum_{i=1}^{N} p_i \log_2\left(\frac{1}{p_i}\right) \tag{3.32}$$

Entropy shows the amount of randomness in a data set and varies from 0 to 1. If there is no amount of uncertainty in the data, then the entropy is 0. For example, this can happen if one value has probability 1 and the others have probability 0. If all values in the dataset are equally probable, then the amount of randomness in the data is maximized, and entropy becomes 1. In this case, the amount of *information* in the data set is maximized. Here is the main idea of the *ID3* algorithm: *A feature is chosen as the next level of the tree if its splitting produces the most information gain.*

Let us look at the *Heartex* clinical trial classification problem using the ID3 approach. The entropy of the starting set is

$$\frac{3}{12}\log\left(\frac{12}{3}\right) + \frac{2}{12}\log\left(\frac{12}{2}\right) + \frac{7}{12}\log\left(\frac{12}{7}\right) = 1.38 \tag{3.33}$$

If we choose age as the splitting attribute, then the entropy of the subset that is $<= 70$ is

$$\frac{1}{3}\times\log\left(\frac{3}{1}\right) + \frac{2}{3}\times\log\left(\frac{3}{2}\right) = 0.906 \tag{3.34}$$

The entropy of the subset that is $70 < a < 80$ is

$$\frac{4}{5}\log\left(\frac{5}{4}\right) + \frac{1}{5}\log(5) = 0.72 \qquad (3.35)$$

The entropy of the subset that is $> = 80$ is

$$\frac{1}{4} \times \log\left(\frac{4}{1}\right) + \frac{1}{4} \times \log\left(\frac{4}{1}\right) + \frac{2}{4} \times \log\left(\frac{4}{2}\right) = 1.5 \qquad (3.36)$$

The weighted sum of these entropies is

$$\frac{5}{12} \times 0.72 + \frac{3}{12} \times 0.906 + \frac{4}{12} \times 1.5 = 1.02 \qquad (3.37)$$

The gain in entropy from using the age as a splitting attribute is

$$1.38 - 1.02 = 0.36 \qquad (3.38)$$

If we choose hypertension as the splitting attribute, then the entropy of the subset that has no hypertension is

$$\frac{2}{7}\log\left(\frac{7}{2}\right) + \frac{1}{7}\log\left(\frac{7}{1}\right) + \frac{4}{7}\log\left(\frac{7}{4}\right) = 1.36 \qquad (3.39)$$

The entropy of the subset that has hypertension is

$$\frac{1}{5}\log\left(\frac{5}{1}\right) + \frac{1}{5}\log\left(\frac{5}{1}\right) + \frac{3}{5}\log\left(\frac{5}{3}\right) = 1.36 \qquad (3.40)$$

The weighted sum of these entropies is

$$\frac{7}{12}1.36 + \frac{5}{12}1.36 = 1.36 \qquad (3.41)$$

The gain in entropy by using the *hypertension* as a splitting attribute is

$$1.38 - 1.36 = 0.02 \qquad (3.42)$$

If we choose the presence of diabetes as the splitting attribute, then the entropy of the subset that has no diabetes is 0, because all cases belong to class 3.

The entropy of the subset that has diabetes is

$$\frac{3}{7}\log\left(\frac{7}{3}\right) + \frac{2}{7}\log\left(\frac{7}{2}\right) + \frac{2}{7}\log\left(\frac{7}{2}\right) = 1.54 \qquad (3.43)$$

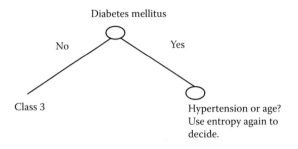

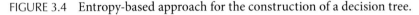

FIGURE 3.4 Entropy-based approach for the construction of a decision tree.

The weighted sum of these entropies is

$$\frac{7}{12} \times 1.54 + \frac{5}{12} \times 0 = 0.89 \tag{3.44}$$

The gain in entropy by using diabetes as a splitting attribute is

$$1.38 - 0.89 = 0.48 \tag{3.45}$$

Please note that in all above equations log actually denotes $\log_2$.

As we can see, *the gain in information is greatest when diabetes is chosen as a splitting attribute.* The resulting tree is shown in Figure 3.4. As we see, the next decision is whether to choose hypertension or age as the splitting attribute of the *yes diabetes mellitus branch.* Again, the same procedure as above can be applied.

Other popular decision tree induction algorithms are *CART* [Bre84] and *SPRINT* [Sha96], the latter of which can be applied to large amounts of data. *CART* is a binary recursive algorithm that produces a sequence of nested pruned trees, all of which are candidate optimal trees. The optimal tree is identified by computing the predictive power of each tree. The importance of each feature is calculated from the sum of improvements that occur in all nodes in which this feature appears as a splitting criterion. *CART* has very interesting features, such as the following:

- *Automatic missing value handling. CART* uses substitute splitting criteria at each node where there are missing values.

- *Enabling of cost-sensitive learning,* where there is a cost assigned to misclassified objects. Cost assignment influences both the tree growing and pruning phases.

- *Enabling of dynamic feature construction*, where new features can be created at a node [Wu08].

3.1.4 Support Vector Machines in Classification

Support Vector Machines (SVMs) constitute one of the most accurate and robust methods in machine learning [Wu08]. An excellent tutorial on SVMs is [Bur98]. *SVMs* are insensitive to the number of dimensions and only a number of examples are required for training. A disadvantage is computational complexity; however, approaches have been developed where the optimization process is significantly faster by breaking it down into smaller problems [Wu08]. In an N-dimensional space and linearly separable data, *Support Vector Machines* construct a separating hyperplane that maximizes the margin between two data sets that belong to two different classes. In a *2-D* plane, the separating hyperplane is defined as

$$<w, x> + a = 0 \qquad (3.46)$$

where *w* is normal to the separating hyperplane, and <*w, x*> is an inner product.

The bigger the distance of the separating hyperplane to the closest point of the two classes, the better the quality of the separation. Two parallel hyperplanes are constructed on each side of the separating hyperplane, which are then brought as close as possible to the two data sets to calculate and maximize the margin. As described in [Mar02], we can visualize the two parallel planes as the borders of a *fat plane* instead of a linear hyperplane.

Support Vectors are defined as the points closest to the separating hyperplane, and the weighted sum of the *Support Vectors* is the normal vector to the hyperplane [Mar03]. What makes the *Support Vector Machine* computationally tractable is that instead of computing a maximal margin hyperplane with inner products in the feature space, we can compute a kernel function in the original space, where the kernel can be computed as a polynomial for example. As discussed in [Ead05], *SVMs* have the reputation of avoiding over-fitting. Another interesting note is made in the same article. Although in classification, one of the well-known problems is the *curse of dimensionality* and many algorithms aim at reducing the dimensionality of the feature space, SVMs increase the dimensionality of the feature space. Their way of working efficiently in the high dimensionality space is margin maximization.

In [Fun02], an incremental *SVM* classifier is proposed that utilizes the proximity of each class point to one of the two parallel planes. The advantages of the proposed algorithm are as follows:

- Ability to add new data very easily

- Ability to compress large datasets with a high compression rate

- Applicability in handling massive data sets effectively

In [Aio05], a multiclass classifier that is able to support a number of prototypes per class is built using *SVM*. This is a nonconvex problem and a greedy optimization algorithm is used that is able to find local optimal solutions. The experimental results showed that using a few linear models per class instead of a single kernel per class results in higher computational efficiency.

3.1.5 Neural Networks in Classification

An *artificial neural network* (*ANN*) is a computational model that mimics the human brain in the sense that it consists of a set of connected nodes, similar to neurons. The nodes are connected with weighted arcs. The system is adaptive because it can change its structure based on information that flows through it. In addition to nodes and arcs, neural networks consist of an input layer, hidden layers, and an output layer. The simplest form of *ANN* is the *feed-forward ANN*, where information flows in one direction and the output of each neuron is calculated by summing the weighted signals from incoming neurons and passing the sum through *an activation function*. A commonly used activation function is the *sigmoid function*:

$$A(x) = \frac{1}{1 + e^{-x}} \tag{3.47}$$

A common training process for feed forward neural networks is the *back-propagation* process, where we go back in the layers and modify the weights. The weight of each neuron is adjusted such that its *error* is reduced, where a neuron's error is the difference between its expected and actual outputs. The most well-known feedforward *ANN* is the *perceptron*, which consists of only two layers (no hidden layers) and works as a binary classifier. When three or more layers exist in the *ANN* (at least one hidden layer), then the perceptron is known as the *multilayer perceptron*.

Another type of widely used feedforward *ANN* is the *radial-basis function ANN*, which consists of three layers and the activation function is a *radial-basis function (RBF)*. This type of function, as the name implies, has radial symmetry such as a Gaussian function and allows a neuron to respond to a local region of the feature space. In other words, the activation of a neuron depends on its distance from a center vector. In the training phase, the *RBF* centers are chosen to match the training samples.

Neural network classification is becoming very popular and one of its advantages is that it is resistant to noise. The input layer consists of the attributes used in the classification [Goh03], and the output nodes correspond to the classes. Regarding hidden nodes, too many nodes lead to over-fitting and too few can lead to reduced classification accuracy [Goh03]. Originally, each arc is assigned a random weight, which can then be modified in the learning process. Classification algorithms using neural network techniques can be found in [Bro88], [Rum86], [Hay99], and [Wit00]. A method for the time series classification of motion data using neural networks is described in [Sho06]. The data are preprocessed using a feature extraction algorithm that computes finite differences between sequences. Then two different neural networks are used for testing and training.

In [Keh97], a predictive modular neural network (*PREMONN*) with a hierarchical structure for time series classification is proposed. The bottom level consists of a bank of predictor modules and the top level is a decision module that processes the prediction errors at the bottom level. Convergence to correct classification with probability 1 has been proven mathematically and can be implemented in a parallel way. Finally, in [Neu97], a combination of *Hidden Markov Models* and *ANNs* is used for time series classification.

3.1.6 Classification Issues

3.1.6.1 Classification Error Types

There are four different types of outcomes to classification, two of which are errors:

1. *False positive.* This means that although object O does not belong to class C, it is classified as a member of class C.

2. *False negative.* Although object O belongs to class C, it is not classified as a member of class C.

TABLE 3.3   Confusion Matrix

	Actual Value: Positive	Actual Value: Negative
Prediction Outcome: Positive	True positive	False positive
Prediction Outcome: Negative	False negative	True negative

3. *True positive.* Object O belongs to class C and is classified as such.

4. *True negative.* Object O does not belong to class C and it is not classified as a member of class C.

The four outcomes are better depicted in a table (Table 3.3) known as the *confusion matrix*.

3.1.6.2 Classifier Success Measures
The following measures can be employed to judge the success of a classifier:

- *Sensitivity:* This measures the proportion of positives that are correctly classified as such. In database queries, this is known as *recall*.

$$\text{Sensitivity} = \frac{\text{Number of true Positives}}{\text{Number of true Positives} + \text{Number of false Negatives}} \quad (3.48)$$

- *Specificity.* This measures the proportion of negatives that are correctly classified as such.

$$\text{Specificity} = \frac{\text{Number of true Negatives}}{\text{Number of true Negatives} + \text{Number of false Positives}} \quad (3.49)$$

- *Accuracy.* This measures all correctly classified samples and is defined by

$$\text{Accuracy} = \frac{\text{Num. of true Pos.} + \text{Num. of true Neg.}}{\text{Num. of true Pos.} + \text{Num. of true Neg.} + \text{Num. of false Pos.} + \text{Num. of false Negs.}}$$

$$(3.50)$$

To increase the success of classification, we can apply more than one classification method. There are two ways in which we can combine results from different classification methods. In one method, we can classify a sample using an ensemble of classification techniques, Cl_{t1}, Cl_{t2}, and so on. Each classification technique classifies the sample to a class C_{Cl}. The class that is chosen the most among the different classifiers is chosen as the class to which the sample belongs. In the second method, the classifier is applied *sequentially*, where one classifier learns from the mistakes of the previous one by varying the weights of the samples based on the success of their classification. In the final step, the information about the classes chosen by each classifier is gathered in a weighted way, where the weight of each classifier method increases with its accuracy.

3.1.6.3 Generation of the Testing and Training Sets

A popular way to generate the training and testing sets is known as the *leave-one-out method* or the *jackknife procedure*. In this method, if we have N samples, testing and training is performed N times. Each time, a sample is left out, which becomes the testing set. All other samples constitute the training set.

3.1.6.4 Comparison of Classification Approaches

Decision trees are easy to understand, and once constructed, they can be easily applied to classify new samples. For this reason, they scale well to large databases. However, they usually apply only to categorical data. Comparing the *Bayes* and *nearest neighbor classifiers*, it is interesting to note that the probability of error of the *Bayes* classifier is always less than or equal to the nearest neighbor classifier. The reason is that the *Bayes* classifier is *optimal* in a probabilistic way, in the sense that it always chooses the most probable class. On the other hand, a basic assumption of the naïve *Bayes* classifier is that the features are independent, which sometimes might not be true and might lead to inaccurate results.

3.1.6.5 Feature Processing

Central to the *K-NN neighbors* algorithm is the computation of the similarity between the unknown class sample and the samples of the different classes, which is usually computed as a *Euclidean* distance. Features with a wide range of values will have a larger effect on the computed similarity. To alleviate the effect of values range on the similarity measure, features should be normalized using either the *min–max* or the *z-score* approaches. If there is noise present in the data, the min and max values might have been corrupted, and in this case, it is preferable to use the *z-score* approach that scales the data according to their mean and standard deviation.

3.1.6.6 Feature Selection

When the sample is represented using a multidimensional feature vector where there is a large number of dimensions, it is quite often desirable to reduce the number of features. This can be according to the usefulness of the different features in discriminating between the different classes. A measure that can be used for this purpose is the divergence measure, whose equation is shown below [Upp02]:

$$Divergence(C_1, C_2 i) = \frac{(\sigma_{1,i} - \sigma_{2,i})^2 + (\sigma_{1,i} + \sigma_{2,i})(\mu_{1,i} - \mu_{2,i})^2}{2\sigma_{1,i}\,\sigma_{2,i}} \quad (3.51)$$

In the above equation, C_1 and C_2 are two classes in our classification problem, and i is the ith feature. The terms $\sigma_{1,i}$ and $\sigma_{2,i}$ denote the standard deviation of the ith feature within classes C_1 and C_2, respectively. The terms μ_1 and μ_2 denote the means of the ith feature within classes C_1 and C_2, respectively. The higher the divergence, the better the feature can discriminate between C_1 and C_2. In case we have more than two classes, we can compute the paired divergences for each feature and add them. Another way we can eliminate redundant features is to compute the correlation between the different features and eliminate one of two highly correlated features (correlation > user specified threshold).

3.2 CLUSTERING

A *cluster* is a set of similar objects, where similarity is defined by some distance measure. Any of the distance measures discussed in the previous chapter can be used for the computation of similarity in clustering. The clustering problem is a challenging one because

1. The attributes, also called features, and their values that differentiate one cluster from another are not known.

2. There are no labeled data, as is the case in the classification. In other words, we do not have data to tell us what features differentiate objects that belong to different clusters. The only thing that might be of use in coarsely defining the clusters is some *a priori* knowledge provided by a domain expert. This lack of guidance in forming the clusters has led to giving the name *unsupervised learning* to clustering in the field of machine learning.

3. As the number of data increases, the number of clusters as well as the number and type of differentiating factors might change.

4. Because there is no guide as to what constitutes a cluster, the success of the clustering algorithms is influenced by the presence of noise in the data, missing data, and outliers. The last refers to data with unusual feature values. In addition, most clustering algorithms are sensitive to the number of features and are most successful in clustering data with a small number of features.

As discussed in the previous section, the algorithms that will be discussed in this section have been developed for nontemporal data, but as the examples will demonstrate, they can easily be extended to temporal data.

3.2.1 Clustering via Partitioning

The main idea in this class of clustering algorithms is to create K clusters of the data, where the number K is entered by the user. These algorithms are suitable mainly for numerical data. The original clustering, also known as *partitioning*, is performed randomly and then objects are moved in and out of clusters, using as guide a criterion of "closeness." Partitioning algorithms are very popular because of their ease of implementation and low computational cost; however, they have these disadvantages: (1) they are sensitive to the presence of noise and outliers, (2) they can discover only clusters with convex shapes, and (3) the number of clusters needs to be specified.

3.2.1.1 K-Means Clustering

The *K-Means clustering* algorithm was first introduced in [Mac67] and is one of the most popular clustering techniques. The main idea in this algorithm is to use the *means* to represent the clusters and use them as a guide to assign objects to clusters.

K-means is a *hill-climbing* algorithm [Lin04], which guarantees convergence to a local optimum, but not necessarily a global optimum. As K increases, the cost of finding the optimal solution decreases, and it reaches its minimum when K is equal to the number of objects [Wu08]. Specifically, the procedure is shown below:

1. Enter the objects to be clustered and also the number K of clusters.

2. Randomly choose K objects to be the original cluster centers.

3. Assign each object to the cluster with the closest mean.

4. Calculate the new mean of each cluster.

5. Repeat from step 3.

6. Stop when the convergence criterion is satisfied. The convergence criterion can be that no object gets assigned to a new cluster. Another criterion, more frequently used, is the minimization of the squared error E of all the objects in the database:

$$E = \sum_{i=1}^{K} \sum_{o \varepsilon C_i} |o - \mu_i|^2 \tag{3.52}$$

where o is an object in the database that belongs to cluster C_i, μ_i is the mean of the cluster C_i, and K is the number of clusters.

In *nearest-neighbor clustering*, a variation of the *k-means* algorithm, each object is assigned to the cluster with the closest neighbor (not the closest mean) at step 3. Despite its shortcomings, the *k-means* algorithm is the most widely used partition-based clustering algorithm [Wu08]. To enable *k-means* to deal with very large data sets, *kd-trees* are employed [Wu08].

Example: We have six time series data sets that consist of the weekly stock return data of various companies, which are represented using a *2-D* vector consisting of the standard deviation and fractal dimension of the time series data. The data, which have been normalized, are shown below and also in Figure 3.5:

S1 = [0.8, 0.1] , S2 = [0.1, 0.6], S3 = [0.2, 0.7], S4 = [0.6, 0.2], S5 = [0.7, 0.3], S6 = [0.3, 0.9]

As Figure 3.5 shows, there are two distinct clusters. A new stock vector at [0.3, 0.6] will be classified to cluster 1.

3.2.1.2 K-Medoids Clustering

The K-Medoids algorithm was first introduced in [Kau90] and is not as sensitive to outliers as is the *k-means*. In this algorithm, each cluster is represented by the most centrally located object, known as *medoid*. Figure 3.6 shows the medoids of two clusters.

The general procedure for the algorithm is as follows:

1. Randomly choose k objects as the initial medoids.

2. Assign each one of the remaining objects to the cluster that has the closest medoid.

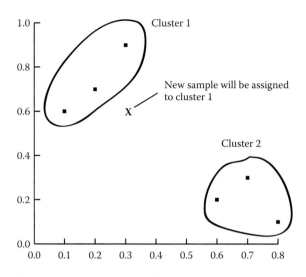

FIGURE 3.5   *K-means* clustering example.

3. In a cluster, randomly select a nonmedoid object, which will be referred to as $O_{nonmedoid}$.

4. Compute the cost of replacing the medoid with $O_{nonmedoid}$. This cost is the difference in the square error if the current medoid is replaced by $O_{nonmedoid}$. If it is negative, then make $O_{nonmedoid}$ the medoid of the

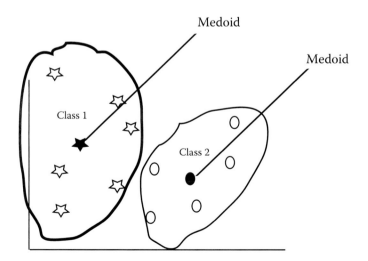

FIGURE 3.6   *K-medoids* clustering example.

cluster. The square error is again the summed error of all objects in the database:

$$E = \sum_{i=1}^{K} \sum_{o \varepsilon C_i} |o - O_{\text{medoid}(i)}|^2 \qquad (3.53)$$

where $O_{\text{medoid}(i)}$ is the medoid of the ith cluster.

5. Repeat from step 2 until there is no change.

3.2.2 Hierarchical Clustering

As the name implies, in this class of algorithms the objects are placed in a *hierarchy* which is traversed in either a *bottom-up* or *top-down* fashion to create the clusters. The advantage of this type of clustering is that it does not require any knowledge about the number of clusters, and its disadvantage is its computational complexity [Lin04]. Quite often a tree-like structure, a *dendrogram*, is used to represented the nested hierarchy levels.

Most *agglomerative* hierarchical algorithms follow a bottom-up approach and start with each object forming its own cluster. Then we merge these clusters into progressively larger clusters until a prespecified criterion is met, such as the number of clusters to be formed.

There are three different variations of the algorithm, based on how clusters are merged:

Single link: In this approach, two clusters are merged if the minimum distance between two objects, one from each cluster, is less than or equal a prespecified threshold distance. An example is shown in Figure 3.7.

Average link: Here, two clusters are merged if the average distance between objects in the two clusters is less than a prespecified threshold.

Complete link: In this approach, two clusters are merged if the maximum distance between points in the two clusters is less than or equal to a prespecified threshold. An example is shown in Figure 3.8.

Single link and complete link agglomerative clustering techniques are proposed in [Sib73] and [Def77], respectively, and they are the most popular hierarchical agglomerative techniques.

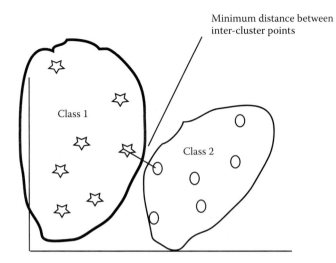

FIGURE 3.7   Agglomerative clustering, single link example.

In contrast to the agglomerative clustering algorithms, in *divisive* clustering we follow a top-down approach. Here we start with all objects belonging in one cluster, and we progressively divide them into smaller and smaller clusters, based on a prespecified criterion such as the number of clusters to be formed. Important hierarchical clustering techniques, such as *AGNES* (agglomerative clustering) and *DIANA* (divisive clustering), are discussed in [Kau90].

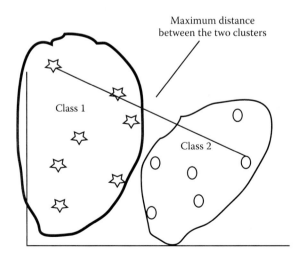

FIGURE 3.8   Agglomerative clustering, complete link example.

3.2.2.1 The COBWEB Algorithm

This algorithm is described in [Fis87] and is an incremental, hierarchical, conceptual clustering algorithm, in which each node in the hierarchy represents a concept and describes probabilistically the objects that belong to this concept. *COBWEB* is incremental, in the sense that objects are added incrementally to the hierarchical tree in a top-down approach looking for the best match. While descending the nodes, an object can be assigned to a cluster (using the *incorporate* operator), a new cluster can be created (with the *create* operator), a cluster can be divided (with the *split* operator), or two clusters can be merged in one (with the *merge* operator). Which operation will be performed depends on the value of a *category utility function* described in [Fis87].

Let us say that the concept at a node is *lung disease*, which is described by three attributes: *has_cough, has_difficulty_breathing, has_blood_in_sputum.* Let us say that there are 3 patients that belong to this category. Each patient's attributes are described by a binary number. The first patient is described as [1 0 0], the second as [0 1 1], and the third as [1 1 1], which means, for examples, that the first patient has a cough but does not have difficulty breathing or blood in his sputum. Then the vector [3 2 2] is stored at the node.

3.2.2.2 The BIRCH Algorithm

In [Zha96], a hierarchical clustering algorithm called *BIRCH* is described. *BIRCH* stands for *Balanced Iterative Reducing and Clustering using Hierarchies* and is an algorithm suitable for clustering in very large databases. The advantages of *BIRCH* over previous approaches are as follows:

- *BIRCH* works in a local fashion. This is achieved by using measurements that indicate the natural closeness of points, such that each clustering decision can be made without scanning all data points or existing clusters.

- It takes into account the structure of the data space. It treats points in a dense region as a single cluster, while points in a sparse region are characterized as outliers and can be optionally removed.

- The algorithm makes full use of available memory while minimizing I/O costs.

The core concepts of *BIRCH,* which works in an incremental way, are the *Clustering Feature* (*CF*) and a *CF* tree, which is a height-balanced tree.

The *Clustering Feature* consists of all the information that needs to be maintained about a cluster. These are the number of points in the cluster, the sum of the data points, and the squared sum of the data points.

The authors note that the *BIRCH* algorithm's performance depends significantly on proper parameter settings, such as the definition of outliers. It is also the first clustering algorithm that deals with noise in an effective way. As noted in [Hal01], *BIRCH* can find a decent clustering with the first database scan; however, different ordering of the input data can result in different clusters.

3.2.2.3 The CURE Algorithm

In [Guh98], a novel hierarchical algorithm is proposed that can detect clusters that are not necessarily convex. *CURE*'s clusters are represented by a fixed number of well-scattered points that are shrunk toward the center of the cluster by a certain fraction. *CURE* differs from the *BIRCH* algorithm in two ways:

- *CURE* begins by drawing a random sample instead of preclustering all data points as in the case of *BIRCH*.

- *CURE* first partitions the random sample and then in each partition the data are partially clustered. Then outliers are eliminated and the preclustered data in each partition are clustered to generate the final clusters.

Experimental results in the article show that *CURE*'s execution time is always less than *BIRCH*'s. More importantly, the results show that, as the size of the database increases, the execution time of *BIRCH* increases rapidly while the execution time of *CURE* increases very little. The reason is that *BIRCH* scans the entire database and uses all points for preclustering, while CURE uses just a random sample.

3.2.3 Density-Based Clustering

In this class of algorithms, the main idea is to keep growing clusters as long as their density is above a certain threshold. The advantage of density-based algorithms, in comparison with partitioning algorithms that are distance-based, is that they can *detect clusters of arbitrary shape*. On the other hand, distance-based algorithms detect only clusters of convex shape.

3.2.3.1 *The* DBSCAN *Algorithm*

The *DBSCAN* algorithm was introduced in [Est96]. *DBSCAN* stands for *Density-Based Spatial Clustering with Noise,* and its main idea is to create clusters that have a high enough density, *high enough* being specified by the user. Objects which are found not to belong to any cluster are characterized as noise. *DBSCAN* can be used on large databases effectively, and as it will be explained, it needs only one input parameter. In addition, distance can be measured using any of the distance measures discussed in Chapter 2. To understand the main idea of *DBSCAN* we need some new concepts, as defined in [Est96]. These concepts are discussed below and are demonstrated in Figure 3.9, which shows the number of *asthma attacks of a patient between January and March.*

1. *Eps neighborhood of an object:* This is the neighborhood that has a radius of size *eps* (specified by the user) around an object.

2. *Core object:* This is an object whose *eps-neighborhood* contains a pre-specified number of objects, *Nmin.* This number could be defined by the user; however, this is not necessary. It was found in [Est96] that the number 4 works well for all database sizes.

3. *Directly density-reachable:* An object *a* is *directly density-reachable* from an object *b,* if *b* is a core object and *a* is in the eps-neighborhood of *b.*

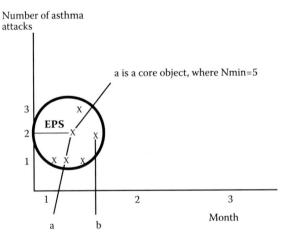

Object *b* is directly reachable from *a*

FIGURE 3.9 Demonstration of density-related concepts in the DBSCAN algorithm.

4. *Density-reachable*: An object *a* is *density-reachable* from an object *b* if a chain of directly reachable objects exists between them.

5. *Density-connected*: An object *a* and an object *b* are *density-connected*, if they are both density-reachable from a core object *c*.

6. *Cluster*: This is defined as a maximal set (in terms of density-reachability) of density-connected objects.

The way *DBSCAN* works is that it checks the *eps neighborhood* of every object in the database and if it finds that this neighborhood contains more than the prespecified number of objects *Nmin* then it defines this object as a *core* object of a new cluster. The process continues iteratively and at some point some density-based clusters might be merged, if they are found to be "close" to each other. An example of density-reachability is shown in Figure 3.9. This figure depicts the number of asthma attacks for six patients in the month of January. Besides being able to create clusters of arbitrary shape, *DBSCAN* can handle noise and outliers well, and can also be implemented in incremental fashion [Est98]. A disadvantage of the algorithm is its sensitivity to the parameters *eps* and *Nmin*, that the user must specify.

3.2.4 Fuzzy C-Means Clustering

Fuzzy C-Means clustering is a clustering technique that allows an object to belong to more than one cluster, by using the concept of a membership function [Dun73], [Bec81]. The cluster center, c_j, of the jth cluster is the weighted mean of all objects that belong to the cluster, where the weight depends on the membership of the object to the cluster, and it is defined as

$$c_j = \frac{\sum_{i=1}^{N} m_{ij}^D x_i}{\sum_{i=1}^{N} m_{ij}^D} \qquad (3.54)$$

where $x_1, x_2, ..., x_N$ are objects that belong to the jth cluster, and m_{ij} is the membership function. The latter is proportional to the inverse of the distance from the cluster center and is defined as

$$m_{ij} = \frac{1}{\sum_{k=1}^{c} \left| \frac{x_i - c_j}{x_i - c_k} \right|^{\frac{2}{D-1}}} \qquad (3.55)$$

where D is a number greater than 1. The goal of the fuzzy *c-means* clustering algorithm is the minimization of the following function:

$$F_D = \sum_{i=1}^{N} \sum_{j=1}^{C} m_{ij}^D |x_i - c_j|^2 \tag{3.56}$$

A disadvantage of the algorithm is that the quality of the results depends on how well the membership function models the actual data.

3.2.5 Clustering via the *EM* Algorithm

The *EM* (*Expectation Maximization*) *Clustering* algorithm is a model-based iterative clustering algorithm that has an *expectation* step and a *maximization* step. In the general *EM* algorithm, the value of an unknown parameter is assumed to belong to a parameterized probability distribution and the goal of the *EM* algorithm is to find the maximum likelihood estimate of the distribution's parameters. In clustering algorithms, the unknown parameter value for each object is the cluster to which it belongs. Assuming K clusters, the aforementioned parameter can be modeled as a binary variable where 1 indicates that the object belongs to one of the K clusters and 0 indicates the object does not belong to a cluster. The initial choice for the number of clusters is an important one, as it can affect the speed and the accuracy of the results.

The expectation and maximization steps shown below are repeated until the algorithm converges:

- *Expectation step*: Estimate the expected value of the parameter given a parameter estimate.

- *Maximization step*: Reestimate the parameters of the probability distribution, where the reestimation criterion is the maximization of the data likelihood, given the expected values computed in the previous step.

A scalable *EM* clustering algorithm that can be applied to very large databases is described in [Bra99].

3.3 OUTLIER ANALYSIS AND MEASURES OF CLUSTER VALIDITY

Outliers are objects that have different characteristics than the other objects in the database, and in general, we are interested in excluding them from our classification or cluster analysis. Outlier detection techniques are discussed in [Kno98], [Arn96], and [Bar94]. There are different ways to detect outliers:

Statistical: In this approach, we assume a probability distribution D for the given data set and then we can use hypothesis testing as described in Appendix A. In this hypothesis testing, the null hypothesis is that the object comes from the D distribution and the alternative hypothesis is that the object comes from a different distribution.

Distance-based outlier detection: In this approach, an object is characterized as an outlier if it is at a distance greater than d from a K number of objects, where d and K are defined by the user.

Deviation-based outlier detection: As the name implies, the criterion to characterize an object as an outlier is its deviation from the general population characteristics of a cluster. For example, one dissimilarity criterion is the difference of the object from the mean of the cluster.

A seminal article regarding measures of cluster validity is [Hal01]. Two criteria of cluster validity are defined by Halkidi et al.:

- *Cluster compactness.* A quantitative measure of compactness is variance, which should be minimized.

- *Cluster separation.* Distance between clusters can be measured in various ways: (1) distance between closest members of the clusters, (2) distance between the most distant members, and (3) distance between centroids.

Several quantitative measures of cluster validity are thoroughly described and tested in [Hal01]. Here, we will briefly mention some:

- *Partition–cluster comparison.* A comparison of the resulting cluster structure with a partition of the data is performed to measure how well the two match.

- *Cophenetic correlation,* which measures the validity of the clusters created by a *hierarchical* clustering method. The *cophenetic matrix* represents the hierarchy diagram, which is the result of the clustering algorithm.

- The *Modified Hubert's Γ index*, which detects crisp clusters. The measure is defined as follows:

$$MH\Gamma = \frac{1}{M}\sum_{i=1}^{N-1}\sum_{j=i+1}^{N} P_{ij}Q_{ij} \qquad (3.57)$$

where $M = N(N-1)/2$, P_{ij} is the proximity matrix, and Q_{ij} is the distance between the centroids of the clusters to which i and j belong. The elements of the proximity matrix have the value 1 if i and j belong to the same cluster and 0 if they do not.

- The *Dunn Index*, which measures compact and well-separated clusters:

$$D_{nc} = \min_{i=1,\dots,nc}\left(\min_{j=i+1,\dots,nc}\left(\frac{d(c_i,c_j)}{\max_{k=1,\dots,nc} diam(c_k)} \right) \right) \qquad (3.58)$$

where $diam(C)$ is the *diameter* of the cluster C, and d is the dissimilarity function of clusters i and j, which is defined as the minimum distance of points that belong to cluster i from points that belong to cluster j.

- The *SD validity index*. This is based on the *average scattering* for clusters, *Scat*, and the *total separation* between clusters, *Dis*, and is defined as

$$SD = a\,Scat(nc) + Dis(nc) \qquad (3.59)$$

where

$$a = Dis(c_{max}) \qquad (3.60)$$

and c_{max} is the maximum number of clusters. The definitions for the scattering and separation functions are given in [Hal01].

- The *Partition Coefficient for fuzzy clusters*. This metric is based on membership functions and is defined by

$$PC = \frac{1}{N}\sum_{i=1}^{N}\sum_{j=1}^{nc} m_{ij}^2 \qquad (3.61)$$

where m_{ij} is the membership function, and nc is the number of clusters. The more PC approaches 1, the more crisp the clustering becomes.

- *Total Variation* of a data set for fuzzy clusters:

$$TV = \sum_{i=1}^{nc} \sum_{j=1}^{N} (m_{ij}^2 \|x_{ij} - c_i\|)^2 \tag{3.62}$$

where x_{ij} is the jth object of the ith cluster, and c_i is the centroid of the ith cluster.

- *The XB Index for fuzzy clusters*, which measures the compactness and degree of separation of fuzzy clusters:

$$XB = \frac{\pi}{N} d_{\min} \tag{3.63}$$

where $d_{\min}$ is the minimum distance between cluster centers:

$$d_{\min} = \min\|c_i - c_j\| \tag{3.64}$$

3.4 TIME SERIES CLASSIFICATION AND CLUSTERING TECHNIQUES

3.4.1 1-NN Time Series Classification

As described in [Xi06], *1-NN Time Series Classification combined with Dynamic Time Warping* is tough to beat. The authors substantiate their argument, by comparing *1-NN* time series classification with a number of other techniques. The results are shown below:

- In [Che05a], a static minimization–maximization approach yields a maximum error of 7.2%. With *1NN-DTW*, the error is 0.33% with the same dataset as in the original article.

- In [Che05b], a multiscale histogram approach yields a maximum error of 6%. With *1NN-DTW*, the error (on the same data set) is 0.33%.

- In [Ead05], a grammar-guided feature extraction algorithm yields a maximum error of 13.22%. With *1NN-DTW*, the error was 9.09%.

- In [Hay05], time series are embedded in a lower dimensional space using a *Laplacian eigenmap* and *DTW* distances. The authors achieved an impressive 100% accuracy; however, the *1NN-DTW* also achieved 100% accuracy.

- In [Kim04], *Hidden Markov Models* achieve 98% accuracy, while *1NN-DTW* achieves 100% accuracy.

- In [Nan01], a *multilayer perceptron neural network* achieves the best performance of 1.9% error rate. On the same data set, *1NN-DTW's* rate was 0.33%.

- In [Rod00], first-order logic with boosting gives an error rate of 3.6%. On the same dataset, *1NN-DTW's* error rate was 0.33%.

- In [Rod04], a *DTW*-based decision tree gives an error rate of 4.9%. On the same dataset, *1NN-DTW* gives 0.0% error.

- In [Wu04], a super-kernel fusion set gives an error rate of 0.79%, while on the same data set, *1NN-DTW* gives 0.33%.

3.4.2 Improvement to the *1NN-DTW* Algorithm Using Numerosity Reduction

In [Xi06], an improvement to the *1NN-DTW* is proposed that is based on *numerosity reduction*, that is, the reduction of the data set without significantly affecting data set integrity, by removing some of the samples used for classification. The authors base their approach on the numerosity reduction discussed in [Wil97], where the novel idea of *nearest enemies* and *associates* is introduced.

A *nearest enemy* of an object p is the closest object in another class, while those objects (in the same class) that have p as one of their closest k neighbors are defined as its nearest *associates*. In this algorithm, which due to its simplicity, is called the *Naïve Numerosity Reduction*, only the nearest enemy and the k associates are stored. An improvement to the *Naïve Numerosity Reduction* is proposed in [Xi06], called the *AWARD* (*Adaptive WARping Window*) algorithm. The main idea of this algorithm is to dynamically adjust the warping window *during* numerosity reduction. Specifically, as AWARD is looking for samples to remove, it is also examining the possibility of expanding the warping window by 1. In other words, if the current window size is r, the classification accuracy using a size of $r+1$ is examined, and, if it is better, then $r+1$ is used as the warping window size.

3.4.3 Semi-Supervised Time Series Classification

In [Wei06], the important problem of *semi-supervised time series classification* is addressed, where both *labeled* and *unlabeled* data are used for classification. This is an important problem because of the large number of unlabeled data that exist in databases today. As noted in the article, semi-supervised learning techniques fall into the following categories:

- *Generative model–based.*

- *Low density–based,* where the decision boundary is pushed away from the unlabeled data.

- *Graph-based,* where prior knowledge is encoded in trees.

- *Co-training methods,* where the features are divided into two disjoint sets, with each data set being used to train a classifier. The predictions of one classifier are used to enlarge the training set of the other.

- *Self-training methods,* where a classifier is first trained using the labeled data. Then the unlabeled data are classified and the datasets, about which we are most confident, are added in the training set.

In [Wei06], a new classification method is proposed which utilizes the *1NN nearest neighbor* classifier with *Euclidean distance* and follows the procedure shown below:

- The classifier is trained using the initial training set. Here, labeled objects are considered positive, and all unlabeled sets are considered negative.

- Then, if the closest neighbor of an unlabeled object is classified as positive, the unlabeled object is also classified as positive. On the other hand, if the closest neighbor of an unlabeled object has not been classified, the unlabeled object is classified as negative.

- Finally, the object in the unlabeled set that is closest to the positive objects is added to the training set. The procedure repeats from the first step until a stopping criterion, such as the minimal nearest neighbor distance, is reached.

The experimental results show that high precision and recall can be achieved with only a small number of labeled objects, and therefore there is a lot of potential value in semi-supervised classification.

3.4.4 Time Series Classification Using Learned Constraints

In [Rat04], a novel method for improving time series classification that learns warping paths for *Dynamic Time Warping* from the data is discussed. The warping path is called the *R-K Band* and it can be of an arbitrary size and shape. To find the warping path, one needs to specify the direction of the search (for example, forward), the initial shape of the

band, heuristic function to evaluate the quality of the operation, operators, and terminal test.

In experimental results, the *Euclidean distance*, a uniform band (the
Sakoe–Chiba band [Sak78]) of size 1 to 100, and *DTW* with an *R-K* band
were tried on a variety of data sets. The *DTW* with *R-K* band always produced higher accuracy than the other methods, and because of the typical
smaller *R-K* band size, it also has a computationally lower cost. Specifically,
the authors demonstrated that their method reduces, by an order of magnitude, both the CPU time of DTW and the error rate. Finally, an important conclusion derived from the authors' experiments is that a wider
bandwidth does not always mean increased accuracy.

3.4.5 Entropy-Based Time Series Classification

In [Che07], an algorithm for time series classification is proposed that is
based on the discretization of a time series using an *entropy measure*. The
goal of the discretization process is to divide a time series into k intervals
according to a *splitting criterion*. The splitting criterion here is the minimization of the entropy within each interval. The entropy of an ith *interval*
is a measure of the information contained in an interval and is defined as

$$e(S_i) = -\sum_{i=0}^{m} p_{ij} \log_2 p_{ij} \tag{3.65}$$

where p_{ij} is the probability of class *j* in interval *i*. The entropy of an
interval is 0, if all values in the interval belong to the same class. On the
other hand, if all classes appear equally often in an interval, then the
entropy of the interval reaches its maximum. In this approach, every point
is considered as a potential breakpoint and the splitting occurs at points
where the entropy has the smallest value. Following entropy discretization, the time series is transformed into a series of strings. Then *k–nearest
neighbor* classification is employed, where the *Euclidean distance* is used
as a distance measure.

The algorithm was compared with various algorithms and was found
to have superior performance. For example, comparing it with *K-NN* that
utilizes *Dynamic Time Warping* and *Euclidean* as the distance measures,
the results were as follows: (1) For the entropy-based approach, the error
rate and execution time were, respectively, 0.02 and 2 sec. (2) For the *K-NN*
approach using Euclidean distance, the error rate and execution time were,
respectively, 0.05 and 2.5 sec. (3) For the *K-NN* approach using the *DTW*,

the error rate and execution time were, respectively, 0.045 and 5430 sec, which are significantly larger than the aforementioned ones.

3.4.6 Incremental Iterative Clustering of Time Series

In [Lin04], a novel algorithm that implements *anytime clustering* of time series is presented. *Anytime* algorithms are algorithms that trade-off quality for execution speed. As the execution time is allowed to increase, the quality of the results also improves. As discussed in the article, time series classification is a difficult problem, because of the high dimensionality, high feature correlation, and presence of noisiness in the data.

The main idea of the proposed algorithm is to utilize the multiresolution property of wavelets to present gradually refined results to the user. Then the *k-means clustering algorithm* is applied starting at the second level of the coefficients and then moving on to finer levels. This algorithm is called the *I-k-Means* algorithm and it has two advantages over the traditional *k-means* algorithm: (1) It is not affected by the choice of the initial clusters, and (2) it can converge to a global optimum as the execution time is allowed to increase. The final centers at the end of each resolution are used as the initial centers for the next level of resolution. The *I-K-Means* algorithm achieves both higher clustering quality and lower computational cost than the traditional *k-means* algorithm.

3.4.7 Motion Time Series Clustering Using Hidden Markov Models (HMMs)

In [Alo03], a method for clustering time series data based on *HMMs* is described. The method is applied to time series data resulting from the motion of objects, such as humans, animals, and vehicles.

As described in [Alo03], when one utilizes *HMMs* to perform clustering, four issues need to be addressed:

- Estimating parameter of each *HMM* cluster
- Selecting the number of the objects that belong to each *HMM* cluster
- Estimating the *HMM* cluster topology and size
- Deciding which sequence to assign to which cluster

In [Alo03], two iterative methods are used to estimate the parameters of the likelihood function: *K-means*, and the *EM algorithm*. Their experimental results show that the two methods have generally similar performance. [Smy01] also utilizes *HMMs* to cluster sequences. The author uses the *Baum–Welch* algorithm to maximize the likelihood, but proposes a novel initialization procedure for the parameters and a *Monte Carlo* cross-validation approach to estimate the number of clusters.

3.4.8 Distance Measures for Effective Clustering of ARIMA Time Series

As described in [Kal01], a time series consists of four components (*trend, cycle, stochastic component,* and a *random component*) which can be described using an *Auto-Regressive Integrated Moving Average (ARIMA) model*. In this article, the *Euclidean distance* between the *Linear Predictive Coding (LPC) cepstrum* of two time series is used, where the *cepstrum* is defined as the inverse *Fourier* transform of the short time logarithmic amplitude spectrum.

The experimental results performed on four different types of real data sets, such as a temperature and an ECG recording dataset, show that the *Euclidean distance* of *LPC cepstrums* results to better clustering results than traditional distance measures such as *Euclidean, DFT, DWT,* and *Principal Component Analysis.*

Another article that deals with preprocessing of a time series to achieve better classification results is [Fu08]. In this article, it is noted that a combination of uniform scaling and *Dynamic Time Warping (DTW)* provides better results for some real world problems than simple *DTW.*

3.4.9 Clustering of Time Series Subsequences

A large number of time series bibliographies are dedicated to time series subsequence clustering, such as [Das98], [Jin02], and [Mor01]. However, [Keo03] makes the surprising argument that such clustering is *meaningless*. This claim is proven in this article using a clustering meaningfulness measure defined as

$$cl_{meaningfulness(X,Y)} = \frac{\text{within–set–}X\text{–distance}}{\text{between–set–}X\text{–and–}Y\text{–distance}} \quad (3.66)$$

This measure should be close to 0 when the data are taken from the centers of clusters of different datasets. The authors used a sliding window to derive the subsequences and they also used randomly extracted subsequences. The astonishing result was that the cluster centers found using time series subsequence clustering are not more similar than the ones found from random walk data.

In [Sim06], the above argument that clustering of subsequences is meaningless is challenged. Instead, it is proven that if appropriate preprocessing of the time series data is done, then the clustering can be meaningful. This preprocessing involves *unfolding* of the regressors of the time series, where a *regressor* is an independent variable in a regression equation. The main idea is to measure the mean distance of the data set to the principal diagonal of its space and the criterion for meaningfulness is based on the prototype positions in the regressor space.

The argument that subsequence clustering is meaningless is also challenged in [Gol06]. The authors argue that the answer to the following question is YES: *Given several time series sequences and a set of time series subsequence centroids from one them is it possible to determine in a reliable way which one of the sequences produced these centroids?*

A salient feature of the work described in [Gol06] is the definition of the *shape* of a set of clusters. This is defined as the ordered list of the pair-wise Euclidean distances among the centroids. The authors propose two variations of an algorithm to match a cluster set to the time series that produced it. Both variations have two phases: in the first phase all the time series sequences are preprocessed and in the second phase the matching is performed, by providing as an answer the sequence that has the smallest distance.

3.4.10 Clustering of Time Series Data Streams

[Rod08] presents an algorithm for the hierarchical incremental clustering of time series data streams, known as *ODAC (Online Divisive Agglomerative Clustering)*. *ODAC* can process thousands of high rate data streams. Most of the data stream incremental clustering algorithms have focused on clustering of samples, while *ODAC* focuses on the clustering of *variables*. In *ODAC*, a hierarchical tree-shaped structure of clusters is constructed using a *top-down* strategy (divisive approach).

The clusters reside on leaves of the tree, where each leaf corresponds to a group of variables. The distance measure that was used is based on the *Pearson* correlation coefficient. Assuming we are given two time series, $X = \{x_1, x_2, \ldots x_N\}$ and $Y = \{y_1, y_2, \ldots, y_N\}$, an incremental form of the Pearson correlation coefficient is defined as

$$corr(X,Y) = \frac{P - \frac{XY}{N}}{\sqrt{X_2 - \frac{X^2}{N}}\sqrt{Y_2 - \frac{Y^2}{N}}} \tag{3.67}$$

where

$$X = \sum x_i \qquad (3.68)$$

$$Y = \sum y_i \qquad (3.69)$$

$$P = \sum x_i y_i \qquad (3.70)$$

$$X_2 = \sum x_i^2 \qquad (3.71)$$

and

$$Y_2 = \sum y_i^2 \qquad (3.72)$$

This is an incremental form of the Pearson correlation coefficient because it is updated every time a new data point arrives.

The algorithm evolves the clusters by constantly monitoring the diameter of each cluster, which is measured as the maximum distance between variables of that cluster. When certain conditions are met, the cluster is split and each of the aforementioned variables is assigned to a new cluster. If no cluster should be divided, then the algorithm looks for possible merging of clusters. Another novel distribution of this work is the use of the *Hoeffding bound* [Hoe63] as a splitting criterion. Specifically, a cluster will be split when the *Hoeffding bound* will give us the statistical confidence that the true diameter of the cluster is known.

Another work that deals with the clustering of data streams is [Agg03]. In this article, the authors propose an approach, called *CluStream*, that has an offline and online component and clusters large evolving data streams. Its main idea is to view the data stream as an *evolving process over time*. The online component stores summary statistics of the streams periodically and the offline component works only with the summary statistics to provide an understanding of the underlying clusters.

In [Guh03], Guha et al. cluster data streams using the *K-Median* idea, where the goal is to find *k* cluster centers, such that the distance of each point to its nearest center is minimized. Their algorithm makes one pass over the data, uses a divide-and-conquer approach, and can offer a reasonably good starting estimate of the centers. The authors note that there is a trade-off between running time and cluster quality. When compared with *BIRCH* on experimental data, the authors found that their algorithm took a somewhat longer time to run, but produced results of higher quality.

3.4.11 Model-Based Time Series Clustering

In [Xi02], a method for the clustering of time series is described that utilizes *mixtures of autoregressive moving average (ARMA) models*. *ARMA* models are described in more detail in the next chapter, and as the name implies they combine autoregression and moving average models.

The value of a time series at time t depends on weighted past observations and weighted past observation errors. In the case of mixture *ARMA* models, it is assumed that the time series data are generated by M $ARMA$ models that correspond to the M clusters. Then an *expectation minimization (EM) algorithm*, which is an iterative form of *Maximum Likelihood Estimation*, is used to learn the mixing coefficients and the parameters of the component models. One advantage of the proposed method is its ability to determine the number of clusters automatically, while a disadvantage is that the clustering performance degrades sharply when the clusters get really close. The method was applied on four real data sets and synthetics sets and it was overall effective in performing time series clustering.

3.4.12 Time Series Clustering Using Global Characteristics

Wang et al. in [Wan05] and [Wan06] propose a method for time series clustering that utilizes summarized information from the time series. In other words, they create a feature vector to represent a time series consisting of global time series characteristics, such as trend and periodicity. Then, feature vectors are used to cluster the time series. Specifically, the following global characteristics were examined by the authors:

- *Trend*. This is an indicator of a long-term change in the data mean (increasing, for example).

- *Seasonality*. This means that the time series is influenced by seasonal factors, such as increased sales in the summer.

- *Serial Correlation*. The *Box–Pierce statistic* [Mak98] is used to calculate the correlation.

- *Nonlinearity*. *Teräesvirta's neural network* test [Ter96] is used to detect nonlinearity.

- *Skewness.* This is a measure of symmetry; its computation was discussed in the previous chapter.

- *Kurtosis.* This is a measure of peakedness, and its computation was also discussed in the previous chapter.

- *Self-similarity.* The *Hurst exponent* [Wil96] is used to measure self-similarity.

- *Chaotic.* The *Lyapunov* exponent [Wol85] is used to measure chaotic behavior.

- *Periodicity.*

The best features for clustering were selected using a *greedy Forward Search* method [Atk00], which is a method for detecting multiple influences in a model. The algorithm was implemented on benchmark data sets and was found to yield meaningful clusters.

3.5 ADDITIONAL BIBLIOGRAPHY

3.5.1 General Classification and Clustering

Further detailed discussion on the topics of data classification and clustering can be found in the books [Dud73], [Gos96], and [Dun03] and in the articles [Rod02], [Han01], [Jai99], and [Hin99]. An impressive study on the comparison of thirty-three classification algorithms can be found in [Lim00]. In [Han93], an algorithm for data driven discovery of quantitative rules in a database is described, where a quantitative rule can serve as a classification rule. Information about concept hierarchies and data relevance is used in the algorithm. A robust hierarchical clustering algorithm, *ROCK,* is proposed in [Guh99]. It uses an object's neighbors and links to measure the similarity between objects.

A discussion of *Support Vector Machines* and their applicability in classification and regression is described in [Gun98]. A popular way to perform clustering operations is via neural networks; such algorithms are discussed in [Joa97], [Koh82], and [Wan03]. There are also several articles on clustering via genetic algorithms, such as [Mau00], [Bhu91], and [Jon91].

3.5.2 Time Series/Sequence Classification

In [Kum05], a time series is divided into subsequences using a sliding window and the *k-NN algorithm*. Various similarity measures were tried such as *Jaccard similarity, Cosine similarity* and *Euclidean distance.* In [Fab08] various similarity measures were combined (after being weighted) in a *Nearest Neighborhood Classifier*. This combination yielded better results than the popular *Nearest Neighbor* with *Dynamic Time Warping* as the similarity measure. In [Yam03], decision trees are used to classify data, where *each data object is a time series*. The algorithm showed promising results when applied to medical time series.

In [Zha06a], a novel feature extraction algorithm is proposed to reduce the dimensionality of the time series classification problem. The algorithm utilizes the relevant information between features and classes to select the features with the most discriminatory power. In [Les99], an algorithm for sequence classification (not just time series) is described that is called *FeatureMine*. *FeatureMine* is a feature extraction algorithm that extracts the most useful features for classification. The criteria for the selection of the features are (1) features have to be frequent, (2) features should not be redundant within a feature set, and (3) they should be distinctive in at least one class.

In [Zha06b], a time series classification method that uses wavelets for feature extraction is proposed to increase classification accuracy. The authors propose a method that combines a novel classification stopping criterion with the usage of *Dynamic Time Warping* for the selection of training data. [Eru07] proposes a method for time series classification that uses *Dynamic Time Warping* for the transformation of the time axis, feature selection to remove features that are not useful for classification, and decision trees for classification. In [Fun06], a method that describes *key sequences* in time series is described. A key sequence is defined as nonredundant and indicative of a time series. These key sequences are then used for time series classification.

In [Zha08a], a method for multivariate time series classification is proposed that utilizes a triangle function for a similarity measure and a *Gini* function to extract meaningful patterns that are used in the classification process. In [Len05], a time series is modeled as being created by a generator of a distinct number of states. It is assumed that for each point in time, one state is active and generates the next value in the time series. This is the main idea behind the *Probable Series Classifier*, which performs classification by finding the state that maximizes the probability that at time i, the state j is active.

In [Pet97], a system that utilizes a predictive modular fuzzy system (*PREMOFS*) to perform a time series classification is described. In this system, time series classification is based on a membership function of the time series to a prespecified set of classes in which the membership function is computed using the predictive accuracy of each class for the observed time series. It is important to note that the membership is assigned in a competitive fashion; in other words, what matters is not how well an observed time series is predicted by a class, but how much better it is than the other classes. The algorithm also has the advantage of being parallelizable.

In [Ead05], an algorithm for time series classification is proposed that employs a genetic algorithm for feature selection and a *Support Vector Machine* for the classification task. In [Pov04], time series classification using *Gaussian mixture models* of *reconstructed phase spaces* is proposed. The [Pov04] method was compared against a *time-delayed neural network,* which it outperformed significantly on three different datasets.

Finally, given a time series consisting of K segments, where each segment is iteratively selected from a collection of classes, [Wan04] addresses the problem of finding K, the breakpoints, and the underlying class of each segment. The authors make the following assumptions about the time series: (1) The K segments are statistically independent, (2) each class model is sufficiently characterized by a set of features, and (3) the probability distribution function (*PDF*) of the features for each class is known. Their proposed approach consists of two phases: In the first phase, a rough segmentation is performed using a piecewise generalized likelihood ratio, and in the second phase the results of the segmentation are refined. This is not an optimal solution to the problem; however, it has the significant advantage of reduced computational complexity compared to an optimal solution, such as *maximum likelihood estimation* realized via *dynamic programming.*

In [Au04], a method that utilizes fuzzy rules to achieve time series classification is employed to handle the *noisiness* in data.

3.5.3 Time Series Clustering

In [Abo03], a clustering approach for time series is described that utilizes fuzzy logic. The reasoning for the use of fuzzy sets in [Abo03] is that changes in time series variables do not have crisp boundaries, instead they have vague boundaries.

In [Kou00], *EasyMiner* is described, which is a temporal data clustering technique that is based on a decision tree approach. In [Rat08], the problem of semi-supervised clustering is addressed, where a large number of *unlabeled* data is available, while only a small number of *labeled* data is available. A method to specifically detect outliers in time series databases is proposed in [Jag99]. In [Yam02], a framework is presented that detects outliers and change points in nonstationary time series data. The unique feature of this work is that the framework detects both change points and outliers using on-line learning of the probabilistic model of the time series. The algorithm has two parts: (1) a data modeling part where the probabilistic data model is learned and (2) a scoring part where each point on the time series data is given a score based on the learned model.

In [Ala05], a method for clustering of financial time series data is described using *fuzzy c-means clustering*. Another article [Mol03] utilizes fuzzy c-means clustering for the purpose of clustering short time series and particularly those with unevenly spaced sampled points. A novel short time distance is also described in the article. In [Li06], an algorithm for the clustering of consumption patterns from imprecise electric load time series data is described. The algorithm first preprocesses the data, then it computes a similarity metric that utilizes a semantic interval-based measure. Then the *k-means* clustering method is utilized for clustering.

In [Rat05], the problem of scaling data mining algorithms is addressed and it is shown that choosing clipping (discussed in Chapter 2) as the time series representation scheme allows time series clustering algorithms to scale to much larger data sets. In [Lin05], an iterative algorithm for the clustering of streaming time series is proposed. This article has two novel contributions. The first one is the introduction of a metric to evaluate clustering in streaming time series data. The second one is the utilization of a *Multiresolution Piecewise Aggregate Approximation Representation* in combination with a *Hoeffding bound* to achieve faster response. The clustering algorithm works incrementally by utilizing information from nearest neighbors of the incoming time series.

In [Ros04], *Self-Organizing Maps* (*SOMs*) are applied to clustering of functional data, such as weather and spectrometric data. *SOMs* reduce the space dimensionality through projection and were introduced by Kohonen [Koh82]. *SOMs* are discussed in more depth in Chapter 4. In [Chi06], an evolutionary algorithm for hierarchical clustering is presented. In [Ama08], environmental data acquired by distributed sensors are reconstructed into time series using decision trees. It is shown that decision trees outperform other reconstruction methods, such as mean value and polynomial interpolation. In [Zhu07], ensemble *hidden Markov models* learners are used to train data for the recognition of 3-D motion of human joints.

In [Nie06], multimedia data, such as images, are clustered by converting them to time series data. This results in significant compression of the data storage requirements. The shapes of objects in the image, such as a face or a leaf, are extracted and then transformed into a time series. The time series are clustered using a *K-Medoids* algorithm and the *Euclidean* and *Dynamic Time Warping* distance measures are utilized. The images are also clustered using the *K-Medoids* algorithm on the color histograms of the images. The results showed that for most images, the time series representation yields a higher maximum accuracy. [Hin98] proposes a density-based clustering algorithm, called *DENCLUE*, which is appropriate for the clustering of large multimedia databases. The main idea of the algorithm is to define a function, *influence function*, that models the effect of an object in its neighborhood. One of the main advantages of *DENCLUE* is its resistance to noise; however, the results are dependent on the choice of its two parameters. One of these parameters is a number that determines the influence of an object in its neighborhood.

REFERENCES

[Abo03] Abonyi, J. et al., Fuzzy Clustering-Based Segmentation of Time Series, *Lecture Notes in Computer Science,* vol. 2811, Springer, pp. 275–285, 2003.

[Agg03] Aggarwal, C.C, J. Han, and P.S. Yu, A Framework for Clustering Evolving Data Streams, *Proceedings of VLDB*, pp. 81–92, 2003.

[Aio05] Aiolli, F. and A. Sperduti, Multiclass Classification with Multi-Prototype Support Vector Machines, *Journal of Machine Learning Research*, vol. 6, pp. 817–850, 2005.

[Ala05] Alanzado, A.C. and S. Miyamoto, Fuzzy c-Means Clustering in the Presence of Noise Cluster for Time Series Analysis, *Lecture Notes in Artificial Intelligence 3558,* pp. 156–163, Springer, 2005.

[Alo03] Alon, J. et al., Discovering Clusters in Motion Time Series Data, *Proceedings of the IEEE International Conference on Computer Vision and Pattern Recognition*, Madison, WI, vol. 1, pp. 375, 2003.

[Ama08] Amato, A., M. Calabrese, and V. Di Lecce, Decision Trees in Time Series Reconstruction Problems, *Proceedings of Instrumentation and Measurement Technology Conference*, pp. 895–899, 2008.

[Arn96] Arning, A., R. Agrawal, and P. Raghavan, A Linear Method for Deviation Detection in Large Databases, *Proceedings 1996, Int. Conf. Data Mining and Knowledge Discovery,* Portland, OR, pp. 164–169, 1996.

[Atk00] Atkinson, A.C. and M. Riani, *Robust Diagnostic Regression Analysis,* Springer, 2000.

[Au04] Au, W.H. and K.C.C. Chan, Mining Fuzzy Rules for Time Series Classification, *IEEE Conference on Fuzzy Systems*, Budapest, vol. 1, pp. 239–244, 25–29 July, 2004.

[Bar94] Barnett, V. and T. Lewis, *Outliers in Statistical Data*, John Wiley and Sons, 1994.

[Bec81] Bezdek, J. C., *Pattern Recognition with Fuzzy Objective Function Algorithms*, Plenum Press, 1981.

[Bhu91] Bhuyan, J. N., V. V. Raghavan, and V. K. Elayavalli, Genetic Algorithms for Clustering with an Ordered Representation, *Proceedings of the 4th Int. Conference on Genetic Algorithms*, pp. 408–415, 1991.

[Bra99] Bradley, P.S., U.M. Fayyad, and C. A. Reina, Scaling EM Clustering to Large Databases, Microsoft Research Technical Report, http://www.lans.ece.utexas.edu/course/ee380l/2000sp/papers/tr-98-35.pdf, 1999.

[Bre84] Breiman, L. et al., *Classification and Regression Techniques*, Wadsworth International Group, 1984.

[Bro88] Broomhead, D.S. and D. Lowe, Multivariate Functional Interpolation and Adaptive Networks, *Complex Systems*, vol. 2, pp. 321–355, 1988.

[Bur98] Burges, C.J.C., A Tutorial on Support Vector Machines for Pattern Recognition, *Data Mining and Knowledge Discovery*, vol. 2, pp. 121–167, 1998.

[Che05a] Chen, L. and M.S. Kamel, Design of Multiple Classifier Systems for Time Series Data, *Multiple Classifier Systems*, pp. 216–225, 2005.

[Che05b] Chen, L., M.T. Ozsu, and V. Oria, Using Multi-scale Histograms to Answer Pattern Existence and Shape Match Queries, *Proceedings of Scientific and Statistical Database Management Conference (SSDBM)*, pp. 217–226, 2005.

[Che07] Chen, X., D. Ye, and X.L. Hu, Entropy-Based Symbolic Representation for Time Series Classification, *Proceedings of the 4th International Conference on Fuzzy Systems and Knowledge Discovery*, pp. 754–760, 2007.

[Chi06] Chis, M. and C. Grosan, Evolutionary Hierarchical Time Series Clustering, *Proceedings of the 6th International Conference on Intelligent Systems and Applications*, pp. 451–455, 2006.

[Das98] Das, G. et al., Rule Discovery From Time Series, *Proceedings of the KDD'98 Conference*, pp.16–22, 1998.

[Def77] Defays, D., An Efficient Algorithm for a Complete Link Method, *The Computer Journal*, vol. 20, no. 4, pp. 364–366, 1977.

[Dom96] Domingos, P. and M. Pazzani, Beyond Independence: Conditions for the Optimality of the Simple Bayesian Classifier, *Proceedings of 13th International Conference Machine Learning*, pp. 105–112, 1996.

[Dud73] Duda, R. and P. Hart, *Pattern Classification and Scene Analysis*, John Wiley and Sons, 1973.

[Dun03] Dunham, M., *Data Mining*, Pearson Education, New Jersey, 2003.

[Dun73] Dunn, J.C., A Fuzzy Relative of ISODATA Process and Its Use in Detecting Compact Well-Separated Clusters, *Journal of Cybernetics*, vol. 3, pp. 32–57, 1973.

[Ead02] Eads, D. et al., Genetic Algorithms and Support Vector Machines for Time Series Classification, *SPIE Proceedings on Applications of Neural Networks, Fuzzy Systems*, vol. 4787, pp. 74–85, 2002.

[Ead05] Eads, D. et al., Grammar-Guided Feature Extraction for Time Series Classification, *Proceedings of the Neural Information Processing Systems Conference (NIPS)*, 2005.

[Eru07] Eruhimov, V., V. Martyanov, and E. Tuv, Constructing High Dimensional Feature Space for Time Series Classification, *Lectures Notes on Artificial Intelligence 4702*, pp. 414–421, Springer, 2007.

[Est96] Ester, M. et al., A Density-Base Algorithm for Discovering Clusters in Large Spatial Databases with Noises, *Proceedings of the Int. Conference on Knowledge Discovery and Data Mining*, pp. 226–231, 1996.

[Est98] Ester, M. et al., Incremental Clustering for Mining in a Data Warehousing Environment, *Proceedings of the 24th VLDB Conference*, New York, pp. 323–333, 1998.

[Fab08] Fabris, F., I. Drago, and F. M. Varejao, A Multi-measure Nearest Neighbor Algorithm for Time Series Classification, *Lecture Notes on Artificial Intelligence 5290*, pp. 153–162, Springer, 2008.

[Fis87] Fisher, D.H., Knowledge Acquisition via Incremental Conceptual Clustering, *Machine Learning*, vol. 2, pp. 139–172, 1987.

[Fu08] Fu, A.W.C. et al., Scaling and Time Warping in Time Series Querying, *The VLDB Journal*, vol. 17, pp. 899–921, 2008.

[Fun02] Fung, G. and O.L. Mangasarian, Incremental Support Vector Machine Classification, *2nd SIAM International Conference on Data Mining*, April 11–13, 2002.

[Fun06] Funk, P. and N. Xiong, Discovering Key Sequences in Time Series Data for Pattern Classification, *Lecture Notes on Artificial Intelligence 4065*, pp. 492–505, Springer, 2006.

[Goh03] Goharian, N. and D. Grossman, Neural Network Classification, http://www.ir.iit.edu/~nazli/cs422/CS422-Slides/DM-NeuralNetwork.pdf, 2003.

[Gol06] Goldin, D., R. Mardales, and G. Nagy, In Search of Meaning for Time Series Subsequence Clustering: Matching Algorithms Based on a New Distance Measure, *Proceedings of ACM CIKM*, pp. 347–356, 2006.

[Gos96] Gose, E., R. Johnsonbaugh, and S. Jost, *Pattern Recognition and Image Analysis*, Prentice Hall, 1996.

[Guh98] Guha, S., P. Medas, and P. Rodrigues, CURE: An Efficient Clustering Algorithm for Large Databases, *Proceedings of the ACM SIGMOD Conference*, *ACM Press*, pp. 73–84, 1998.

[Guh99] Guha, S., R. Rastogi, and K. Shim, ROCK: A Robust Clustering Algorithm for Large Databases, *Proceedings of the IEEE Conference on Data Engineering*, 1999.

[Guh03] Guha, S. et al., Clustering Data Streams: Theory and Practice, *IEEE TKDE*, vol. 15, no. 3, pp. 515–528, 2003.

[Gun98] Gunn, S., Support Vector Machines for Classification and Regression, Technical Report, University of Southampton, 1998.

[Hal01] Halkidi, M., Y. Batistakis, and M. Varzigiannis, On Clustering Validation Techniques, *Journal of Intelligent Information Systems*, vol. 17, no. 2–3, pp. 107–145, 2001.

[Han93] Han, J., Y. Cai, and N. Cercone, Data-Driven Discovery of Quantitative Rules in Relational Databases, *IEEE Transactions on Knowledge and Data Engineering,* vol. 5, no. 1, pp. 29–40, 1993.

[Han01] Han, J., M. Kamber, and A. K. H. Tung, *Spatial Clustering Methods in Data Mining: A Survey,* Taylor & Francis, 2001.

[Har68] Hart, P., The Condensed Nearest Neighbor Rule, *IEEE Transactions Information Theory,* vol. 14, pp. 515–516, 1968.

[Hay99] Haykin, S., *Neural Networks—A Comprehensive Foundation,* 2nd ed., Prentice Hall, 1999.

[Hay05] Hayashi, A., Y. Mizuhara, and N. Suematsu, Embedding Time Series Data for Classification, *Machine Learning and Data Mining in Pattern Recognition,* pp. 356–365, 2005.

[Hin98] Hinneburg, A. and D. A. Keim, An Efficient Approach to Clustering in Large Multimedia Databases with Noise, *Proceedings of KDD Conference,* pp. 58–65, 1998.

[Hin99] Hinneburg, A. and D. A. Keim, Clustering Methods for Large Databases: From the Past to the Future, Technical Report, *ACM SIGMOD Tutorial,* 1999.

[Hoe63] Hoeffding, W., Probability Inequalities for Sums of Bounded Random Variables, *Journal of the American Statistical Association,* vol. 58, no. 301, pp. 13–30, 1963.

[Jag99] Jagadish, H.V., N. Koudas, and S. Muthukrishnan, Mining Deviants in a Time Series Database, *Proceedings of 1999 International Conference on Very Large Databases,* Edinburgh, UK, pp. 102–113, September 1999.

[Jai88] Jain, A.K. and R.C. Dubes, *Algorithms for Clustering Data,* Prentice Hall, 1988.

[Jai99] Jain, A.K., M.N. Murty, and P. J. Flynn, What is a Data Warehouse? *ACM Computing Surveys,* pp. 264–323, September 1999.

[Jin02] Jin, X., Y. Lu, and C. Shi, Distribution Discovery: Local Analysis of Temporal Rules, *Proceedings of the 6th Pacific-Asia Conference on Knowledge Discovery and Data Mining,* pp. 469–480, May 6–8, 2002.

[Joa97] Joachims, T., D. Freitag, and T. Mitchell, WebWatcher: A Tour Guide for the World Wide Web, *Proceedings of the 15th International Joint Conference on Artificial Intelligence,* pp. 770–775, 1997.

[Jon91] Jones, D.R. and M.A. Beltramo, Solving Partitioning Problems with Genetic Algorithms, *Proceedings of the 4th International Conference on Genetic Algorithms,* pp. 442–449, 1991.

[Kal01] Kalpakis, K. and D.G. Puttagunta, Distance Measures for Effective Clustering of ARIMA Time Series, *2001 IEEE International Conference on Data Mining,* San Jose, CA, pp. 273–280, 2001.

[Kau90] Kaufmann, L. and P. J. Rousseeuw, *Finding Groups in Data: An Introduction to Cluster Analysis,* John Wiley and Sons, 1990.

[Keh97] Kehagias, A. and V. Petridis, Predictive Modular Neural Networks for Time Series Classification, *Neural Networks,* vol. 10, pp. 31–49, 1997.

[Keo03] Keogh, E. and J. Lin, Clustering of Time-Series Subsequences is Meaningless: Implications for Previous and Future Research, *Knowledge and Information Systems,* vol. 8, pp. 154–177, Springer, 2003.

[Kim04] Kim, S., P. Smyth, and S. Luther, Modeling Waveform Shapes with Random Effects Segmental Hidden Markov Models, Technical Report, *UCI-ICS*, 2004.

[Kno98] Knorr, E.M. and R.T. Ng, Algorithms for Mining Distance-Based Outliers in Large Datasets, *Proceedings of the International Very Large Databases Conference*, pp. 392–403, 1998.

[Koh82] Kohonen, T., Self-Organized Formation of Topological Correct Feature Maps, *Biological Cybernetics*, vol. 43, pp. 59–69, 1982.

[Kou00] Koundourakis, G., M. Saraee, and B. Theodoulidis, Data Mining in Temporal Databases, *Citeseer*, 2000.

[Kum05] Kumar, P. et al., Using Sub-sequence Information with KNN for Classification of Sequential Data, *Lecture Notes on Computer Science 3816*, pp. 536–546, 2005.

[Len05] Lenser, S. and M. Veloso, Non Parametric Time Series Classification, *Proceedings of the International Conference on Automation and Robotics (ICRA)*, pp. 3918–3923, 2005.

[Les99] Leash, N., M.J. Zaki, and M. Ogihara, Mining Features for Sequence Classification, *Proceedings of the KDD Conference*, pp. 342–346, 1999.

[Li06] Li, Q., S.S. Liao, and D. Li, A Clustering Model for Mining Consumption Patterns from Imprecise Electric Load Time Series Data, *Lecture Notes on Artificial Intelligence 4223*, pp. 1217–1220, Springer, 2006.

[Lim00] Lim, T.S., W.H. Loh, and Y.S. Shih, A Comparison of Prediction Accuracy, Complexity and Training Time of Thirty-Three Old and New Classification Algorithms, *Machine Learning*, vol. 40, pp. 203–209, 2000.

[Lin04] Lin, J. et al., Iterative Incremental Clustering of Time Series, *Proceedings of the Conference on Extending Database Technology (EDBT)*, pp. 106–122, 2004.

[Lin05] Lin, J. et al., A MPAA-Based Iterative Clustering Algorithm Augmented by Nearest Neighbors Search for Time-Series Data Streams, *Lecture Notes on Artificial Intelligence 3518*, Springer, pp. 333–342, 2005.

[Mac67] MacQueen, J., Some Methods for Classification and Analysis of Multivariate Observations, *Proceedings of 5th Berkeley Symposium on Mathematics Statistics Prob.*, vol. 1, pp. 281–297, 1967.

[Mak98] Makridakis, S., S.C. Wheelwright, and R.J. Hyndman, *Forecasting Methods and Applications*, John Wiley & Sons, 1998.

[Mar03] Markowetz, F., Classification by Support Vector Machines, http://www.bioconductor.org/workshops/2003/NGFN03/svm.pdf, 2003.

[Mau00] Maulik, U. and S. Bandyopadhyay, Genetic Algorithm-Based Clustering Technique, *Pattern Recognition*, vol. 33, pp. 1455–1465, 2000.

[Mol03] Möller-Levett, C.S. et al., Fuzzy Clustering of Short Time-Series and Unevenly Distributed Sampling Points, *Lecture Notes on Computer Science 2810*, pp. 330–340, Springer, 2003.

[Mor01] Mori, T. and K. Uehara, Extraction of Primitive Motion and Discovery of Association Rules from Human Motion, *Proceedings of the 10th IEEE International Workshop on Robot and Human Communication*, pp. 200–206, 2001.

[Nan01] Nanopoulos, A., R. Alcock, and Y. Manolopoulos, Feature-Based Classification of Time Series-Data, *International Journal of Computer Research*, pp. 49–61, 2001.

[Neu97] Neukirchen, C. and G. Rigoll, Time Series Classification Using Hidden Markov Models and Neural Networks, *Proceedings of the IAR Annual Meeting*, 1997.

[Nie06] Niennattrakul, V. and C.A. Ratanamanahatana, Clustering Multimedia Data Using Time Series, *IEEE International Conference on Hybrid Information Technology*, pp. 372–379, 2006.

[Pet97] Petridis, V. and A. Kehagias, Predictive Modular Fuzzy Systems for Time-Series Classification, *IEEE Transactions on Fuzzy Systems*, vol. 5, no. 3, pp. 141–144, August 1997.

[Pov04] Povinelli, R.J. et al., Time Series Classification Using Gaussian Mixture Models of Reconstructed Phase Spaces, *IEEE Transactions on Knowledge and Data Engineering*, vol. 16, no. 6, pp. 779–783, 2004.

[Qui86] Quinlan, J.R., Induction of Decision Trees, *Machine Learning*, vol. 11, no. 1, pp. 81–106, 1986.

[Qui93] Quinlan, J. R., *C4.5: Programs for Machine Learning*, Morgan Kaufmann, 1993.

[Rat04] Ratanamahatana, C.A. and E. Keogh, Making Time Series Classification More Accurate Using Learned Constraints, *Proceedings of the SIAM International Conference on Data Mining*, pp. 11–22, 2004.

[Rat05] Ratanamahatana, C. et al., A Novel Bit Level Time Series Representation with Implication of Similarity Search and Clustering, *Lecture Notes on Artificial Intelligence 3518*, pp. 771–775, Springer, 2005.

[Rat08] Ratanamahatana, C.A. and D. Wanichsan, Stopping Criterion Selection for Efficient Semi-Supervised Time Series Classification, in *Software Engineering, Artificial Intelligence, Networking, and Parallel/Distributed Computing*, Roger Lee ed., vol. 149, pp. 1–14, Springer, 2008.

[Rod00] Rodriguez, J.J., C.J. Alonso, and H. Bostrom, Learning First-Order Logic Time Series Classifiers: Rules and Boosting, *Proceedings of PKDD Conference*, pp. 299–308, 2000.

[Rod02] Roddick, J. F. and M. Spiliopoulou, A Survey of Temporal Knowledge Discovery Paradigms and Methods, *IEEE Transactions on Knowledge and Data Engineering*, vol. 14, no. 4, pp. 750–767, July/August 2002.

[Rod04] Rodriguez, J.J. and C.J. Alonso, 2004. Interval and Dynamic Time Warping-Based Decision Trees, *Proceedings of the 2004 ACM Symposium on Applied Computing*, pp. 548–552, 2004.

[Rod08] Rodrigues, P.P., J. Gama, and J.P. Pedroso, Hierarchical Clustering of Time Series Data Streams, *IEEE Transactions on Knowledge and Data Engineering*, vol. 20, no. 5, pp. 615–627, 2008.

[Ros04] Rossi, F., B.C. Guez, and A. E. Golli, Clustering Functional Data with the SOM Algorithm, *Proceedings of the European Symposium on Artificial Neural Networks Conference*, pp. 305–312, 2004.

[Rum86] Rumelhart, D.E. and J. McClelland, Eds, *Parallel Distributed Processing*, vol. 1, The MIT Press, 1986.

[Sak78] Sakoe, H. and S. Chiba, Dynamic Programming Algorithm Optimization, for Spoken Word Recognition, *IEEE Transactions on Acoustics, Speech, and Signal Processing*, vol. ASSP-26, pp. 43–49, 1978.

[Sha96] Shafer, J., R. Agrawal, and M. Mehta, SPRINT: A Scalable Parallel Classifier for Data Mining, *Proceedings 1996 International Conference on Very Large Data Bases*, Bombay, India, pp. 544–555, Sept. 1996.

[Sho06] Shou, L. et al., Classifying Motion Time Series Using Neural Networks, *Lecture Notes on Computer Science 4261*, pp. 606–614, Springer, 2006.

[Sib73] Sibson, R., Slink: An Optimally Efficient Algorithm for the Single Link Cluster Methods, *The Computer Journal*, vol. 161, no. 1, pp. 30–34, 1973.

[Sim06] Simon, G., J.A. Lee, and M. Varleysen, Unfolding Preprocessing for Meaningful Time Series Clustering, *Neural Networks*, vol. 19, pp. 876–888, Elsevier, 2006.

[Smy01] Smyth, P., Clustering Sequences with Hidden Markov Models, *Adv. Neural Inf. Process*, vol. 9, pp. 648–655, 2001.

[Ter96] Teräesvirta, T., Power Properties of Linearities for Time Series, *Studies in Nonlinear Dynamics & Econometrics*, vol. 1, no. 1, pp. 3–10, 1996.

[Tou02] Toussaint, G.T., Proximity Graphs for Nearest Neighbor Decision Rules; Recent Progress, *34th Symposium on Computing and Statistics, Interface*, pp. 17–20, 2002.

[Upp02] Uppalun, R., T. Mitsa, E. Hoffman, M. Sonka, and G. Mclen Method and Appara tvs for Analyzing CT Images to Determine the Presence of Pulmonary Tissue Pathology, U.S. Patent #US 6466687, Issue date: 10/15/2002.

[Wan03] Wan, J. H., J. D. Rau, and W. J. Liu, Two-Stage Clustering via Neural Networks, *IEEE Transactions on Neural Networks*, vol. 14, no. 3, pp. 606–615, May 2003.

[Wan04] Wang, Z.J. and P. Willett, Joint Segmentation and Classification of Time Series Using Class-Specific Features, *IEEE Transactions on Systems, Man, and Cybernetics*, vol. 34, no. 2, pp. 1056–1067, April 2004.

[Wan05] Wang, X., K.A. Smith, and R.J. Hyndman, Dimension Reduction for Clustering Time Series Using Global Characteristics, *Lecture Notes on Computer Science 3516*, pp. 792–795, 2005.

[Wan06] Wang, X., K.A. Smith, and R.J. Hyndman, Characteristic-Based Clustering for Time Series Data, *Data Mining and Knowledge Discovery*, vol. 13, pp. 335–364, 2006.

[Wei06] Wei, L. and E. Keogh, Semi-Supervised Time Series Classification, *Proceedings of the SIGKDD Conference*, pp. 748–753, 2006.

[Wil96] Willinger, W., V. Paxon, and M.S. Taqqu, Self-Similarity and Heavy Tails: Structural Modeling of Network Traffic, *A Practical Guide to Heavy Tails: Statistical Techniques and Applications*, pp. 27–53, 1996.

[Wil97] Wilson, D.R. and T.R. Martinez, Instance Pruning Techniques, *Proceedings of the International Conference on Machine Learning (ICML)*, pp. 403–411, Morgan-Kaufmann, 1997.

[Wit00] Witten, I. H. and E. Frank, *Data Mining Practical Machine Learning Tools and Techniques*, Morgan Kaufmann, 2000.

[Wol85] Wolf, A. et al., Determining Lyapunov Exponents from a Time Series, *PHYSICA D*, vol. 16, pp. 285–315, 1985.

[Wu04] Wu, Y. and E.Y. Chang, Distance-Function Design and Fusion for Sequence Data, *Proceedings of the ACM Conference on Information and Knowledge Management (CIKM)*, pp. 324–333, 2004.

[Wu08] Wu, X. et al., Top 10 Algorithms in Data Mining, *Knowledge Information Systems*, vol. 14, pp. 1–37, 2008.

[Xi02] Xiong, Y. and D.Y. Yeung, Mixtures of ARMA Models for Model-Based Time Series Clustering, *IEEE International Conference on Data Mining*, pp. 717–720, 2002.

[Xi06] Xi, X. et al., Fast Time Series Classification Using Numerosity Reduction, *Proceedings of the 23rd International Conference on Machine Learning*, Pittsburgh, PA, pp. 1033–1044, 2006.

[Yam02] Yamanishi, K. and J. Takeuchi, A Unifying Framework for Detecting Outliers and Change Points from Non-Stationary Point Data, *Proceedings of SIGKDD*, pp. 676–681, 2002.

[Yam03] Yamada, Y. et al., Decision-Tree Induction from Time-Series Data Based on a Standard-Example Split Test, *Proceedings of the 20th International Conference on Machine Learning*, pp. 840–847, 2003.

[Zha96] Zhang, T., R. Ramakrishnan, and M. Livny, BIRCH: An Efficient Data Clustering Method for Very Large Databases, *Proceedings of the ACM SIGMOD Conference*, pp. 103–114, 1996.

[Zha06a] Zhang, H. et al., Multi-scale Feature Extraction for Time Series Classification with Hybrid Feature Selection, *Lecture Notes in Control and Information Sciences, Springer*, vol. 344, pp. 933–944, 2006.

[Zha06b] Zhang, H. et al., Feature Extraction for Time Series Coefficients Classification Using Discriminating Wavelet Coefficients, *Lecture Notes in Computer Science 3971*, Springer, pp. 1394–1399, 2006.

[Zha08a] Zhang, X. et al., Extracting Meaningful Patterns for Time Series Classification, *IEEE Congress on Evolutionary Computation*, pp. 2513–2516, 2008.

[Zha08b] Zhang X. et al., A Novel Pattern Extraction Method for Time Series Classification, *Optimization and Engineering*, vol. 10, no. 2, pp. 253–271, 2008.

[Zhu07] Zhu, H.L., P.Y. Du, and J. Xiang, 3D Motion Recognition Based on Ensemble Learning, *8th International Workshop on Image Analysis for Multimedia Interactive Service*, pp. 28, 6–8 June 2007.

CHAPTER 4

Prediction

Prediction is quite often the ultimate goal of temporal data mining. It is because of this important and intimate relationship with data mining that a whole chapter is dedicated to the subject of temporal prediction. It has a variety of applications in such diverse areas as financial forecasting, meteorology, seismology, and medical disease detection. For example, a company is interested in predicting its average closing price for next month. In another example, a doctor would like to predict the reaction of his patients to a new diabetes medication, particularly the duration of hypoglycemic episodes. The first example falls into the area of *time series prediction,* which is also known as *time series forecasting.* Note that the terms *prediction* and *forecasting* will be used interchangeably in the remainder of the chapter. The second example falls in the area of *event prediction.* The differences between time series prediction and event prediction are summarized below:

- *Problem statement*: In univariate *time series forecasting* the problem is to predict the value of a variable at a multiple of a time interval. For example, a company would like to predict its sales for next month. In *event prediction*, we would like to predict the *occurrence* of an event or the *number of occurences* of an event or the *duration* of an event given the existence of certain conditions. For example, a doctor just put one of his epileptic patients on a new drug that is very effective, but upon therapy initiation might cause a severe migraine attack. The doctor would like to predict the duration of this attack given knowledge about the age of the patient and the number and duration of epileptic attacks in the last year. In this book, we will only examine event prediction problems that deal with the prediction of an event's duration. The reason is that an event's duration can be modeled using a continuous variable and we can use linear regression for its prediction, where *linear regression* is one of the most widely available regression methods. Prediction of the occurence of an event can be modeled with a categorical dichotomous variable and the appropriate type of regression is *logistic regression* [Fel09]. Prediction of the number of occurence's of an event can be modeled with a count variable and the appropriate type of regression is *Poisson regression* [Orm09].

- *Data:* In *time series forecasting,* the data are historical data obtained at regular time intervals. Information about past patterns can be used to predict future patterns. For example, in the company example, the data could be sales for the previous 12 months. In *event prediction,* the data are population data about variables that are related to the occurrence of an event or series of events. In the migraine example, the data could be durations of migraine attacks of other patients on the drug along with information about their ages, preexisting medical conditions, and severity of epilepsy.

We will first examine the use of regression in *temporal event prediction* and then we will discuss the use of various methods for *time series prediction.*

4.1 FORECASTING MODEL AND ERROR MEASURES

The forecasting model, whether it is about the prediction of an event's duration or the values of a time series, can be expressed as follows:

$$Actual\ Value\ =\ Forecasted\ Value\ +\ Residual \tag{4.1}$$

It is desirable that the residuals do not follow a specific distribution or exhibit any patterns, instead they should resemble white noise. If there are any patterns in the residuals, these patterns were not captured in the forecasting model, and therefore our model is inadequate.

Most of the techniques developed for time series analysis and prediction assume *stationary* data. *Stationarity* means that the mean, variance, and autocorrelation of the time series do not change over time. Differencing the data is often useful in removing nonstationarity. Ways to deal with other types of nonstationarity, such as changing variance and the existence of a trend, are discussed later in the chapter.

The most common error measures of the success of our forecasting model are *Mean Square Error (MSE), Mean Absolute Error (MAE)*, and *Mean Absolute Percentage Error (MAPE)*. These measures are all based on the residual values, so it is assumed that the actual value for each forecasted point is also known. Given that the actual values in practice are not known, it might initially seem that these measures are useless. However, they can be used to test the success of a forecasting algorithm on a particular data set by reserving part of the known data points and their values as a testing set. The error measures are defined as follows:

$$MSE = \sum_{i=1}^{N} R_i^2 / N \qquad (4.2)$$

$$MAE = \sum_{i=1}^{N} |R_i| / N \qquad (4.3)$$

$$MAPE = \frac{1}{N} \times \sum_{i=1}^{N} \left| \frac{R_i}{A_i} \right| \qquad (4.4)$$

where R_i is the residual at point i, A_i is the actual value, and N is the number of points on the time series that are being forecasted.

Three more interesting measures are defined in [Jun08a]. They are Normalized Mean Square Error (*NMSE*), Prediction on Change of Direction (*POCID*), and Average Relative Variance (*ARV*):

$$NMSE = \frac{\sum_{j=1}^{N} R_i^2}{\sum_{j=1}^{N} (A_j - A_{j+1})^2} \qquad (4.5)$$

$$POCID = 100 \times \frac{\sum_{j=1}^{N} D_j}{N} \qquad (4.6)$$

$$ARV = \frac{\sum_{j=1}^{N} R_i^2}{\sum_{j=1}^{N} (F_j - A_{mean})^2} \qquad (4.7)$$

In *POCID*, the direction D_j is defined as

$$D_j = 1 \ if \ (A_j - A_{j-1}) \times (F_j - F_{j-1}) > 0 \qquad (4.8)$$
$$D_j = 0 \ \text{otherwise.}$$

In the above equation, F_j and F_{j-1} are forecasted values.

4.2 EVENT PREDICTION

Regression examines whether a variable (the dependent variable) can be predicted using one or more variables, known as the independent variables. If there is just one independent variable, then the regression is called *simple regression*. If there are two or more independent variables, the regression is called *multiple regression*. Another categorization of regression is *linear* versus *nonlinear* regression, depending on the kind of curve that describes the relationship between the dependent and independent variables. In this book, we will focus on linear regression for two reasons: (1) A large number of variable relationships can be described using linear regression; (2) by appropriate transformations of the independent and/or dependent variable, nonlinear relationships can be transformed to linear ones. Once we have a regression equation that models the behavior of a dependent variable given some independent variables, we can use the equation to *predict* the value of the dependent variable for new observed values of the independent variables.

4.2.1 Simple Linear Regression

The equation of the fitted line is

$$Y = b + aX \qquad (4.9)$$

where Y is the dependent variable and X is the independent variable. The estimated (predicted) value of the dependent variable is known as the *fitted value*. The difference between the actual value and the fitted value is known as the *residual*. In other words: *Actual Value = Fitted Value + Residual*.

A popular method for the estimation of the a and b coefficients in the equation above is *least squares estimation*. In this method, the line is chosen such that it minimizes the sum of the residuals. The equation for the computation of the slope is

$$a = \frac{\sum(X_i - X_{\text{mean}})(Y_i - Y_{\text{mean}})}{\sum(X_i - X_{\text{mean}})^2} \qquad (4.10)$$

The equation for the computation of the constant is

$$b = Y_{\text{mean}} - aX_{\text{mean}} \qquad (4.11)$$

Let us look at an example. Assume that a team of researchers is experimenting with a new drug for the treatment of diabetes. They are currently conducting a clinical trial and one of the issues being investigated is whether there is a linear relationship between the daily dosage (in mgs) and the total monthly duration of hypoglycemic episodes (hypoglycemia: drop of blood sugar below a certain level). The data are shown in Table 4.1, where the total duration is measured in hours.

Using the least square estimation formulas above, we derive the following equation:

$$Dur\_Hypogl\_Episodes = 0.669 + 0.008 \times mgs \qquad (4.12)$$

TABLE 4.1   Example Simple Linear Regression Data

Daily Medication Dosage (mg)	Total Duration of Hypoglycemic Episodes
100	1.7
100	0.5
100	1.2
100	3.1
200	1.4
200	2.5
200	3.8
200	2.5
200	0.7
300	3.5
300	4.4
300	2.3
400	3.8
400	4.5
400	4.1
400	3.3
400	5.4
400	3.2

The quality of fit can be evaluated using two metrics, the *standard error of estimate* and R^2. The standard error of estimate is defined as

$$error_{st} = \sqrt{\frac{\Sigma residual_i^2}{N-2}} \qquad (4.13)$$

where $residual_i$ is the residual of the ith point and N is the number of observations. R^2, also known as the *coefficient of determination*, determines the amount of variation of the dependent variable explained by the regression equation. An interesting note is that R^2 is the square of the Pearson correlation coefficient [Fie09]. The equation for R^2 is

$$R^2 = 1 - \frac{\sum residual^2}{\sum (Y_i - Y_{mean})^2} \qquad (4.14)$$

It is desirable to have a value of R^2 as close to 1 as possible. In the example above, the values of the standard error estimate and R^2 are

$$error_{st} = 0.97 \qquad (4.15)$$

$$R^2 = 0.53 \qquad (4.16)$$

The standard error indicates that on average our estimate of the total duration of hypoglycemic episodes will be off by approximately 1 hour. In addition, as R^2 shows, about 53% of the variation in the duration of hypoglycemic episodes is explained by the daily medication dosage. Let us see now how adding independent variables can improve the regression model. This brings us to the topic of *multiple regression*.

4.2.2 Linear Multiple Regression

In multiple regression, there are more than one independent variable. In this case, we are no longer talking about a fitted line. In the case of 3 independent variables for example, we are talking about a *fitted plane*. The equation below shows the multiple regression equation:

$$Y = a + c_1 X_1 + c_2 X_2 + \dots \qquad (4.17)$$

Let us now consider the additional independent variables in the above presented example. So far we have examined only the drug dosage as a

TABLE 4.2   Data for the Prediction of the Total Monthly Duration of Hypoglycemic Episodes

History	Age	Daily Medication Dosage (mg)	Duration of Hypoglycemic Episodes
0	60	100	1.7
0	55	100	0.5
0	65	100	1.2
1	80	100	3.1
0	55	200	1.4
0	85	200	2.5
1	90	200	3.8
0	75	200	2.5
0	60	200	0.7
1	63	300	3.5
0	80	300	4.4
0	62	300	2.3
0	73	400	3.8
0	78	400	4.5
1	65	400	4.1
0	62	400	3.3
1	55	400	5.4
0	60	400	3.2

potential factor for predicting the total monthly duration of hypoglycemic episodes. Let us now add two more factors, the *age of the patient* and whether the *patient has a history of hypoglycemia* (See Table 4.2). Hypoglycemia history is a categorical variable that will be modeled as a binary variable where 1 indicates that there has been such a history and 0 that there has not.

The resulting multiple regression equation is

$$Dur\_Hypogl\_Episodes = -2.673 + 1.134 \times Previous + 0.045 \times Age + 0.008 \times mgs \qquad (4.18)$$

R^2 has now increased to 0.81 which indicates that 81% of the variability in the duration of hypoglycemic episodes is explained by the independent variables. The standard error has also been reduced to 0.59. The above equation can be interpreted as follows by the clinical researchers: With everything else held constant, for every 10 years increase in age, the duration of hypoglycemic episodes increases by 0.45. With everything else held constant, for every 100 mgs increase in the medication dosage the increase of the duration of hypoglycemic episodes is close to 1 hour

(0.8 specifically). Finally, with everything else held constant, existence of hypoglycemic history increases the duration of hypoglycemic episodes by approximately 1 hour. Now that we have the regression equation, let us do a prediction for a patient with no hypoglycemia history, 77 years old, who is on 200 mgs of the drug:

$$Dur\_Hypogl\_Episodes = -2.673 + 0.045 \times 77 + 0.008 \times 200 \\ = 2.39 \text{ hours} \tag{4.19}$$

Although the coefficients of simple linear regression can be computed by hand, for the computation of the coefficients of multiple regression, a statistical software package needs to be used. A plethora of such packages exists today. In this book, Stat Tools by *Palisade*™ *Decision Tools* has been used.

Let us now look at another example, where the temporal prediction deals with the *duration* of an event. Specifically, a financial firm has introduced a new type of financial investment choice for its existing investors, and a financial advisor is trying to estimate the time needed to explain the new product to an investor. The specific investor has $100K to invest, has no experience with this product and already has $90K invested in the firm. The advisor constructs a regression analysis where the dependent variable is the duration of the consultation (in minutes) and the independent variables are the amount already invested by the investor in the firm (in thousands), the amount to be invested in the new investment (in thousands), and whether the investor has any experience with the type of the new financial investment. The latter is modeled as a categorical variable (0 indicates no experience; 1 indicates experience). The data are shown in Table 4.3, where the dollar amounts are in thousands:

The regression equation is

$$Consult\_time = 16.466 - 0.024 \times Amount\_invested + 0.267 \\ \times Amount\_to\_be\_invested - 8.484 \times \text{experience} \tag{4.20}$$

The R^2 and standard error of estimate are, respectively, 0.90 and 6.39. The high value of R^2 indicates that 90% of the variation in the consultation time is explained by the amount the investor already has invested in the firm, the amount he/she plans to invest, and the experience of the investor with similar products. The standard error indicates that using the above regression equation, we are off on average by approximately 6 min. For

TABLE 4.3   Data for the Prediction of the Duration of a
Temporal Event Using Regression

Amount Already Invested	Amount to Be Invested	Relevant Experience	Consultation Time
80	70	0	30
28	90	0	50
150	20	1	15
32	150	1	45
45	100	0	40
30	90	0	33
200	20	1	10
88	30	0	20
30	150	0	60
150	50	1	13

the specific investor that the financial advisor was wondering about, the estimated consultation time is

$$16.466 - 0.024 \times 100 + 0.267 \times 90 \approx 38 \; minutes \qquad (4.21)$$

4.2.3 Other Regression Issues

Regarding the quality of fit of the regression model, so far we have examined only R^2 and the standard estimate of error. Statistical software packages also provide another very useful number, the *p-value* for each regression coefficient. The smaller the *p-value*, the more certain we are about the significance of having this coefficient in our regression equation. The *p-value* becomes meaningful in light of a *significance level, a.* Let us assume that in our first example, the clinical researchers set the significance level to 95%; therefore, our alpha is $\alpha = 0.05$. The meaning of the *p-value* is as follows: If *p-value* $< \alpha$, then we can reject the null hypothesis that the coefficient is 0. In other words, if *p-value* $< \alpha$, we are certain that the corresponding variable is statistically significant, in the sense that it should be in our regression equation. In our example, the *p-values* of the regression coefficients are as shown in Table 4.4. From Table 4.4, we see that the *p-values* of all

TABLE 4.4   Regression Coefficient *p*-Values

Coefficient	*p*-Value
Presence of Hypogl. History	0.0033
Age	0.0047
Mgs	<0.0001

regression coefficients are less than our α (0.05), and therefore, they are all statistical significant in our regression equation.

It is important to note that the variables in a linear regression equation do not have to be the original variables, but may be some transformations of them. This could happen if the original independent variables and the dependent variable had a nonlinear relationship, which we wanted to *linearize*. *Linearization* can be achieved by transforming the independent variables and/or the dependent variable using a transformation, such as *log*. Examples of linearization will also be presented in the time series forecasting section.

A good idea before creating a regression model is to examine whether there are linear relationships between some of the independent variables. This can be examined by creating *scatterplots* of pairs of variables. If any such relationship is detected, then we need to keep only one of the involved variables in our regression equation. Having an independent variable that is linearly related with another is not the only reason to remove a variable from a regression equation. If the *p-value* of the corresponding coefficient in the regression analysis is significantly larger than our set α level, then this indicates that the variable is not statistically significant in our regression equation and could be removed.

One of the assumptions of linear regression is that the variance of the errors is the same for all levels of the independent variable. This is an assumption that is quite often violated. To test whether this happens in our data set, we create a scatterplot of the dependent variable and each one of the independent variables. If one or more of the scatterplots has a fan shape (a widening distribution) of values as the independent variable values increase, then this is an indication of an assumption violation. An example is shown in Figure 4.1.

It is a simulated graph, where the dependent variable is the monthly dollar amount (per household) spent on imported luxury food items, such as wine and cheese, and the independent variable is the square footage of the house owned or rented by the people of the household. As the graph shows, as the square footage increases the variability of the dollar amount also increases. A common way to remedy this is to perform a logarithmic transformation of the dependent variable.

Another assumption of linear regression is that the errors are normally distributed. We can check whether this assumption holds by plotting the residuals. If nonnormality is detected, then a logarithmic transformation of the dependent variable could correct the situation. Finally, in linear regression it is assumed that the errors are independent of each other. This assumption holds for most data except time-series, which are quite often

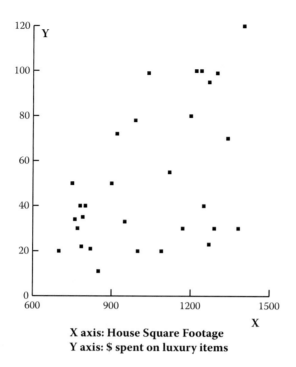

X axis: House Square Footage
Y axis: $ spent on luxury items

FIGURE 4.1   Example of an increasing variance.

positively autocorrelated. Time series prediction will be discussed in the next section. It is interesting to note, that given the above assumptions, regression using least-squares estimation produces a maximum likelihood estimate of the regression equation's parameters [Pre02].

Finally, an interesting question that often arises is how to compare two regression lines. The comparison can be done in three different ways: (a) compare the slopes (ignoring the intercept); (b) compare the intercepts (ignoring the slopes); (c) perform an overall test of coincidence of the two lines. For further discussion of this subject, see [Gla05].

4.2.4 Learning to Predict Rare Events in Event Sequences

In [Wei98], a genetic algorithm called *TIMEWEAVER* that predicts rare events in event sequences is discussed. Important concepts in the algorithm are the following:

- *Event sequence*: A time-ordered sequence of events

- *Target event*: The event to be predicted, which is described as a set of feature value pairs

- *Warning time, W:* The lead time necessary for a prediction to be useful

- *M:* Maximum amount of time before the occurence of the target event for which a prediction is considered correct

The algorithm consists of the following two steps:

- *Identifying prediction patterns.* A genetic algorithm is employed to find a set of patterns. A steady-state genetic algorithm is employed, where only few individuals are changed in each *iteration*. The population algorithm is initialized with patterns that create a single event. The genetic algorithm employed has a novel approach regarding the algorithm's *change and replacement strategies* because the genetic algorithm must satisfy two conflicting criteria: *finding the optimum areas in the search space* and also *maintaining a population diversion*. To find the best patterns a fitness criterion is used:

$$fitness = \frac{(\beta^2 + 1)\,precision \times recall}{\beta^2\,precision + recall} \tag{4.22}$$

where *recall* and *precision* are defined as

$$recall = \frac{number\ of\ target\ events\ predicted}{total\ target\ events} \tag{4.23}$$

$$precision = \frac{Number\ of\ target\ events\ predicted}{Number\ of\ target\ events\ predicted + disc.\ false\ positive\ predictions} \tag{4.24}$$

where this definition of precision counts correct predictions only once. It also takes into account the fact that a prediction is "active" during its monitoring time and therefore adjusts for false predictions that have overlapping monitoring periods.

To encourage diversity, a *niching measure* is employed which is shown below:

$$niche\ count = \sum_{j=1}^{n}(1 - distance(i,j))^3 \tag{4.25}$$

This measure is an indicator of the similarity of individual i to the entire population, where the distance is the fraction of target events that are predicted differently in i and j. Increased similarity of an individual to the rest of population means smaller distance and increased niche value.

- *Generating prediction rules.* A greedy algorithm is used to form a list of prediction rules from a set of candidate patterns.

4.3 TIME SERIES FORECASTING

4.3.1 Moving Averages

The simplest form of forecasting is to use an average of past patterns to forecast future patterns. This is the main idea behind *moving averages*. To forecast the time series value at time t_1, t_{1+1}, t_{1+2}, we average the past values using a sliding window with width τ, where τ has to be carefully chosen such that it tracks the original series closely, while discarding small and probably random variations.

The moving average forecasting method was applied to part of the *S&P 500*™ *time series (weekly data)*, to forecast the *S&P 500* closing price for the next week. The moving average was applied using two sliding windows of size 5 and 11. Figure 4.2 shows the application of the moving average with size 5 and size 11.

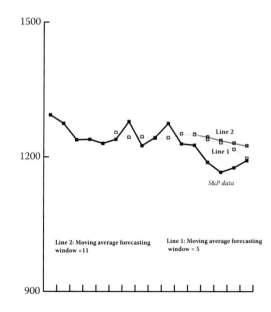

FIGURE 4.2 Moving average forecasting with window sizes 5 and 11 for *S&P 500* data .

TABLE 4.5   Moving Average Forecasting for *S&P 500* Data

Time Period	Actual Values	Forecasted Values Using a Window of Size 5	Forecasted Values Using a Window of Size 11
1	1294.59		
2	1275.47		
3	1238.33		
4	1239.22		
5	1230.13		
6	1239.40	1255.55	
7	1279.64	1244.51	
8	1225.19	1245.34	
9	1243.26	1242.72	
10	1275.09	1243.52	
11	1229.23	1252.52	
12	1226.27	1250.48	1251.78
13	1188.03	1239.81	1245.57
14	1166.46	1232.38	1237.62
15	1176.74	1217.02	1231.08
16	1192.33	1197.35	1225.40

Using a sliding window of size 5 yields a significantly smaller mean square error than the window of size 11 (1154 versus 2614). Table 4.5 shows the forecasted values and the actual values for the *S&P 500* time series. The smaller MSE for window size 5 can be attributed to the fact that, for window size 11, local peaks and valleys disappear and the actual shape of the time series gets distorted.

4.3.2 Exponential Smoothing

In exponential smoothing, via the use of a smoothing constant, we can weigh *differently recent observations* and *differently past observations*. This is a clear advantage in comparison with the moving average method, where the only choice we have is the size of the sliding window. The formula for simple exponential smoothing is shown below, where F_{t+i} is the forecasted value, a is the smoothing constant, and Y_t, Y_{t-1}, and so on are past observations:

$$F_{t+i} = aY_t + a(1-a)Y_{t-1} + a(1-a)^2 Y_{t-2} + a(1-a)^3 Y_{t-3} + \ldots \quad (4.26)$$

A question that arises in exponential smoothing is what value to give to the smoothing constant. In general, the smoothing constant always has a value between 0 and 1. A value close to 0 means that observations from the distant past play an important role. A value close to 1 means that recent observations play an important role in forecasting. For the *S&P 500* time series, two values for the smoothing were tried, 0.3 and 0.7. Simple exponential smoothing with $a = 0.7$ produced the smallest mean square error of 791.8 (versus 1109.5 for $a = 0.3$). Table 4.6 shows the exponential smoothing results. The results are also shown in Figures 4.3 and 4.4. The fact that $a = 0.7$ produces the smaller mean square error means that in the series, the most recent observations are the more important ones in forecasting.

Exponential smoothing can be improved, in the case of data with a trend, by using *Holt's formulas* shown below [Alb06]. The term L_t represents the level of the series (in the absence of noise), the term T_t represents the trend of the series, and the new constant b is usually set equal to a and deals with changes in the trend.

$$F_{t+i} = L_t + iT_t \tag{4.27}$$

where

$$T_t = b(L_t - L_{t-1}) + (1-b)T_{t-1} \tag{4.28}$$

TABLE 4.6   Exponential Smoothing Forecasting for *S&P 500* Data

Time Period	Actual Values	Forecasted Values Using $a = 0.3$	Forecasted Values Using $a = 0.7$
1	1294.59		
2	1275.47	1294.59	1294.59
3	1238.33	1288.85	1281.21
4	1239.22	1273.70	1251.19
5	1230.13	1263.35	1242.81
6	1239.40	1253.39	1233.93
7	1279.64	1249.19	1237.76
8	1225.19	1258.33	1267.08
9	1243.26	1248.38	1237.76
10	1275.09	1246.85	1241.61
11	1229.23	1255.32	1265.06
12	1226.27	1247.49	1239.97
13	1188.03	1241.13	1230.38
14	1166.46	1225.20	1200.74
15	1176.74	1207.58	1176.74
16	1192.33	1198.33	1176.74

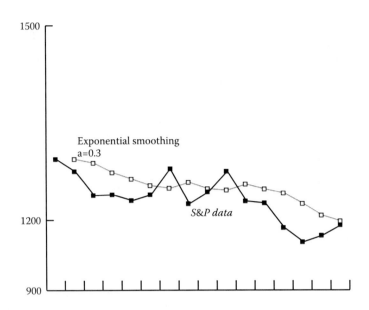

FIGURE 4.3 Simple exponential smoothing forecasting with $a = 0.3$ for *S&P 500* data.

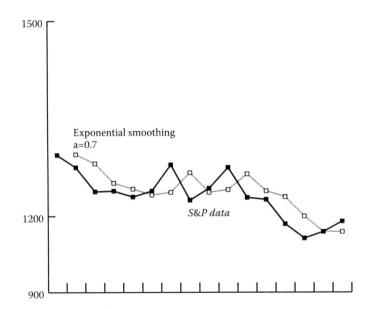

FIGURE 4.4 Simple exponential smoothing forecasting with $a = 0.7$ for *S&P 500* data.

and

$$L_t = aY_t + (1-a)(L_{t-1} + T_{t-1}) \qquad (4.29)$$

4.3.3 Time Series Forecasting via Regression

For data with a linear trend, regression can be used for forecasting where the independent variable is time. Let us use linear regression to forecast the last *S&P* closing weekly price shown in Table 4.6. That price is 1,192.33.

The data and the fitted regression line are shown in Figure 4.5.

The computed regression equation is

$$Closing\_price = 1285.13 - 6.25 \times week \qquad (4.30)$$

The R^2 is 0.5629 and the standard error of estimate is 25.56. The forecasted closing price for the last value of the table is 1185.13.

For data with nonlinear trends, such as exponential, we can still use linear regression to perform forecasting, but we must first perform a logarithmic transformation of the dependent variable.

4.3.4 Forecasting Seasonal Data via Regression

Regression can be used to forecast time series data that exhibit seasonality by modeling seasonality via a dummy variable. As before, once we have the regression equation, we can forecast the observed value for a time period

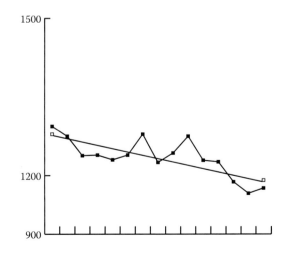

FIGURE 4.5   Regression line for *S&P500* data.

TABLE 4.7   Seasonal Data for
Number of Hotel Guests

Time Period in Months	Guests	Season
1	350	0
2	320	0
3	280	1
4	250	1
5	300	1
6	319	1
7	344	0
8	350	0
9	299	1
10	288	1
11	346	0
12	350	0
13	300	0
14	280	0
15	260	1
16	276	1
17	288	1
18	300	1
19	310	0

by plugging the time period and the other independent variables into the regression equation. Let us consider an example where a hotel with capacity of 350 beds has the largest number of guests during two peak seasons: November to February and July to August. We use a dummy variable to model seasonality, where 0 indicates a peak season and 1 a slow season. The data are shown in Table 4.7.

The regression equation is shown below:

$$Number\_of\_guests = 338 - 1.09 \times timeperiod - 41 \times season \quad (4.31)$$

R^2 is 0.51 and the standard error of estimate is 23.16. We are therefore, on average, 23 guests off in our estimate of hotel reservations.

4.3.5 Random Walk

In this forecasting method, the main idea is that the difference between successive observations is random (*a random walk*). If the mean difference is positive, then there is an upward trend in the series, while if the mean difference is negative, there is a downward trend in the series. An observation, k periods away from the last known observation Y_t, can be forecasted as follows:

$$F_{k+t} = Y_t + (k \times mean\ difference) \tag{4.32}$$

Let us go back to the *S&P 500* example. The 15 weeks of *S&P 500* data and their differences are shown in Table 4.8. The differences are also shown in Figure 4.6.

As can be seen from the table and the figure, the differences do not follow any pattern, instead they appear random. Therefore, the random walk model is applicable.

As in previous examples, let us now forecast the closing price of the very last week in the actual value column of Table 4.6, using the random walk method. Again, that value is 1192.33. The mean difference of the *S&P* values in Table 4.8 is −8.41. Using the last known value, 1176.74, the forecasted value is

$$Forecasted\_value = 1176.74 + (1 \times (-8.41)) = 1168.33 \tag{4.33}$$

TABLE 4.8   Successive Value
Differences in *S&P 500* Data

S&P500 Values	Difference
1294.59	
1275.47	−19.12
1238.33	−37.14
1239.22	0.89
1230.13	−9.09
1239.40	9.27
1279.64	40.24
1225.19	−54.45
1243.26	18.07
1275.09	31.83
1229.23	−45.86
1226.27	−2.96
1188.03	−38.24
1166.46	−21.57
1176.74	10.28

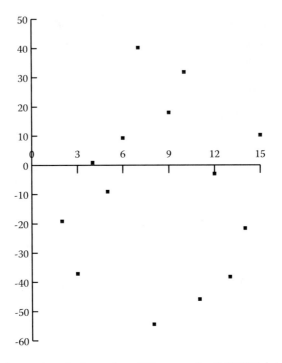

FIGURE 4.6   Successive closing prices differences for *S&P 500* data.

4.3.6 Autocorrelation

Quite often, successive observations of time series are correlated. Most often this is a positive correlation, in the sense that large values follow large values and small values follow small values. The autocorrelation formula is shown below:

$$AC_m = \frac{\sum_{i=1}^{N-m}(Y_i - Y_{mean})(Y_{i+m} - Y_{mean})}{\sum_{i=1}^{N}(Y_i - Y_{mean})^2} \tag{4.34}$$

where $Y_1, Y_2,..., Y_N$ is a series of observations, Y_{mean} is the mean of the observations, and m is the *lag* we are computing.

Computing the autocorrelation for a time series can be used for two different purposes:

1. To test whether the time series is *random*. In this case, the autocorrelation between successive observations will be very small.

TABLE 4.9   Autocorrelations for
the *S&P 500* Data

Lags	Autocorrelation
Lag 1	0.513
Lag 2	0.169
Lag 3	0.058

2. To find the most important *lags.* These are lagged versions of the time series with the highest autocorrelation.

These, as will be explained next, can be used to compute the *autoregression equation* of the time series. Table 4.9 shows the autocorrelation for the first three lags of the *S&P 500* data shown in Table 4.8. As we see from the table, the first lag has the highest value.

4.3.7 Autoregression

In autoregression, the main idea is to perform time series forecasting via a regression equation where the independent variables are *lagged values of the dependent variable.* A general form of the autoregression equation is

$$Y(t) = constant + C_1 \times Y(t-1) + C_2 \times Y(t-2) + \ldots \qquad (4.35)$$

The difficult question is how many lags to include in the regression equation. Autocorrelation could be a guide as to how many lags to include: high magnitude of autocorrelation indicates a lag that should be included in the regression equation. Let us compute the autoregression equation for the *S&P 500* data above, using Lag 1, because it has the highest value.

The autoregression equation using Lag 1 is

$$Closing\_price = 428.45 + 0.6 \times lag1 \qquad (4.36)$$

The R^2 and standard error estimate are

$$R^2 = 0.42 \qquad (4.37)$$

and

$$err_{st} = 27.43 \qquad (4.38)$$

The R^2 value is small, which means that *lag1* is not enough to explain the variability in the *S&P 500* value. As in previous examples, let us forecast the closing price of the last week in Table 4.6 (1192.33). Using the above autoregression equation, the value is 1181.56.

4.3.8 ARMA Models

As the name implies, *ARMA* (*AutoRegressive Moving Average*) models combine autoregression and moving average models. Sometimes they are also referred to as *ARIMA* (*AutoRegressive Integrated Moving Average*) models. *ARMA* modeling is a very powerful modeling technique that can be used to model a large number of time series data. For detailed reading on the subject, a popular source is [Box94]. Generally, the *ARMA* model equation for a time series is given by

$$Y(t) = \sum_{i=1}^{M} a_i Y(t-i) + \sum_{i=1}^{N} b_i E(t-i) + \varepsilon(t) \qquad (4.39)$$

where the first summation is the *autoregressive part* and the second summation is the *moving average part*. $\varepsilon(t)$ is an error term that is assumed to be a random variable sampled from a normal distribution. The autoregressive part consists of weighted past observations, and the moving average part consists of weighted past estimation errors. In other words, $E(t-1)$ is the difference between the actual value and the forecasted value in the previous observation.

ARMA models are designed for stationary time series, although they can also be applied to nonstationary ones, after the latter are transformed to stationary time series. As described in [Tra01], there are generally three steps in fitting an *ARIMA* model to a time series:

- *Model Identification:* Here, the mean, autocorrelation, and partial autocorrelation based on the time series samples are computed. Then, these can be used to estimate M, N in the autoregressive equation, that is, the orders of the autoregressive and moving average parts. A way to do that is described in [Ciz05]. An alternative method to estimate M, N is described in [Den97]. This method is based on *AIC*, which is an *entropy-based selection* criterion described in [Aka76]. Specifically, various combinations are tried and the combination that yields the smallest *AIC* value is chosen. The *AIC* has been widely used for a number of applications in engineering and statistics and consists of two terms, where the first term shows the lack-of-fit of the model and the second term is a model of the unreliability of the model [Mut94].

- *Parameter Estimation:* The parameters of the model are derived using either *Maximum Likelihood* or *Least Squares estimation*. For example, in the *Least Squares* approach, assuming only an autoregressive part, the autoregression coefficients are estimated by minimizing the sum of squares of the prediction errors.

- *Forecasting:* This is done to predict future values and can be either short- or long-term forecasting.

The process above describes an *offline* approach to constructing the *ARIMA* model. An *online* approach is described in [Tra01], where an iterative least-squares parameter estimator method (*ELS*) is used.

4.4 ADVANCED TIME SERIES FORECASTING TECHNIQUES

In this section, we will discuss advanced techniques for time series forecasting, such as *neural networks* and *genetic algorithms*. It is still a matter of debate as to whether complex techniques perform forecasting better than simple techniques, such as autoregression and moving average. However, a considerable amount of work has been done in advanced time series forecasting, and it is worthwhile discussing it because for some data sets, such as nonlinear and chaotic, these techniques do indeed outperform conventional techniques.

4.4.1 Neural Networks and Genetic Algorithms in Time Series Forecasting

The advent of *neural network* (*NN*) research has fueled a significant amount of research dedicated to the use of *NNs* in time series forecasting. A review of such research can be found in [Dor96], [Fra01], [Zha04], and [Pal05]. The main reason for this research is that *NNs* can be used to model and predict *nonlinear time series*, an area of considerable difficulty. In the simplest form of these algorithms, time-delayed samples of the time series are fed to the *input layer* of a neural network. As the complexity of the time series increases, so does the size of the input layer and the number of the corresponding weights.

Quite often a *genetic algorithm* is used to *fine-tune* important aspects of the neural network. *Genetic algorithms* are heuristic algorithms based on biological concepts, specifically organism evolution. Their main operations are *inheritance, mutation, selection,* and *crossover.* A *fitness* function is used to guide the solution process, specifically to select which subjects from the population will breed the next generation.

For example, in [Fer04], a hybrid intelligent method is presented that employs a genetic algorithm to select the best training algorithm and also the best structure for the neural network. In addition, the genetic algorithm selects the best number of lags to represent the time series. The neural network in this algorithm is a *multilayer perceptron.* In [Fer05], a *Time-Delay–Added Evolutionary Forecasting method* is used for time series forecasting. Specifically, the method performs a search using evolutionary techniques to find the minimum number of embedded dimensions to determine the characteristic phase space of the time series. Time phase distortions are adjusted using a behavioral statistical test.

Given the widespread application of neural networks in time series forecasting, Berardi and Zhang in [Ber03] investigated their success in terms of bias and variance. *NNs* have the advantage of being able to model any kind of nonlinear complex data, with the downside, however, of being so data-dependent that over fitting becomes a problem. The most important findings of their study can be summarized as follows:

- The learning capability of the *NN* and its success in time forecasting are affected by both hidden and input nodes.

- Overspecifying the number of input lags has a significant effect on the model variance.

- Overspecifying the number of hidden nodes has a significant effect on the model bias.

- Underspecifying the number of hidden nodes or input lags can have a severe negative effect on the model bias.

In [Teo01], the authors describe a method that utilizes wavelets to initialize neural networks used in time series forecasting. A *multilayer perceptron neural network (MLP)* is trained using back propagation. The weights in back propagation are usually initially chosen as random values. The choice of the initial values is important as it can lead to getting trapped in local optimums. The novelty of the proposed algorithm is that it initializes the weights using wavelets and clustering concepts.

The specific neural network used in the article is the *WP-MLP* network, which utilizes wavelets as a feature extraction method. Its input is a tapped delay line followed by wavelet decomposition and the *MLP*. The hierarchical and counter propagation clustering algorithms are employed for the neural network weight initialization. The number of clusters is chosen to be the number of neurons in the hidden layer of the *WP-MLP*.

In [Bar07], a review is presented of time series prediction with *Self-Organizing Maps* (*SOMs*). A *SOM* is an unsupervised type of neural network and it is not very widely used in time series prediction. This is in contrast to *MLP* and radial basis function (*RBF*) neural networks, which are supervised architectures and are very popular in time series prediction. The main idea in an *SOM* is to project a high dimensional continuous space onto a lower dimension discrete space. The author notes the following advantages of *SOMs* compared to *MLP* and *RBF*:

- Simple growing architecture.

- Local function approximation. *SOMs* generate the output by mathematical operations on localized space, which allows a better understanding of the underlying process.

- No need to prespecify the number of prototypes and, therefore, the number of neurons.

4.4.2 Application of Clustering in Time Series Forecasting

An interesting method that clusters temporal patterns in overlapping sliding windows of time series is described in [Gev99]. In contrast to traditional *NN* and autoregressive methods, a prediction model is created for each *separate* cluster, where the cluster consists of a set of temporal patterns. A *hierarchical top-down fuzzy clustering technique* is used, where temporal patterns are assigned to the clusters using a degree of membership. A prediction model is created for each cluster using the set of members that have the maximum degree of membership to the cluster. Finally the prediction of a new sample's value is performed by mixing the prediction models that apply to the last sample using its degree of membership to the existing fuzzy clusters as a weight in the prediction results.

Another article that utilizes clustering as a means to perform time series forecasting is [Sfe04]. Specifically, a hybrid clustering algorithm is proposed in the article that creates data clusters, where the data within each cluster are described by the same linear relationship. The clustering algorithm utilizes the correlation coefficient of the time series to find the most important lags of the time series. Then, a data vector is formed from the lagged versions. The clusters are initialized using *k-means*, where the number of clusters is defined by the user. For each cluster, a linear regression model is developed and data are assigned to a cluster based on their distance from the corresponding hyperplane. The algorithm returns a forecast value for each cluster. A pattern recognition algorithm, based on *k-means clustering*, is used to choose a cluster and its corresponding forecast value as the final output. Experimental results on three different data sets exhibited a reduction in the forecasting error by 9%, compared to other methods, including an *autoregressive* model, an *ARIMA* model, and a neural network.

4.4.3 Characterization and Prediction of Complex Time Series Events Using Time-Delayed Embedding

In [Pov03], the authors address the challenging problem of finding *events* in complex time series, such as chaotic time series. In the article, an *event* is defined as an important occurrence in a time series, where importance is context-dependent. The method proposed searches for an optimal pattern by defining a characteristic function to model the concept of *eventness*. Then, the search for the optimal pattern is performed with the help of a genetic algorithm in a reconstructed phase space using time-delayed embedding. A greedy search is used to find an optimal collection of temporal pattern clusters. The method can also be used for *event prediction*. Specifically, in the article, the proposed algorithm is used to predict the release of metal droplets from a welder.

To reconstruct a phase space from a time series signal, we need to choose a time delay and determine an embedding dimension. Overall, the steps of the algorithm in [Pov03] are as follows:

1. Training phase:

  • Define an *event characterization function* and an *objective function* and formulate the problem as a *constrained optimization problem*. An example of a characterization function is

$$g(x(t)) = x(t+1) \qquad\qquad (4.40)$$

which allows one-step prediction. The objective function will be used to order temporal pattern clusters according to their ability to statistically differentiate between events and nonevents.

- Assign a time delay τ and determine an embedding dimension. The time delay was set to 1 in the article. The dimension of the embedding space, which is also the length of the temporal pattern, is determined as $Dim_E = 2m + 1$, where m is the original state space dimension. This definition comes from Takens' theorem [Tak80], which states that if the embedding dimension is given by the above equation, then the reconstructed phase space is topologically equivalent to the original state space. In the article, Dim_E was chosen to be 2.

- Unfold the time series in the reconstructed space; that is, each pair of sequential points generates a 2-D phase space point.

- Form the augmented phase space, by adding the extra dimension of the characteristic function.

- Search for the optimal collection of temporal pattern clusters.

- Evaluate the training stage results and repeat if necessary.

2. Testing phase:

- Embed the time series used for testing into the phase space.

- Identify/predict events, using the optimal collection of temporal pattern clusters.

- Evaluate testing stage results.

The proposed algorithm was compared against a *Time-Delay Neural Network* and *C4.5*. As experimental results showed, the proposed algorithm outperformed the other two methods in the testing phase.

4.5 ADDITIONAL BIBLIOGRAPHY

A popular reference on regression, data analysis in general, and time series forecasting is [Alb06]. Two books on the subject of regression analysis are

[Gel06] and [Men03]. Statistics books with useful information regarding the topics discussed in the chapter are [Fie09] and [Gla05]. There is a large number of books examining the topic of time series analysis and forecasting in detail, such as [Abr83], [Box94], [Bro02], [Cha96], [Del98], [Mak83], [Shu06], and [Yaf00]. A book that addresses the interesting issue of developing foresight in today's knowledge economy is [Tsou04].

Additional methods for time series forecasting using neural networks can be found in [Abb05], [Jan04], and [Zha02]. In [Jun08a], a *NN*-genetic hybrid method is presented that utilizes a neural network for time series forecasting, where the parameters and structure of the *NN* are tuned using evolutionary strategies and a greedy adaptive search procedure. Also, in [Jun08b], a hybrid method, called *GRAPES,* is used for time series forecasting, which is a combination of a randomized adaptive search procedure and evolutionary concepts. Other neural-network hybrid methods are discussed in [Wen02] and [Kon04]. A number of them will also be discussed in Chapter 7, applied to *financial series forecasting.*

Regarding genetic algorithms, in [Luq07] an evolutionary algorithm is described that uses a *Michigan* approach, where the time series forecasting solution is based on the *total population* instead of the most fitted individual. The algorithm allows the evolution of common behaviors, but uncommon situations are also allowed and not discarded as noise. This way, the algorithm has the advantage of taking into account local time series behaviors. In [Nun05], *artificial immune system (AIS)* techniques, that are traditionally used in anomaly detection, are extended to time series forecasting.

In [Pop03], time series forecasting is performed using *independent component analysis*, which is an extension of *principal component analysis.* Specifically, forecasting is performed in the independent component domain and then transformed back in the time domain. In [Pop06], autocorrelation is used to understand a time series to select the best algorithm to preprocess it, such as choosing an algorithm for seasonality removal. In [Cam07], *Support Vector Machines* are used for time series prediction, where nonlinear dynamics determine the *model order,* which is the number of past samples needed to predict the future values of the time series.

In [Han06], a time series forecasting method is proposed that exploits the fact that most time series, such as financial, are controlled by both *macroscopic* and *microscopic laws.* The proposed algorithm decomposes the time series into simpler ones that have different smoothness and then samples them in a multiscale fashion. Each simple time series is modeled and processed separately and the results are integrated at the end. In [Zha08], the authors propose a method for the forecasting of nonlinear and non-Gaussian time series. The method utilizes a *radial basis function (RBF) neural network model* and it makes the assumption that the measurement noise is a *hidden Markov model (HMM)*. The forecasting algorithm is based on the *RBF-HMM model* combined with a *sequential Monte Carlo* method. In [Kim08], a method is proposed for predicting the number of *on-demand video requests* using time series to model the video request data. This is useful in choosing which videos to cache based on their request popularity.

In [Ban08], a method to predict nonstationary or *chaotic time series* is described. The method utilizes data preprocessing based on correlation analysis and multiple fuzzy predictors. To improve the prediction accuracy, an error compensation procedure was used based on correlation analysis. In [Che08], the authors address the problem of *long-term time series forecasting*, which is important for many areas such as climate prediction and urban growth planning. As the authors note, because a huge amount of historical data would be needed to develop long-term forecasts, an alternative method is to use a supervised method where past observations (that could be from simulated data) are used to predict future values. The novel idea in their article is that they use a *semi*-supervised way based on *HMM Regression*. Because there might be inconsistencies between historical and simulation data, the authors also develop a data calibration method. The efficiency of the methods was demonstrated on a variety of data sets.

In [Oh03], a *mixture of locally linear models* is used to perform time series prediction. As the authors note, the attractiveness of the idea lies in that locally linear models have better performance and flexibility. In this method, only nearby states are used to perform the nonlinear mapping. In addition, *Principal Component Analysis* is used to choose parameters of the prediction system.

REFERENCES

[Abb05] Abbas, S.R., and M. Arif, Competitive Neural Network Based Algorithm for Long Range Time Series Forecasting Case Study; Electric Load Forecasting, *9th International IEEE Multitopic Conference (INMIC)*, pp. 1–6, December 2005.

[Abr83] Abraham, B. and J. Ledolter, *Statistical Methods for Forecasting*, Wiley, 1983.

[Aka76] Akaike, H., Canonical Correlation Analysis of Time Series and the Use of an Information Criterion, *System Identification: Advances and Case Studies*, R. K. Mehra and D.G. Lainiotis, eds., Academic Press, 1976.

[Alb06] Albright, S.C., W.L. Winston, and C. Zappe, *Data Analysis & Decision Making*, Thomson Higher Education, 2006.

[Ban08] Bang, Y.K. and C.H. Lee, Fuzzy Time Series Prediction with Data Preprocessing and Error Compensation Based on Correlation Analysis, *Proceedings of the 3rd International Conference on Convergence and Hybrid Information Technology*, pp. 714–721, 2008.

[Bar07] Barreto, G.A., Time Series Prediction with the Self-Organizing Map: A Review, Perspectives of Neural-Symbolic Integration, in *Studies in Computational Intelligence*, Springer, vol. 77, pp. 135–158, 2007.

[Ber03] Berardi, V.L. and G.P. Zhang, An Empirical Investigation of Bias and Investigation in Time Series Forecasting: Modeling Considerations and Error Evaluation, *IEEE Transactions on Neural Networks*, vol. 14, no. 3, pp. 668–679, May 2003.

[Box94] Box, G.E.P., G.M. Jenkins, and G.C. Reinsel, *Time Series Analysis, Forecasting, and Control*, 3rd ed., Prentice Hall, 1994.

[Bro02] Brockwell, P.J. and R.A. Davis, *Introduction to Time Series and Forecasting*, 2nd ed., Springer-Verlag, 2002.

[Cam07] Camastra, F. and M. Filippone, SVM-Based Time Series Prediction with Nonlinear Dynamic Methods, *Proceedings of Knowledge-Based & Intelligent Information & Engineering Systems Conference (KES)*, vol. 3, pp. 300–307, 2007.

[Cha96] Chatfield, C., *The Analysis of Time Series*, 5th ed., Chapman & Hall, 1996.

[Che08] Cheng, H. and P.N. Tan, Semi-Supervised Learning with Data Calibration for Long-Term Time Series Forecasting, *Proceedings of ACM SIGKDD Conference*, pp. 133–141, 2008.

[Ciz05] Cizek, P., W. Hardle, and R. Weron, Statistical Tools for Finance and Insurance, http://fedc.wiwi.hu-berlin.de/xplore/tutorials/xegbohtmlnode39.html, 2005.

[Del98] DeLurgio, S.A., *Forecasting Principles and Applications*, Irwin McGraw-Hill, Boston, MA, 1998.

[Den97] Deng, K., A.W. Moore, and M.C. Nechyba, Learning to Recognize Time Series: Combining ARMA Models with Memory-Based Learning, *IEEE International Symposium on Computational Intelligence in Robotics and Automation*, pp. 246–251, 1997.

[Dor96] Dorffner, G., Neural Networks for Time Series Processing, *Neural Network World*, vol. 6, no. 4, pp. 447–468, 1996.

[Fer04] Ferreira, T.A.E., G.C. Vasconselos, and P.J.L. Adeodato, A Hybrid Intelligent System Approach for Improving the Prediction of Real World Time Series, *Congress on Evolutionary Computation*, vol. 1, pp. 736–743, 2004.

[Fer05] Ferreira, T.A.E., G.C. Vasconcelos, and P.J.L. Adeodato, A New Evolutionary Method for Time Series Forecasting, *Proceedings of the 2005 Annual Conference on Genetic and Evolutionary Computing*, pp. 2221–2222, 2005.

[Fie09] Field, A., *Discovering Statistics Using SPSS*, 3rd ed., Sage Publishing, 2009.

[Fra01] Frank, R.J., N. Davey, and S.P. Hunt, Time Series Prediction and Neural Networks, *Journal of Intelligent and Robotic Systems*, vol. 31, pp. 91–103, 2001.

[Gel06] Gelman, A. and J. Hill, *Data Analysis Using Regression and Multilevel/ Hierarchical Models,* Cambridge University Press, 2006.

[Gev99] Geva, A.B., Non-stationary Time Series Prediction Using Fuzzy Clustering, *Proceedings of the 18th International North American Fuzzy Information Processing Society Conference (NAFIPS)*, pp. 413–417, 1999.

[Gla05] Glantz, S., Primer of Biostatistics, 6th ed., McGraw-Hill Medical, 2005.

[Han06] Han, X. et al., A Novel Time Series Forecasting Approach with Multi-level Data Decomposing and Modeling, *The Sixth World Congress on Intelligent Control and Automation,* pp. 1712–1716, 2006.

[Jan04] Jana, P.K., A Neural Network Based Time Series Forecasting, *Proceedings of Intelligent Sensing and Information Processing International Conference,* pp. 329–331, 2004.

[Jun08a] Junior, A., A Study for Multi-Objective Fitness Function for Time Series Forecasting with Intelligent Techniques, *Proceedings of the 10th Annual Conference on Genetic and Evolutionary Computation (GECCO'08)*, Atlanta, Georgia, vol. 1, pp. 1843–1846, July 12–16, 2008.

[Jun08b] Junior, A. and T.A. Ferreira, A Hybrid Method for Tuning Neural Network for Time Series Forecasting, *Proceedings of the 10th Annual Conference on Genetic and Evolutionary Computation (GECCO'08)*, Atlanta, Georgia, pp. 531–532, July 12–16, 2008.

[Kim08] Kimiyama, H. and S. Itoh, Method of Predicting Number of On-Demand Video Requests Using Time Series Data for Video Cache System, *Proceedings of the 6th International Conference on Advances in Mobile Computing and Multimedia*, Linz, Austria, pp. 200–205, 2008.

[Kon04] Kong, S.G., Time Series Prediction with Evolvable Block-Based Neural Networks, *International Joint Conference on Neural Networks,* vol. 2, pp. 1579–1583, 2004.

[Luq07] Luque, C., J.M.V. Ferran, and P.I. Vinuela, Time Series Forecasting by Means of Evolutionary Algorithm, *Parallel and Distributed Processing Symposium*, pp. 1–7, 26–30 March 2007.

[Mak83] Makradakis, S., S.C., Wheelwright, and V.E. McGhee, *Forecasting: Methods and Applications*, 2nd ed., Wiley, 1983.

[Men03] Mendenhall, W. and T.L. Sincich, *A Second Course in Statistics: Regression Analysis* (6th ed.), Prentice Hall, 2003.

[Mut94] Mutua, F.M., The Use of the Akaike Information Criterion in the Indentification of an Optimum Flood Frequency Model, *Hydrological Sciences*, vol. 39, no. 3, pp. 235–244, June 1994.

[Nun05] Nunn, I. and T. White, The Application of Antigenic Search Techniques to Time Series Forecasting, *Proceedings of GECCO'05 Conference*, Washington, DC, June 25–29, pp. 353–360, 2005.

[Oh03] Oh, S.K., K.H. Seo, and J.J. Lee, Time Series Prediction by Mixture of Linear Local Models, *The 29th International Conference of the IEEE Industrial Electronics Society,* vol. 5, pp. 1905–1908, 2003.

[Orm09] Orme, J.G. and T. Combe-Orme, *Multiple Regression with Discrete Dependent Variables*, Oxford University Press, 2009.

[Pal05] Palit, A.K. and D. Popovic, *Computational Intelligence in Time Series Forecasting: Theory and Engineering Applications*, 1st ed., Springer, 2005.

[Pop03] Popescu, T.D., Multivariate Time Series Forecasting Using Independent Component Analysis, *Proceedings of IEEE Conference on Emerging Technologies and Factory Automation*, vol. 2, pp. 366–371, 5–8 December 2004.

[Pop06] Popoola, A. and K. Ahmad, Fuzzy Models for Time Series Analysis: Towards Systematic Data Preprocessing, *IEEE International Conference on Engineering of Intelligent Systems*, pp. 1–5, April 2006.

[Pov03] Povinelli, R.J. and X. Fing, A New Temporal Pattern Identification Method for Characterization and Prediction of Complex Time Series Events, *IEEE Transactions on Knowledge and Data Engineering*, vol. 15, no. 2, pp. 339–352, March/April 2003.

[Pre02] Press, W.H., S.A. Teukolsky, W.T. Ve Herling, B.P. Flannery, *Numerical Recipes in C*, 2nd ed., Cambridge University Press, 2002.

[Sfe04] Sfetsos, A. and C. Siriopoulos, Time Series Forecasting with a Hybrid Clustering Scheme and Pattern Recognition, *IEEE Transactions on Systems, Man, and Cybernetics*, vol. 34, no. 3, pp. 399–405, May 2004.

[Shu06] Shumway. R.H. and D.S. Stoffer, *Time Series Analysis and Its Applications: With R Examples*, Springer, 2006.

[Tak80] Takens, F., Detecting Strange Attractors in Turbulence, *Proceedings of Dynamic Systems and Turbulence*, pp. 366–381, 1980.

[Teo01] Teo, K.K., L. Wang, and Z. Lin, Wavelet Packet Multi-layer Perceptron for Chaotic Time Series Prediction: Effects of Weight Initialization, *Lecture Notes in Computer Science*, vol. 2074, pp. 310–317, Springer-Verlag, 2001.

[Tra01] Tran, N. and D.A. Reed, ARIMA Time Series Modeling and Forecasting for Adaptive I/O Prefetching, *Proceedings of ACM International Conference on Supercomputing*, pp. 473–484, 2001.

[Tsou04] Tsoukas, H. and J. Sheperd, Eds. *Managing the Future: Foresight in the Knowledge Economy*, Wiley–Blackwell, 2004.

[Wei98] Weiss, G.M. and H. Hirsch, Learning to Predict Rare Events in Event Sequences, *Proceedings of the 4th International Conference on Knowledge Discovery and Data Mining*, pp. 359–363, AAAI Press, 1998.

[Wen02] Wen, C.H. and M. Yao, A Combination of Traditional Time Series Forecasting with Fuzzy Learning Neural Networks, *International Conference on Machine Learning and Cybernetics*, vol. 1, pp. 21–23, 4–5 November 2002.

[Yaf00] Yaffee, R.A. and M. McGee, *Time Series Analysis and Forecasting*, Academic Press, 2000.

[Zha02] Zhang, T. and A. Fukushinge, Forecasting Time Series by Bayesian Neural Networks, *Proceedings of the 2002 International Conference on Networks*, vol. 1, pp. 382–387, 12–17 May 2002.

[Zha04] Zhang, G., B.E. Patuwo, and M.Y. Hu, Forecasting with Artificial Neural Networks: The State of the Art, *Int. J. Forecasting*, vol. 14, pp. 35–62, 1998.

[Zha08] Zhang, D. et al., A Framework for Time Series Forecasts, *International Colloquium on Computing, Communication, Control, and Management*, pp. 52–56, 2008.

CHAPTER 5

Temporal Pattern Discovery

Temporal pattern discovery deals with the discovery of *temporal patterns of interest in time series or temporal sequences*, where the interest is determined by the domain and the application. Below are some examples of temporal patterns of *interest*:

- Customers who buy laptops on a computer company's site in a transaction, frequently buy printers from the same site on a later transaction. This is an example of a frequent sequence.

- Patients who are on drug X for over a month, sometimes start suffering from severe headaches after a month. This is a temporal association tion rule, but also a potentially *causal* rule.

- A Web site's number of hits has a sharp peak around noon and 7–8 p.m. and a deep valley between 1 a.m. and 6 a.m. This is a time series pattern.

There is a significant amount of work in the literature for discovering temporal patterns of interest in sequence databases and time series databases. In sequence databases, the problem is usually examined under three frameworks: *sequence mining (or frequent sequential pattern discovery), temporal association rule discovery,* and *frequent episode discovery*. The last two terms typically refer to the discovery of sequential patterns in a *single* data sequence, while the first term typically refers to the discovery of sequential patterns in multiple data sequences stored in a transactional database. Alternatively, the term *intertransaction pattern discovery* can be used to refer to the result of sequence mining, and the term *intra-transaction pattern discovery* can be used to describe the result of association rule discovery.

In *time series*, a significant amount of work has been done in the following areas:

- *Streaming data pattern discovery*, because of its application in a variety of areas, such as financial data analysis and security monitoring.

- *Motif and anomaly discovery*, because of their increasing utility in various applications, such as bioinformatics and computer network monitoring.

This chapter is organized as follows: The three frameworks for frequent patterns in sequences are discussed in Section 5.1 (sequence mining), Section 5.2 (frequent episode discovery), and Section 5.3 (temporal association rule discovery). In Section 5.4 we discuss pattern discovery in *time series*, where particular emphasis is placed on *motif* and *anomaly discovery*. Section 5.5 discusses various methods for pattern discovery in *data streams*, while Section 5.6 presents techniques for mining temporal patterns in multimedia. Finally, additional bibliography is discussed in Section 5.7.

5.1 SEQUENCE MINING

5.1.1 *Apriori* Algorithm and Its Extension to Sequence Mining

A *sequence* is a time-ordered list of objects, in which each object consists of an *itemset*, with an itemset consisting of all *items* that appear (or were bought) together in a transaction (or session). Note that the order of items in an itemset does not matter. A sequence database consists of *tuples*, where each tuple consists of a *customer id, transaction id*, and *an itemset*. The purpose of *sequence mining* is to find *frequent* sequences that exceed a user-specified support threshold. Note that the *support* of a sequence is the percentage of tuples in the database that contain the sequence.

An example of a sequence is {(*ABD*), (*CDA*)}, where items *ABD* indicate the items that were purchased in the first transaction of a customer and *CDA* indicate the items that were purchased in a later transaction of the same customer. An example of a transactional database is shown in Table 5.1.

Table 5.1 shows the transaction sequences of different customers during their visits to a Web site. For example, customer 3 visited the Web site two times, where he purchased item F in the first transaction and items CG

TABLE 5.1   Transaction Sequences for Individual Customers

CustomerID	Transactions
1	(AB) (CG)
2	(AB) (F) (CG)
3	(F) (CG)
4	(F) (AB) (CG)

in the second transaction. Regarding containment of sequences, sequence (AB) (CG) is part of the transaction sequences of customers 1, 2, and 4, but not of customer 3. Another example is (A) (CG), which is a subsequence of the transaction sequences of customers 1, 2, and 4.

The *Apriori* algorithm was first proposed by Agrawal in [Agr93], for the discovery of frequent itemsets. It is the most widely used algorithm for the discovery of frequent itemsets and association rules. The main concepts of the *Apriori* algorithm are as follows:

1. Any subset of a frequent itemset is a frequent itemset.

2. The set of itemsets of size k will be called C_k.

3. The set of frequent itemsets that also satisfy the minimum support constraint is known as L_k. This is the seed set used for the next pass over the data.

4. C_k+_1 is generated by joining L_k with itself. The itemsets of each pass have one more element than the itemsets of the previous pass.

5. Similarly, L_k+_1 is then generated by eliminating from C_k those elements that do not satisfy the minimum support rule. Because the candidate sequences are generated by starting with the smaller sequences and progressively increasing the sequence size, *Apriori* is called a *breadth-first* approach. An example is shown below, where we require the minimum support to be two. The left column shows the transaction ID and the right column shows the items that were purchased.

Table 5.2 shows a transaction database, while Table 5.3 shows the candidate itemset for size 1. We see that the item E does not have the minimum support. So the table of frequent items of size 1 (L_1) is shown in Table 5.4.

After joining L_1 with itself, Table 5.5 shows the support of itemsets with size 2. Table 5.6 shows the frequent itemsets of size 2. Finally, Table 5.7 shows the most frequent itemset for size 3.

TABLE 5.2   Itemsets Purchased during Transactions

TID	Itemsets
1	A B C D
2	A B C E
3	C D
4	A B

TABLE 5.3 C_1—Candidate Itemset for Size 1

Item	Count
A	3
B	3
C	3
D	2
E	1

TABLE 5.4 L_1—Frequent Itemsets with Minimum Support for Itemset Size 1

Item	Count
A	3
B	3
C	3
D	2

TABLE 5.5 C_2—Candidate Itemset for Size 2

Itemset	Count
A B	3
B C	2
C D	2
A C	2
A D	1
B D	1

TABLE 5.6 L_2—Frequent Itemsets with Minimum Support of Item Size Size 2

Itemset	Count
A B	3
B C	2
C D	2
A C	2

TABLE 5.7 L_3

Itemset	Count
A B C	2

An extension of the *Apriori* algorithm that deals with *sequence mining* was proposed in [Agr95]. The process for the discovery of *frequent sequences* is similar to that of the *Apriori* algorithm, where the support of a sequence is defined as the fraction of tuples in which the sequence is contained. Multiple passes are made over the database to generate candidate sequences where *nonmaximal* sequences are pruned out. Going back to Table 5.1, we see that the sequence (AB) (CG) has support 75%. The sequence (F) (CG) also has support 75%. Note that both sequences are *maximal*, because they are not contained in any longer frequent sequences.

As noted in [Lin07], the problem with the *Apriori* approach is that it can generate a very large number of candidate sequences. For example, to generate a frequent sequence of length 100, it must generate 2^{100} candidates, because these are its subsequences, which are also frequent.

5.1.2 The *GSP* Algorithm

GSP was proposed by the same team as the *Apriori* algorithm in [Agr96], and it can be 20 times faster than the *Apriori* algorithm because it has a more intelligent selection of candidates for each step and introduces time constraints in the search space.

Agrawal and Srikant formulated *GSP* to address three cases:

1. The presence of time constraints that specify a maximum time between adjacent elements in a pattern. The time window can apply to items that are bought in the same transaction or in different transactions.

2. The relaxation of the restriction that the items in an element of a sequential pattern must come from the same transaction.

3. Allowing elements in a sequence to consist of items that belong to a taxonomy.

The algorithm has very good scalability as the size of the data increases.

The *GSP* algorithm is similar to the *Apriori* algorithm. It makes multiple passes over the data as is shown below:

- In the first pass, it finds the frequent sequences. In other words, it finds the sequences that have minimum support. This becomes the next seed set for the next iteration.

- At each next iteration, each candidate sequence has one more item than the seed sequence.

Two key innovations in the *GSP* algorithm are how candidates are generated and how candidates are counted. Regarding *candidate counting*, a hash table is used. Regarding *candidate generation*, a novel definition of a contiguous subsequence is used, which is described below.

5.1.2.1 Candidate Generation

The main idea here is the definition of a *contiguous subsequence*. Assume we are given a sequence $S = \{S_1, S_2,.... S_N\}$, then a subsequence C is defined as a *contiguous subsequence* if any of the following constraints are satisfied:

1. C is extracted from S by dropping an item from either the beginning or the end of the sequence (S_1 or S_N).

2. C is extracted from S by dropping an item from a sequence element. The element must have at least 2 items.

3. C is a contiguous subsequence of C' and C' is a contiguous subsequence of S.

Here is an example. Let us assume we have the sequence

$$S = \{(10, 30, 40), (50, 60, 70), (90, 100), (110, 120)\}$$

Then the sequence $S_1 = \{(30, 40), (50, 60, 70), (90, 100), (110)\}$ is defined as a contiguous subsequence. The sequence $S_2 = \{(10, 30, 40), (50, 60, 70), (110, 120)\}$ is NOT a contiguous subsequence. The sequence $S_3 = \{(10, 30, 40), (50, 60, 70), (90), (110, 120)\}$ is a contiguous subsequence.

However, as the authors in [Pei04] note, there are three inherent costs associated with *GSP*:

- When dealing with large sequence databases, a very large number of candidate sequences could be generated.

- Multiple scans of databases are required.

- When dealing with long sequences, the number of candidates could experience exponential growth. This could be the case in *DNA* sequences or stock market data.

Finally, in [Mas98], the authors propose *PSP*, a method for sequential pattern discovery that is similar to *GSP*, but uses a different intermediary data structure that makes it more efficient.

5.1.3 The *SPADE* Algorithm

In [Zak01], Zaki proposes an algorithm that takes a different approach than *Apriori* and *GSP* for the discovery of sequential patterns. Instead of assuming a *horizontal* database, where a customer has a set of transactions associated with him, *SPADE* assumes a *vertical* database, where each *item* is associated with a customer id and a transaction id. The support of each k-sequence is determined by the temporal join of k-1 sequences that share a common suffix [Abr01].

As we saw, both *Apriori* and *GSP* make multiple database scans. *GSP* also uses a hash structure. The innovation of the *SPADE* algorithm is that it decomposes the original problem into smaller problems and uses *lattice search* techniques, to solve the problems independently in memory. The important contribution of this algorithm is that all sequences are discovered in only *three* database scans. This becomes even more important if one considers how large the search space is. As noted in [Mas05], *SPADE* needs to load the database into main memory. This could be a potential disadvantage in memory-demanding applications.

In [Zak98], Zaki et al. propose an algorithm that finds frequent sequences that always precede plan failures, with applications in emergency situations. The algorithm (*PlanMine*) is based on the *SPADE* algorithm and it successively filters out uninteresting sequences. In the first step, sequences that are not related to plan failures are filtered out. In the second step, it filters out frequent sequences that appear frequently in plan failures, but also have high support in plans that did not fail. In the third step, redundant patterns are removed, which are patterns that have subpatterns in both good and bad plans. In the final step, *dominated patterns* are removed. These are patterns that have subsequences that have lower support in good plans and higher support in bad plans than the original sequence in which they are contained.

5.1.4 The *PrefixSpan* and *CloSpan* Algorithms

In [Pei04], the authors propose a technique for sequential patterns mining, *PrefixSpan*, that deviates from the standard ways of generating candidates and testing them, such as *GSP* and *SPADE*. Instead, they propose a method that has the following two key features:

- It is *projection-based*. What is projected is the actual database. Specifically, the sequence database is iteratively projected into smaller projected databases. Han, Pei, et al. first developed the *FreeSpan* algorithm [Han00], using projected databases.

- It grows patterns sequentially in the projected databases, by investigating only locally frequent segments. Specifically, in a database scan, the frequent patterns above a certain support are recorded. Then for each one of the frequent patterns, their ending locations are used as the beginning locations of new scans. Frequent patterns discovered in this scan are then appended to the frequent patterns discovered in the previous scan. As noted in [Wu05], each projected database has the same prefix subsequence.

One of the computational costs of *PrefixSpan* is the creation of the projected databases. The authors propose *pseudoprojection* as a way to reduce the number and size of the projected databases. In this method, the index of the sequence and the starting position of the projected suffix are recorded, instead of performing a physical projection.

In experimental results on different data sets, *PrefixSpan* outperformed *GSP* and *SPADE*. In a comprehensive performance study, the authors explain the reasons *PrefixSpan* outperformed the other methods:

- It grows patterns without candidate generation.

- The projections are an effective way to perform data reduction.

- The memory space utilization is approximately steady.

In [Yan03a], a variation of *PrefixScan*, called *CloSpan*, is proposed that is tailored to detect *closed* subsequences only. A *closed sequential pattern* is one for which there is no superset sequence (with the same support) containing the pattern. Previous work for the discovery of closed patterns includes *MAFIA* [Bur01], *DepthProject* [Aga99], *CLOSET* [Pei00], and *CHARM* [Zak02]. The authors in [Yan03a] note three important advantages of *CloSpan*: (1) It is faster than previous work (*PrefixSpan*) by one order of magnitude, (2) it can mine long sequences that cannot be mined by other techniques, and (3) it returns a smaller number of sequences than other methods that discover sequences that are not constrained to be closed. It does that while keeping its expressive power, since superset sequences can be produced from the returned results. *CloSpan* generates a set of candidate sequences and then performs post pruning by incorporating efficient search space pruning methods. Another work that deals with closed subsequences is [Li08]. Specifically, the authors describe a method for detecting closed subsequences with gap constraints.

In [Che05], the authors propose an extension of the *PrefixSpan* algorithm called *MILE*. The *MILE* algorithm utilizes the knowledge about existing patterns to avoid redundant database scans. The algorithm can be further optimized by utilizing prior knowledge about the data distribution. The authors note that as the number of patterns grows, the faster *MILE* becomes in comparison with *PrefixSpan*. The fact that *MILE* uses more memory than *PrefixSpan* can be addressed by using a memory balancing approach.

5.1.5 The *SPAM* and *I-SPAM* Algorithms

In [Ayr02], the authors propose a sequential pattern mining technique that utilizes a *bitmap representation* called *SPAM*. The algorithm is the first sequential mining method that utilizes a *depth-first* approach to explore the search space. Combining this search strategy with an effective pruning technique that reduces the number of candidates makes the algorithm particularly suitable for very long sequential patterns. However, the algorithm requires that the whole database can be stored in main memory, which is the main drawback of the algorithm.

As sequences are generated traversing the tree, two types of children are generated from each node: sequence-extended sequences (sequence-extension step or S-step) and itemset-extended sequences (item-extension step or I-step). Finally, an efficient representation of the data is used, which is a vertical bitmap representation. The bitmap is created for each item as follows: If item A belongs to transaction j, then a 1 is recorded for A in transaction j. Otherwise a 0 is recorded.

SPAM was compared against *SPADE* and *PrefixSpan* using several small, medium, and large data sets. On small data sets, *SPAM* was about

2.5 times faster than *SPADE*, but *PrefixSpan* was a bit faster than *SPAM* on very small data sets. On large data sets, *SPAM* was faster than *PrefixSpan* by about an order of magnitude. The authors attributed the fast performance on large data sets to its efficient bitmap-based representation.

An improved version of *SPAM* is proposed in [Wu05], called *I-SPAM*. In this version, approximately half of the maximum memory requirement of *SPAM* is needed with no sacrifice in the performance time.

5.1.6 The Frequent Pattern Tree (FP-Tree) Algorithm

In [Han04], the authors propose a pattern discovery method that has the novel feature that it does not use candidate generation, which is a common step in all methods that utilize an *Apriori-like* approach. As noted previously, candidate generation can be a very computationally expensive step, especially for long patterns. There are two important contributions of the work described in [Han04]:

- A novel tree structure (*the FP-tree*) is described, which is an efficient and compact way to store information about the frequent patterns in the database.

- An efficient mining technique is described, called *FP-growth*, which mines the complete set of frequent patterns and stores them in the tree. Experimental results show that *FP-growth* is approximately an order of magnitude faster than the *Apriori* algorithm.

The *FP-tree* top node is labeled as the null node. Each subsequent node corresponds to an item whose count is also noted in the label of the node. The count represents the number of transactions that are present in the path reaching the node. The main idea behind the construction of the tree is to merge transactions that share a common prefix item set. Therefore, more frequent itemsets, which have a higher likelihood to be shared, are placed closer to the root node.

Besides being a highly compact way to represent a database, the *FP-tree* is *complete;* that is, it contains all the information about the frequent patterns in a database. Also only *two scans* of the database are needed: one to find the frequent items and the other to construct the *FP-tree*. Performance comparisons show that *FP-growth* has much better scalability than *Apriori*, when the number of transactions increases from 10k to 100k. In large databases, *FP-growth* is about an order of magnitude faster than

TABLE 5.8   Transaction Database

TID	Purchased Items
1	B, C, D
2	B, A, D
3	B, C, M
4	C, A

Apriori. Table 5.8 shows a transaction database, and Figure 5.1 shows *the FP-tree* for it.

5.1.7 The *Datte* Algorithm

In [Bou07], an algorithm, called the *Datte* algorithm, is described for the discovery of frequent temporal patterns and temporal rule construction. In the first step, which is the discovery of frequent temporal patterns, the algorithm finds patterns that exceed a minimum support. In the second step, it generates temporal association rules that have confidence higher than a minimum prespecified confidence. The input set to the *Datte* algorithm is a set of intervals, and specifically the duration of these intervals, which are also called the *primitives* of the algorithm.

A temporal pattern in *Datte* is of the form *(S, R)* and has a dimension *k*. *S* is the set of the primitive intervals, and *R* is a matrix that denotes the

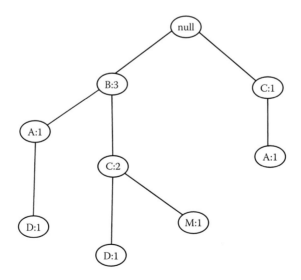

FIGURE 5.1   FP-tree for Table 5.8.

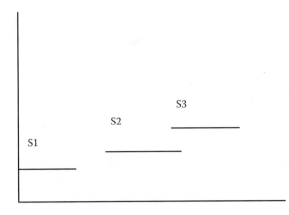

FIGURE 5.2   Intervals used as input to the *Datte* algorithm.

relationships between these intervals. For example, assuming a time series that contains 3 primitive intervals, the matrix R is

$$R = \begin{matrix} r_1^1 & r_1^2 & r_1^3 \\ r_2^1 & r_2^2 & r_2^3 \\ r_3^1 & r_3^2 & r_3^3 \end{matrix} \tag{5.1}$$

These relationships are one of *Allen's relationships* such as *before, after, meets,* or *overlaps*. For example, for the intervals S_1, S_2, S_3 shown in Figure 5.2, the matrix R becomes

$$R = \begin{matrix} eq & b & b \\ a & eq & m \\ a & m & eq \end{matrix} \tag{5.2}$$

where *eq* denotes equals, *b* denotes the *Allen relationship before*, *m* denotes the *Allen relationship meets*, and *a* denotes the *Allen relationship after*. The pattern generation follows an *Apriori-like* approach.

5.1.8 Incremental Mining of Databases for Frequent Sequence Discovery

Given that the state of the database constantly changes and it is possible that new frequent sequences will appear, while older frequent sequences can become infrequent, incremental mining of temporal databases is gaining increasing attention. This means that instead of rescanning the entire database to apply a sequence pattern discovery algorithm, such as *GSP*, algorithms that *reuse* prior information about frequent patterns are developed. Algorithms for the mining of incremental databases, where prior information about frequent patterns is used, are described in [Lin07] (the *DSPID* algorithm) and [Gui05] (the *TSET* algorithm).

In [Mas03], an algorithm called *ISE* is proposed, which reuses minimal information from previously discovered sequences and also reduces the number of candidate sequences. *ISE* allows both the insertion of new sequences and the extension of the suffix of older sequences. In [Par99], the *ISM* algorithm for incremental mining of databases is proposed, which is based on the *SPADE* algorithm and keeps information about *maximally frequent sequences* and *minimally infrequent sequences*. It also provides a user interface that allows the user to specify parameters, such as the minimal support.

[Wan96] describes an algorithm for the incremental mining of temporal patterns, where only the affected part of the data structures and database are scanned. In [Cho08], a framework for quantifying changes in temporally evolving data is presented. The changes are expressed with the following transformations: *create, merge, split, continue,* and *cease.*

5.2 FREQUENT EPISODE DISCOVERY

[Man97] deals with the problems of finding temporal patterns of *interest* under the framework of discovering *frequent episodes* in a temporal data sequence. Note that here we are dealing with a *single* data sequence, while in the previous algorithms we were looking for patterns in *multiple* data sequences. *An episode is defined as a sequence of events appearing in a specific order within a specific time window,* for example, the occurrence of flu followed by pneumonia. The algorithm that is used constructs frequent episodes from simple frequent subepisodes.

An event sequence is defined as a pair of events and corresponding timestamps as follows:

$$(E_1, t_1), (E_2, t_2),\ldots(E_n, t_n)$$

An example of an event sequence with four events is shown below:

$$(A, 1), (B, 5), (C, 10), (D, 15)$$

An episode is said to *belong* to a sequence if the events in the episode appear in the same order in the sequence. An episode can be either serial, where order matters, or parallel, where order does not matter. An example of a serial episode is (A->B->C), while an example of a parallel episode is of the form (ABC). As we see, both the aforementioned serial and parallel episodes belong to the event sequence defined above.

The *frequency* of an episode is defined as the fraction of window width, in which the episode occurs at least once, and the size of the window is user defined. A *frequent episode* is defined as an episode that has a frequency above a user defined threshold. Counting frequencies of parallel episodes is very similar to counting frequent itemsets, and therefore, an *Apriori-style* algorithm can be employed (the *WINEPI* algorithm). First, frequent *1-node* episodes are discovered. They are then joined to find *2-node* episodes, whose frequency is counted to find *2-node frequent* episodes. The process is continued iteratively to find all frequent episodes.

Finding serial episodes, where order of events matters, can be addressed using *finite state automata*. For example, for the episode A->B->C->D, we need a four state automaton. The automaton goes to its first state when it finds an event of type A, then it goes to its second stage when it recognizes an event of type B, then it goes to its third state when it recognizes an event of type C, and so on.

Another way to find frequent episodes, suggested in [Man97], is based on counting the *minimal* occurrences of episodes (the *MINEPI* algorithm). In *MINEPI*, instead of frequency, the concept of *support* is used, where a minimum support threshold is used. A *minimal occurrence* of an episode is defined as a temporal window W of the original time sequence, in which the episode appears. In addition, no subwindow of W contains the episode [Lax06]. In [Man97], the algorithms were applied to different data sets, such as alarms from a computer network and pages from an Internet server log. Comparing the *WINEPI* and *MINEPI* algorithms, the authors state that although *MINEPI* generally outperforms *WINEPI*, the latter algorithm can be more efficient in the first phases of episode discovery.

5.3 TEMPORAL ASSOCIATION RULE DISCOVERY

The third framework under which sequential pattern mining is examined is *temporal association rule discovery*. As discussed in [Che00], a *temporal association rule* can be defined as a pair (R, T) where R is an association rule and T is a temporal feature, such as a period or a calendar. There are three measures of success regarding the discovery of association rules, which can also be applied to sequence mining:

- *Support.* The support of a rule is equal to the probability that this rule will be found in the transactions recorded in the database.

- *Confidence.* The confidence of the rule "*if computer purchased, then printer purchased*" is equal to the conditional probability of the printer purchase, given the computer purchase, that is, P(printer| computer).

- *Informativeness.* In [Smy92], a measure computes the usefulness of a rule in terms of the information it provides. If we denote as *Co* the purchase of a computer and *Pri* the purchase of a printer then this measure, often called the *J-measure*, is defined as

$$J(Co, Pri) = \left[P(Pri|Co) \log\left(\frac{P(Pri|Co)}{P(Pri)} \right) \right.$$
$$\left. + (1 - P(|Pri|Co)) \log\left(\frac{1 - P(Pri|Co)}{1 - P(Pri)} \right) \right] \tag{5.3}$$

5.3.1 Temporal Association Rule Discovery Using Genetic Programming and Specialized Hardware

In [Het02] and [Het05], Hetland and Sætrom propose a method to discover temporal association rules using *evolutionary techniques* and *special purpose hardware*. This hardware is a co-processor designed to detect complex patterns in unstructured data at the rate of 100 MB/s.

Regarding the genetic algorithm, each individual in the population represents a rule and is assigned a *fitness score* that shows how well the individual can solve the problem. The authors make two important points about their algorithm: (1) The main advantage of their technique lies in the freedom it gives its user to choose the quality metric and the rule format. (2) Their approach is very direct in terms of finding temporal patterns; they define the problem as *mining interesting rules* using an interestingness measure.

The algorithm works on discrete symbols, which were produced using the method described in [Keo02]. The authors developed the algorithm to work in both a supervised and an unsupervised learning environment. In the supervised learning setting, the fitness measure is the correlation between the occurrences of a rule and the desired occurrences of a rule. In the unsupervised learning setting, different fitness measures can be used that are based on confidence, support, and interestingness.

The experimental results show that the algorithm has good predictive power and creates rules that have high interestingness.

5.3.2 Meta-Mining of Temporal Data Sets

[Abr99] addresses the issue of discovering *metarules*, that is, rules describing rules in large temporal data sets. Relationships between rules can give unique insights about the characteristics of a system. Given a rule set $R = \{r_1, r_2, \ldots r_n\}$, then a *meta rule* between two rules ri and rj, Mij, describes the relationship between the two rules. The process for the discovery of the metamining rules can be described as follows:

- *Rule set generation.* The rules describing the data set are generated and separated in hierarchies, or rule set layers.

- *Input generation.* The rule set layers are transformed to have a consistent format.

- *Meta rule set base generation.* A separation algorithm produces the following rule categories: new, retired, changed, and unchanged.

- *Processing of categories.*

5.3.3 Other Techniques for the Discovery of Temporal Association Rules

Additional information on temporal association rules can be found in [Blu83], [Che98a], [Che98b], [Ima94], [KouURL], [Win05]. In [Lee04], the authors propose a data mining system for the discovery of *fuzzy temporal association rules*. A fuzzy calendar is developed which allows users to express their temporal queries in a natural and user-friendly way. Five operators are introduced to deal with temporal expression with various granularities.

In [Ale00], a method of discovering temporal association rules is described that utilizes an extension of the *Apriori* algorithm and introduces the notion of *temporal support*. For example, for a product that has been in the market for the last ten months, its temporal support is ten months. In [Ng07], the authors propose a novel method for the *incremental mining* of temporal association rules with numerical attributes and it was tested for crime pattern discoveries. In [Hua07], an efficient algorithm is described, called *SPFA*, for discovering temporal association rules. The main idea of *SPFA* is to segment the database into smaller databases, where all items in a database have either the same starting time or the same ending time.

In [Mör04], the authors describe the *Temporal Data Mining Method*, which is a framework that discovers temporal rules in *multivariate* time series, using a rule language called *UTG (Unification-Based Temporal Grammar)*. UTG is a hierarchical rule language that allows the expression of complex patterns:

- At its base, there are *Primitive Patterns* (single points in time).

- Then we have *Successions,* which model the temporal concepts of *duration* and *persistence*. A time interval is modeled with a set of points that all have the same Primitive Pattern label.

- *Events* represent coincidence and synchronicity. A time interval is represented as the overlap of several successions.

- *Sequences* represent order. A sequence is represented with several Events occurring sequentially.

- The final layer consists of *Temporal Patterns*. A Temporal Pattern consists of several Sequences.

Other relevant work by Mörchen can be found in [Mör06], [Mör07a], and [Mör07b].

In [Das98], the authors address the problem of discovering rules in *multivariate* time series. Their method discretizes the time series data. Subsequences are formed by sliding a window through the time series, and then the subsequences are clustered using a measure of similarity. Finally, in [Höp02], an *Apriori-like* algorithm is described that mines temporal rules using *Allen's temporal relations* and a sliding window to limit the pattern length.

5.4 PATTERN DISCOVERY IN TIME SERIES

5.4.1 Motif Discovery

5.4.1.1 General Concepts

The recent interest in *motif discovery* is mainly triggered by the recent advances in bioinformatics, where over-represented *DNA* sequences indicate biological significance [Dur98], [Pev00]. Here we will discuss some basic concepts related to motif discovery [Chi03]:

- *Motif definition*: This is the approximate repetitions of subsequences within a time series [Chi03].

- *Match*: We are given a time series T containing a subsequence C starting at position p_1 and a subsequence M beginning at location p_2. Given a distance measure $D(C, M)$, a positive number R called the range, then M is called a matching subsequence if $D(C, M) < R$.

- *Trivial match*: This is a subsequence that matches another because of their significant overlap.

- *Most significant motif*: Given a time series T, the most significant motif in T is the subsequence that has the highest count of nontrivial matches.

- *Planted motif problem* [Pev00]: We are given t random strings of length n. An unknown motif y of length w is planted with exactly one approximate occurrence of y. Find the unknown motif y.

- *Random projection algorithm* [Tom01]: A solution to the aforementioned planted motif problem, it works by using random projections to reduce the search space. The algorithm chooses k positions as a type of a *mask*. The mask is then superimposed at all positions. This way, each substring is mapped to a string (of size k) by reading the symbols through the mask, which becomes the projection.

The problem of finding motifs is a difficult one because typically the number of occurrences of motifs, their shape, and duration of occurrences are unknown. In Chapter 6, we describe a number of algorithms for the discovery of motifs in biological sequences (*DNA*, *RNA*, and *protein*).

5.4.1.2 Probabilistic Discovery of Time Series Motifs

In [Chi03], an algorithm for the discovery of motifs in time series is introduced that can discover motifs *in the presence of noise*. The algorithm leverages off the work in [Tom01] that utilizes random projections, by first creating a discrete representation of the time series. The algorithm does the following:

- Discretizes the time series using *PAA*

- Extracts subsequences using a sliding window

- Converts them into symbolic form using *SAX*

- Places them in a matrix

Note that the *PAA* and *SAX* algorithms were discussed in Chapter 2. After the construction of the matrix, random projection begins by selecting two columns at random to act as a mask. A *collision matrix* counts how many words (that correspond to subsequences) are hashed to the same bucket. If the entries in the matrix are uniform, this is in an indication that there are no motifs.

5.4.1.3 Discovering Motifs in Multivariate Time Series

In [Min07], the problem of discovering motifs in *multivariate time series* is addressed as locating regions of high density in the space of all time subsequences and avoiding the problem of trivial matches. Motif occurrences are detected using a *greedy mixture learning algorithm* that allows variable length occurrences. Specifically, the algorithm lets various motif models compete to explain the time series data. Each motif is modeled using an *HMM*. The *HMMs* are fit to the time series data using a generalized *Viterbi* algorithm.

As described in [Min07], the benefits of using *HMMs* for motif discovery are as follows:

- No direct estimation of the neighborhood size is needed.

- To account for variable-length motifs, the motif models are allowed to shrink and stretch.

- The *HMM* models provide a way to measure goodness of fit.

- The algorithm is globally optimal because it finds the state sequence that maximizes the data log likelihood.

- It finds all the motif occurrences.

- It disregards spurious motifs.

5.4.1.4 Activity Discovery

In [Min06], the problem of *activity discovery* is addressed as finding motifs in multivariate time series. Such series are generated by wearable sensors and street video cameras. The algorithm has the following advantages:

- It can handle nonlinear time warping.

- It can handle variable-length motifs.

- It can even find sparse motifs.

The main steps of the algorithm are as follows:

- Time series quantization into strings of discrete symbols

- Analysis of the quantized time series to discover seed motifs, where a generalized suffix tree is used for linear time subsequence searches

- Seed refinement, where seeds can be merged or split

- Motif occurrence detection using probabilistic models trained from the refined seeds

5.4.2 Anomaly Discovery

5.4.2.1 General Concepts

Anomalies in time series are data points that are *significantly different* from the normal pattern of the time series. There are several applications of anomaly detection ranging from disease detection to computer network traffic optimization, biosurveillance, fraud detection, and so on. An intuitive way to solve the problem is to develop a model of normal behavior using a normal training set or expert knowledge and then to characterize any aberrations as anomalies.

In [Gwa03], the authors perform a probabilistic analysis of sequences to estimate the probability that a suspicious subsequence occurs randomly, that is, the *probability of a false alarm*. The specific question addressed in the article is whether a certain number of occurrences of a subsequence can be generated purely by chance. The novel contributions of the article are as follows:

- It quantifies the probability that a subsequence will occur within a larger sequence within a window of size *w*.

- It provides a way to compute the expected number and variance of such occurrences.

- It establishes a limiting threshold that allows the set up of an alarm threshold, such that the probability of false alarms is very small.

5.4.2.2 Time Series Discords

In [Keo05], the authors propose an algorithm for finding unusual time series where the notion of *time discords* is introduced. A time discord is a subsequence of a longer time series that is maximally different from all other subsequences of the series [Keo05]. Discords can be used to detect anomalies in an efficient way. Some interesting points are made in the article:

- Discords are not necessarily found in *sparse space*, which means it would not be effective to project in an *n-space* and use outlier techniques to find discords.

- Discords cannot be discovered by breaking the problem into smaller problems and recombining them to find the solution.

- Discords are a very attractive way to detect anomalies because they require the specification of only one parameter, the length of the subsequence.

In the brute force algorithm for finding discords, for each possible subsequence the distance to the nearest non-self-match is found. The subsequence with the highest distance is identified as the discord. Note that the distance has to be defined by a symmetric function that takes two sequences as input and returns a positive number as the distance. An example of such a distance metric is the Euclidean metric. In [Keo05], a heuristic modification to the brute form algorithm is proposed which improves the algorithm's running time. The algorithm was successfully applied in finding changes in patient's monitoring, such as respiratory patterns and *ECG* signals.

In [Keo02], an algorithm, called *TARZAN*, is described that finds surprising patterns in time series by using a suffix tree to encode the frequencies of the observed patterns and uses a *Markov* model to predict the frequency of previously unseen patterns. In [Yan07], an algorithm that finds unusual time series in terabyte-sized data sets is described. The proposed algorithm requires only two linear scans of the disk and a very small buffer in main memory.

5.4.2.3 VizTree

In [Lin05], a novel visualization framework for massive time series data sets is described that enables very efficient pattern discovery, such as motifs and anomalies. It allows the user to view simultaneously the local and global structure of the time series and interactively find patterns in the data. The main idea behind *VizTree* is to transform the time series into a series of symbols using *SAX* and encode the data in a suffix tree.

Different colors are used to visualize the different frequencies and other properties of the time series. Besides visualization and pattern discovery,

VizTree can be used for monitoring purposes in both an *append* and *amnesic* fashion. In the append fashion, the tree is updated in constant time and some information about the global structure of the data is maintained, while in the latter case, newer data are assigned higher weights and some of information of the older data needs to be stored.

As described in [Lin04], *VizTree* was used by the *Aerospace Corporation* in making engineering assessments to make a go/no-go decision for the launching of a space vehicle. These assessments are based on incoming telemetry data in the prelaunch hours. *VizTree* has the dual capabilities of mining archived data and monitoring live incoming data.

5.4.2.4 Spacecraft Anomaly Detection Using Support Vector Machines

In [Fuj05], a method for the detection of anomalies in spacecrafts is described. Although the method is developed for a very specific application, it is worthwhile discussing because of its potential applicability in other areas. The anomaly detection algorithm operates on telemetry data, which consist of *multidimensional* time series data.

The authors define as *anomaly* any unexpected changes in the causal relationships inside the spacecraft system. The main hypothesis behind the algorithm is that causal relationships will appear in the form of principal component axes in a high dimensional feature vector space. The algorithm requires no *a priori* knowledge provided by experts about dynamic models of the system.

The difficulty of dealing with a high dimensional feature is overcome by utilizing *Support Vector Machines* and a kernel function. The authors use the *Kernel PCA* algorithm [Sch98] to compute the principal components vector in a high dimensional feature space. This algorithm applies a kernel method to linear *PCA*. In summary, the learning phase of the algorithm consists of the following steps:

- Acquire the training data.

- Extract the feature vector.

- Optimize the directional vector. This represents the normal data after mapping into the feature space.

- Predict the directional distribution. This is achieved by optimizing the parameters of a *von Mises–Fisher* distribution [Dhi03], which is considered the most natural distribution for directional data.

5.4.3 Additional Work in Motif and Anomaly Discovery

In [Jen06], *Gemoda*, an algorithm for generic motif discovery is proposed. The algorithm discovers motifs that are maximal in length and can accept any type of similarity measure to be used. The algorithm was applied to the discovery of motifs in amino acid sequences and conserved protein substructures. In [Hil07], two anomaly detection methods that employ *dynamic Bayesian networks* to detect anomalies in sensor data are described. The algorithms can be employed for quality assurance purposes in sensor networks. Dynamic Bayesian networks model the evolution of a system by tracking changes in its states. The algorithm requires no a priori knowledge about the type of anomalies that might take place.

In [Chi05], a comparison of different time series anomaly detection techniques is performed. The authors found that *multiple layer perceptron neural networks* and *radial basis functions* perform better than *principal components analysis*. Also they found that a novel method known as *D-Markov chain* performs better than the other aforementioned techniques. In [Wei05a], an anomaly detection system is described that is based on *time series bitmaps* and can be applied to many domains. This is in contrast to many existing anomaly detection systems that are customized to a particular domain. In [Das96], a novelty detection algorithm derived from immunology is described. Specifically, it utilizes the *negative selection* mechanism of the immune system that can discriminate between what constitutes *self* and *other*. In the article, a set of normal patterns are defined as *self* and everything else is a novelty or anomaly.

5.4.4 Full and Partial Periodicity Detection in Time Series

Periodicity detection is an important task in many disciplines. As noted in [Vla05], examples include oceanography, medicine, and industrial manufacturing. For example, in medicine, disturbances of the heartbeat can indicate arrhythmia.

In [Elf05], the authors present a novel *convolution-based* algorithm for periodicity detection. The algorithm detects the periodicity rate of a time series for two types of periodicity: (1) *segment periodicity* and (2) *symbol periodicity*. Segment periodicity deals with the periodicity of the entire time series. For example, in the sequence *123412341234*, the periodic segment is *1234*. In contrast, symbol periodicity deals with the periodicity of individual symbols. For example, in the sequence 1ab1cd1ef1, the symbol 1 is periodic because it appears periodically at positions 0, 3, 6, 9. The main idea of the algorithm is to use *convolution* to shift and compare the time series for all possible values of the period. The definition of convolution is as follows: Let us assume we have two time series: $A = \{a_1, a_2, \ldots, a_N\}$ and $B = \{b_1, b_2, \ldots, b_N\}$. Then, their convolution is defined as a series of size N, as shown in Equation 5.4:

$$A(conv)B = \sum_{j=0}^{i} a_j b_{i-j} \tag{5.4}$$

The convolution is computed more easily in the Fourier transform domain, as shown in Equation 5.5:

$$A(conv)B = InvFFT(FFT(A)\ FFT(B)) \tag{5.5}$$

where FFT stands for the Fast Fourier transform and InvFFT is the inverse Fast Fourier transform.

In [Vla06], another novel periodicity detection algorithm is described. The basis of this algorithm is the utilization of the *power spectrum* to detect how much of the power of the signal is at each frequency. Two estimators of the power spectral density are the *circular autocorrelation* and the *periodogram*. The periodogram is given by the square of each Fourier coefficient. The circular autocorrelation of a sequence $X = \{x_1, x_2, \ldots, x_N\}$ is given by Equation 5.6:

$$Circ\_Autocorr(a) = \frac{1}{N} \sum_{n=0}^{N-1} x_a x_{n+a} \tag{5.6}$$

The above autocorrelation examines how similar a sequence is to its previous values for time lag a. The method proposed by the authors is known as *AUTOPERIOD* and it consists of two phases. First, the periodogram is used to extract period candidates, which are called *hints*. These period candidates are detected, if their power at a frequency exceeds a predefined threshold. These are coarse candidates of periodicity estimates and could be false, primarily because of spectral leakage. To get a more refined estimate of periodicity, the authors use the autocorrelation. This is done as follows: If the period detected with the periodogram lies on a hill of the autocorrelation function, then this is likely a valid period.

In [Ma01], the problem of *partial periodicity* is addressed, where periodic patterns appear only *intermittently* in the time series. This is, in contrast to fully periodic time series, that were discussed above, where every point in the time series contributes to the periodicity. The authors in [Ma01] refer to these patterns as *p-patterns*. These patterns can be detected despite presence of noise, shifts in phase and/or periods, and imprecise time information. The period of *p-patterns* is found using a *chi-squared test* and then temporal associations are found. The authors developed two versions of their algorithm. In one version, the period of the patterns is detected first and the associations are detected second. This algorithm has good computational efficiency. In the other version, the associations are detected first and the period second, which results in an algorithm that has a higher tolerance to noise.

In [Han99], another algorithm for detecting *partially* periodic patterns is described. The patterns that can be detected from the algorithm can take the form *a* * *b*, where * is the *don't care* character, which can be any character. The authors note that for a single period of a partially periodic pattern, an *Apriori-like* algorithm can be used. The algorithm that is proposed in [Han99] can detect partially periodic patterns not just for a period, but for a set of periods as well. Besides the *Apriori* concept, the algorithm uses a new concept, called the *max-subpattern hit set* property, which allows the counting of all frequent patterns from a small subset of patterns derived from the time series. The method is computationally efficient, in the sense that it needs only two scans of the database to find the partially periodic patterns.

In [Oat02], an algorithm called *PERUSE* is described, which has been developed for the detection of recurring patterns in time series. As is noted in the article, the algorithm was influenced by work in the bioinformatics literature, such as the *MEME* algorithm, which is described in Chapter 6. The author decomposes the pattern discovery problem in two parts:

1. A supervised learning part. A teacher shows exemplars to the learner and labels them according to whether they contain the desired patterns. This problem is addressed using the *EM* algorithm.

2. An unsupervised learning part. Here, time series are used to generate an approximation to the teacher.

Finally, in [She06], the authors propose a method, called *DPMiner*, for the detection of *dense periodic patterns*, that is, short period patterns that do not exist in the entire time series, but only in a certain range of the time series. The authors proved the efficiency of their method on both real and synthetic data sets.

5.4.5 Complex Temporal Pattern Identification

In [Hon02], the authors discuss a method for the extraction of temporal patterns in a signal that is also dominated by *noise* and *nonlinear temporal warping*. The main idea of the algorithm is to use a *hidden Markov model (HMM)*. The proposed approach detects patterns, even if the number and length of patterns are unknown and the sequence contains irregular non-pattern signals. A specific type of *HMM* is used. It is a *threshold HMM*. The thresholds are placed on the population of the segments, the spatial variance of the segments, and the temporal warping thresholds.

In [Pov03], a novel algorithm is described that finds temporal patterns, such as *chaotic* and *nonperiodic*, in complex time series. The patterns that the algorithm is designed to find are patterns that characterize and predict *important events*. A significant difference of this algorithm with other temporal pattern discovery algorithms is that the patterns are not predefined, instead it finds the patterns that match the goal of the defined problem in an optimal fashion.

An *event characteristic function* that is problem specific is defined. The search for the optimum temporal patterns is performed in a phase-space that is created using time-delayed embedding. A set of temporal clusters is created, and then an objective function f is used to order the temporal clusters according to their ability to characterize events. The objective function can be defined in a way to minimize false positives, as shown in Equation 5.7:

$$f = \frac{t_p}{t_p + f_p} \tag{5.7}$$

where t_p is the true positive and f_p is the false positive.

Another way to define the objective function is to maximize the characterization and prediction accuracy, as shown in Equation 5.8:

$$f = \frac{t_p + t_n}{t_p + t_n + f_p + f_n} \tag{5.8}$$

where t_n corresponds to true negative and f_n corresponds to false negative.

There are two forms of search to find the optimal temporal patterns. The first search is done using a genetic approach to find a single optimal cluster. The second search is done using a greedy method to find a collection of optimal clusters. The accuracy of the new algorithm was compared against a *time-delay neural network algorithm (TDNN)* and *the C4.5 decision tree algorithm*. It was shown that the new algorithm gave significantly better results than both the *TDNN* and decision tree algorithms, in regards to the overall prediction accuracy and the positive prediction accuracy. For example, the overall prediction of the *TDNN* in the testing phase was 77.71%, while the corresponding accuracy of the new method was 96.43%. Finally, in [Pol00], the authors use *fuzzy temporal clustering* for *nonstationary time series analysis*. Nonstationary time series are important because they characterize many physical and biological processes.

5.4.6 Retrieval of Relative Temporal Patterns Using Signatures

In [Win08], a novel algorithm is described for content-based queries of temporal patterns, that is, the discovery of previously identified patterns. The patterns are modeled using a set of states and a set of relationships between the states. The similarity of two temporal patterns A and B is computed using the *Jaccard similarity* and is defined as [Win08]:

$$S(A, B) = \frac{|S_c| + |R_c|}{\sqrt{\left(N_A^S + N_A^R\right)\left(N_B^S + N_B^R\right)}} \quad (5.9)$$

S_c is defined as the number of common states between A and B, and R_c is the number of common relationships between A and B. N_A^S is the number of states in pattern A, N_A^R is the number of relationships in pattern A. Similarly, N_B^S is the number of states in pattern B and N_B^R is the number of relationships in B. For a pattern of size k, there are k states and $k(k-1)/2$ relationships.

Each element in a target set is represented by a bit pattern known as the *element signature*. Each pattern is represented by a bit pattern which is the *bitwise union of all element signatures*.

5.4.7 Hidden Markov Models for Temporal Pattern Discovery

In [Ge00], a *hidden Markov model (HMM)* is used to automatically detect patterns in time series. The authors make the important note that much of the work in pattern matching is *distance-based*, that is, based on a dissimilarity metric. The problem with this approach is that the distance measure depends on the application domain and on the type of the waveforms. Instead of looking for a universal distance metric, the authors propose the idea of probabilistic generative modeling, that is, a modeling technique for the waveform. Specifically, in this method, the waveform is decomposed in segments that are linked in a Markov way. Each segment is modeled using a parametric function with additive noise. Then the pattern matching problem becomes the problem of finding the best parameters given the observed waveforms. Experimental results on a variety of waveforms prove the method to have robustness.

5.5 FINDING PATTERNS IN STREAMING TIME SERIES

Streaming time series (or, shortened, *data streams*) constitute an important type of data in many areas such as network, sensor data, and financial data analysis. As noted in [Pap05], idiosyncrasies of these applications are that enormous amounts of data arrive at high rates and that the end users need information to respond immediately, such as sell/buy a stock or identify an alarm. These streams are often correlated and in many cases a few *hidden variables* [Pap05] are enough to compactly identify the situation. As noted in [Che07b], the problem of pattern discovery in streaming time series can be defined as continuously monitoring the data stream to match a stream subsequence against a set of potential query patterns. The match is based on a distance measure D and a threshold δ.

5.5.1 *SPIRIT, BRAID, Statstream,* and Other Stream Pattern Discovery Algorithms

In [Pap05] Papadimitriou et al. describe *SPIRIT*, an algorithm that uses *Principle Component Analysis* to identify hidden variables and trends in multiple streaming time series. *SPIRIT* is adaptive and automatic and it is able to automatically detect changes in the incoming streams and the number of hidden variables. The algorithm has the advantage of scaling linearly with the number of streams.

Other work in data streams is presented in [Sak05], [Zhu02], [Guh03], [Dom01], and [Gan02]. In [Sak05], the authors find pairs of time series streams that exhibit *lag correlations*. Their algorithm is called *BRAID* and they prove that *BRAID* can detect lag correlations with little or no error at all. Their experimental results showed that the maximum error was approximately 1%.

In [Zhu02], the authors propose an algorithm called *StatStream* that finds strong correlations between pairs of time series data streams. The algorithm is based on the *discrete Fourier transform* and utilizes a three-level interval hierarchy. The algorithm is parallelizable and efficient in the sense that it beats the direct computational approach by several orders of magnitude.

In [Guh03], the authors propose an algorithm that computes correlations between data streams utilizing *Singular Value Decomposition (SVD)* and a combination of dimensionality reduction and sampling to make *SVD* appropriate for streaming content. In [Dom01], an algorithm that constructs a decision tree from continuously changing data streams is proposed. In [Gan02], a survey of techniques that perform change detection and incremental mining is performed. Focus is put on techniques that use block evolution, where a data set is updated on a periodic basis by inserting or deleting blocks of records.

In [Pap06], Papadimitriou and Yu propose a novel method to discover optimal multiscale patterns in time series streams. These patterns provide a concise description of the time series and are discovered by examining the time series at multiple time scales. The main idea of the algorithm is that it *learns* an optimal orthonormal transform *from the data*. This is in contrast to using a set of predetermined basis functions such as *Fourier* or *wavelet functions*. The algorithm can be applied in an incremental setting to streams. The authors also propose a criterion to find the best window size for the data.

5.5.2 Multiple Regression of Streaming Data

In [Che02], the authors present a method for *multidimensional regression* of time series data streams. Specifically, they address the very interesting problem of how one can extend data cube technology and perform regression and trend analysis efficiently in the multidimensional space. The main contributions of the article are the following:

- To perform linear and multiple linear regression, one does not need the entire stream.

- To reduce the storage requirements, a partially materialized data cube is constructed with a *tilt time frame* as its time dimension. In this frame, time is recorded at different levels of granularity, such that the most recent time is recorded at the finest granularity and the more distant time is recorded at the coarsest granularity.

- To further reduce memory space, only two layers are computed: (1) *an observation layer,* which is the layer that the analyst uses to make decisions; and (2) the *minimally interesting layer,* which is the minimal layer that the analyst would like to study. Fully materializing only two layers allows a lot of room as to how the cuboids in between can be computed (fully, partially, or not at all).

5.5.3 A Warping Distance for Streaming Time Series

In [Che07b], the authors propose a novel warping distance measure (*SpADe*), which can handle *shifting and scaling both in the amplitude and temporal dimensions.* Because these types of distortions are quite often present in data streams, *SpADe* can be incorporated in a pattern discovery algorithm in streaming time series.

The authors note that traditional warp distances, such as *DTW, LCSS,* and *EDR,* do not handle well shifting and scaling in the amplitude dimension and, therefore, can be problematic in stream data pattern discovery. They also note that a solution that at first might appear suitable for data streams—that is, subsequence segmentation—is also problematic because there are no clear boundaries that can guide data stream segmentation.

The main idea behind *SpADe* is to measure distances between the shapes of two time series. *SpADe* comes from *spatial assembling distance,* and it is based on finding the best combination of local pattern matches, using the shortest path that connects two end points. *SpADe* can be used for both full sequence matching and subsequence matching.

Experimental results show that *SpADe* is more efficient than other measures such as *DTW* and *EDR* for various scales of time shifting. Regarding accuracy, *SpADe* is more accurate than other distance metrics for full sequence matching of smooth curves. For subsequence matching on streaming times series, *SpADe* handles amplitude scaling and shifting much better than *DTW, Euclidean,* and *EDR.* Regarding time shifting, *DTW* and *EDR* perform well, as expected, *Euclidean* does not perform well (also as expected), and *SpADe* performs the best. The authors further increase the speed of the algorithm by using wavelets to find the most important shape coefficients and indexing them in an *R-tree.* Additionally, *SpADe* is calculated incrementally to further speed it up.

5.5.4 Burst Detection in Data Streams

In [Zhu03], the authors propose an efficient technique for the detection of *bursts* in data streams. *Burst detection* is an important activity in many areas, such as astrophysics and stock market data monitoring. Bursts in data streams can be computed from either *points* or *aggregates*. In the aggregate form, a window is used to compute an aggregate of information from points. The window can be one of four types [Zhu03]:

- *Landmark window.* This corresponds to a fixed time period.
- *Sliding window.* The window size is fixed, but the beginning point of the window is movable.
- *Damped window.* Here, the weights decrease exponentially as we move from current time to the past.
- *Elastic window.* The user can recommend a range of sizes for the sliding window.

The main idea of the algorithm in [Zhu03] is to find abnormal aggregates in data streams using an elastic window. Bursts are detected using a *shifted wavelet tree.* The algorithm was tested for the detection of Gamma Ray bursts in astrophysical data and high stock price volatility in stock data. Compared with a brute force approach, it was found to be several orders of magnitude faster. More information about elastic burst detection can also be found in [Sha04].

The authors in [Zha06] propose another algorithm for elastic burst detection that utilizes a family of multidimensional overlapping data structures, called *Shifted Aggregation Trees*, and a heuristic search algorithm. Finally, in [Ara06], a genetic algorithm is employed to detect and track bursts in event sequences.

5.5.5 The *MUSCLES* and *Selective MUSCLES* Algorithms

In [Yi00], a method for the mining of *co-evolving data streams* is presented. The method is called *MUSCLES* and in addition to data mining, it can be used for outlier detection and forecasting of either future, missing, or delayed values. The authors tackle the problems of delayed and missing values as multivariate linear regression, where the incoming information comes from (1) the past values of the sequence whose value we are trying to estimate and (2) all past and present values of other sequences.

The regression coefficients can be used to detect correlation, because if a regression coefficient is high, this indicates a strong correlation between the dependent and the corresponding independent variable.

Regarding its predictive ability, *MUSCLES* was compared against two other popular techniques: (1) using the immediately previous value, as a predictor of the future value, and (2) auto-regression. For various data sets, *MUSCLES* outperformed the other two techniques. Because *MUSCLES* uses an incremental approach, it can handle sequences that are indefinitely long.

To be able to handle a large number of sequences, the authors propose a method called *Selective MUSCLES*. The main idea here is that we do need information from every single sequence, so some preprocessing to select the most informative time series can save us a lot of time. The measure of goodness used to select the best sequence subset is the expected estimation error. Experimental results show that *Selective MUSCLES* can be up to 110 times faster than *MUSCLES* and can, occasionally, have even better prediction accuracy.

5.5.6 The *AWSOM* Algorithm

The authors in [Pap04] propose an algorithm called *AWSOM*, which is developed for sensors operating in hostile and remote environments. The algorithm allows sensors to detect representative patterns and trends and it requires very few resources, such as memory. Specifically, the authors note that an algorithm that processes stream data from sensors should have the following requirements:

- Be able to detect periodic components, filter out noise, and identify simple patterns.

- Use limited memory.

- Be *one pass* algorithm and *online*.

- Be able to detect outliers.

- Should not require human supervision.

AWSOM uses the *wavelet transform* as the time series representation technique, because of its success in compression and the fact that it can be computed quickly and incrementally. Also a wavelet-transformed signal can easily capture periodicities by exhibiting high energy at the scales that correspond to the frequency. Overall, *AWSOM* achieves all of the above requirements by providing a concise linear model with a few coefficients and it can capture periodic components as well as self-similarity. The algorithm operates in an unsupervised fashion and it requires only one pass over the data. Experiments on real and synthetic data have showns that the algorithm can successfully capture periodicities and bursts in the data.

5.6 MINING TEMPORAL PATTERNS IN MULTIMEDIA

In [Tse05], an algorithm is proposed (*SP-Miner*) that discovers complex spatial relationships among objects that appear in multimedia contents captured using a digital camera and another algorithm (*CP-Miner*) that discovers temporal changes in spatial patterns. The discovered spatial relationships are of the form *northeast, southeast, northwest,* and *southwest.* For example, object 1 is to the northeast of object 2. Locations of objects are described using Cartesian coordinates. Images are described by a collection of objects and the time they were captured.

SP-Miner is an extension of the well-known *Apriori* algorithm. A spatial pattern is called a frequent spatial pattern if its support exceeds a certain threshold. Finally the inputs to the *CP-Miner* algorithm that discovers temporal changes in spatial patterns are the t_i and t_i+_1 patterns discovered by *SP-Miner*. *CP-Miner* extracts the patterns that contain the same objects but differ in their spatial relations.

In [Gal01], an *image segmentation* algorithm that is based on spatiotemporal clustering is described. Specifically, the image segmentation is not performed on individual frames, but instead multiple images are utilized to segment objects of interest, such as a beating heart. Features, such as image brightness and motion in the x, y, and z directions, are extracted from the sequence of frames, which are then clustered using the *k-means* clustering algorithm.

In [Hon01], the authors describe a framework that indexes semantically annotated satellite images and extracted association rules, and stores them in a database. Then the user can perform high level queries using *SQL* entered from a user interface. The authors describe the flow of their system as follows:

- Cluster the images using a *self-organizing map*. The accuracy of clustering was measured using precision and recall.

- Transform the time series of images into a sequence of cluster addresses.

- Extract the events and time-dependent rules from the rules.

- Index the events and rules in an *R-tree* and store in the database.

- Search for complex time variation patterns, such as "search for frequent events that occur between two tornadoes."

In [Rad04], an unsupervised pattern discovery is proposed to discover temporal patterns in multimedia. Specifically, in video, the problem is formulated as a time series clustering algorithm. The authors make the following important note: While past work in the area has focused on algorithms to detect *semantic events* (for scripted content, such as news) or *sparse events* (for unscripted content, such as sports), this current work proposes an unsupervised approach. This unsupervised approach utilizes a time series clustering framework to detect patterns of interest in unscripted content. The problem of detecting these patterns is formulated as detecting outliers in a stationary background process consisting of low/mid-level audiovisual features. However, the background process is stationary for a small period of time and the modeling has to be adaptive to the content. An *eigenvector-based* approach is utilized to segment the images. Also the detected events (the outliers) are ranked based on a confidence measure. Experimental results show that the algorithm was able to effectively extract highlight events from sports video.

In [Che07a], the authors propose a hierarchical temporal association mining technique with the purpose of automatic video analysis to detect events. The mining technique utilizes an extension of the *Apriori* algorithm to find frequent patterns in video databases. The confidence of the patterns is used to confirm that these patterns indeed characterize the event of interest. The patterns are then used to create rules that are used in video event detection.

Finally, in [Fle06], a method is proposed to classify video content based on movement temporal patterns. These patterns are placed in a hierarchy, which allows the construction of a lexicon of patterns of movement. Then these patterns are used as features in an *SVM* classification method. Temporal relations between movement events are expressed using *Allen's temporal relations*. Experimental results on real-world data show the method to have better performance than a *hidden Markov model*. A significant advantage of hierarchical patterns over *HMMs* is that the former can capture global information about an event by finding relationships that are not locally apparent, while the latter can only capture local transitions between states. An interesting note is that the real-world video data are from the *Human SpeechHome* project, where the speech development of a single child is observed and modeled in an unprecedented scale.

5.7 ADDITIONAL BIBLIOGRAPHY

5.7.1 Sequential Pattern Mining

In [Mas05] and [Zha03], surveys of sequential pattern mining are presented. Additional work in temporal pattern discovery can be found in [Die85] and [Han98]. For example, in [Han98] an algorithm is described for a fixed-length periodicity detection using a combination of data cube and *Apriori* algorithm technologies. In [Rai00], a visualization tool is described that uses a cyclical graph to create association rules visualization.

In [Bet98], the problem of finding patterns with multiple granularities is addressed. The algorithm uses *timed finite automata* for the discovery of *event structures,* where an event structure consists of events and temporal constraints, such as events occurring in the same day. [Wu07] describes an algorithm (the *TPrefixSpan*) for describing nonambiguous *interval-based* temporal events. In [Mar05], an algorithm for temporal case-based reasoning is described, which allows the abstraction of temporal sequences of cases, known as episodes.

In [Zha07], the authors describe a method for the detection of *follow-up correlation patterns*. These patterns refer to correlated itemsets that do not necessarily belong to the same transaction. In [Cot03], the discovery of temporal meta rules is discussed. In [Tum06], an algorithm for mining frequent subsequences (*ProMFS*) is described. The main idea of the algorithm is the estimation of probabilistic and statistical properties of the appearance of elements in the sequence. These are the *probability of an element in the sequence,* the *average element distance,* and the *probability of an element* to appear *after another.* Then a shorter sequence is built which is analyzed using the *GSP* algorithm.

In [Low99], an algorithm for temporal pattern matching using *fuzzy templates* is described. Fuzzy set theory allows the creation of fuzzy rules from linguistic rules. The algorithm was applied to the monitoring of planning in anesthesia. Fuzzy templates can represent the following types of incomplete knowledge: (1) *events of vague duration,* (2) *intervals of vague duration between events,* (3) *fuzzy signal levels,* and (4) *fuzzy pattern matching,* because pattern matching is itself a fuzzy process.

In [Lin02], a method for the *fast discovery* of sequential patterns is proposed. The method is called *MEMISP* and its computational speed is achieved by using memory indexing. The method does not generate a candidate set and fits the entire database in main memory. This way only one database scan is required. If the database is too large to fit into main memory, then the database is divided into partitions and each partition is mined separately. In this case we need to make a second scan of the database to validate the patterns. Experimental results indicate that *MEMISP* outperforms both *GSP* and *PrefixSpan*.

5.7.2 Time Series Pattern Discovery

In [Li04], an algorithm, called *INFER*, is developed that discovers infrequent episodes in time series databases. This is one of the few works that address the problem of finding patterns with low support, but high confidence. It achieves this by combining *bottom-up* and *top-down* searching techniques. In [Wei05b], the authors address the problem of *on-the-fly* subsequence matching of streaming time series to predefined patterns. Their proposed algorithm takes into account similarity of patterns and merges similar patterns into a wedge. This results in significant computational gains.

The authors in [Ngu08] describe three algorithms for *fast correlation analysis* of time series data sets. Specifically, they present algorithms for *bivariate and multivariate correlation queries* and also correlation queries based on a new correlation measure that utilizes *DTW*. In [Ude04], a method for discovering all frequent trends in a time series is proposed. The main novelty of the method is that it is domain independent. In many pattern discovery algorithms, domain knowledge is utilized in the early stages of the algorithms, to set the window width, for example. The method described in [Ude04] achieves domain independence by creating a modification of the *Shape Language* proposed by Agrawal et al. (described in Chapter 2) and using the relative time series changes to encode the time series in a finite alphabet string. Then the translated time series is represented using a suffix tree.

In [Yun08], the authors propose a method (*WSpan*) for the discovery of weighted sequential patterns. The weight of a pattern depends on its importance, and it plays an early and crucial role in the proposed algorithm. Specifically, during the discovery of the patterns, attention is paid to both the support and the weight of the pattern. Experimental results show that the algorithm detects fewer, yet important patterns, and therefore achieves its goal. In [Bro02], the authors address the problem of *frequency selective time series analysis.* They achieve this by analyzing only sub-bands of the frequency spectrum and specifically by transforming a part of the spectrum in a long sub-band AR model.

In [Jia04], a hybrid method for streaming data pattern discovery is proposed, which combines a *segmental semi-Markov model* and a *distance-based method*. In [Hoc02], a framework for the interactive querying of time series data is presented. The framework allows the user to draw time-boxes, which are rectangular query regions drawn on the temporal data. The *x-axis* shows the time period in which the user is interested, and the *y-axis* allows the user to specify the range of values in which she is interested.

In [Ler04], the authors present techniques for the following four problems:

- Find correlations using a sliding window in time series data from financial, physical, and other applications.

- Find bursts in sensor data.

- Perform hum searching, that is, find recorded music based on similar hums.

- Maintain and investigate temporal data in databases.

In [Yan03b], the authors address the problem of detecting periodic patterns. However, they deviate from what is commonly examined in literature, synchronous periodic patterns, and they focus on *asynchronous periodic patterns*. These are patterns that may occur only within a subsequence or with their occurrences shifted due to noise. The main idea of the algorithm is to employ two parameters:

- *Min_rep.* This is minimum number of required nondisrupted repetitions.

- *Max_dis.* This is the maximum allowed length for a piece of disturbance.

The algorithm consists of two steps: (1) potential periods by distance-based pruning are generated, and (2) an iterative procedure is employed to find the longest valid subsequence.

In [Aβf08], a framework for the *threshold-based* mining of time series is described. The framework, *T-time* allows the user to do the following: (1) identify of potentially useful threshold values, and (2) visually explore and interact with other data analysis parameters and also extract useful knowledge. The framework also includes the *Euclidean distance* and *DTW*. It was successfully applied to gene expression data, in a way that allowed the clustering of patients to different phenotypes, by simply changing the threshold.

In [Ahm07], the authors present *Pulse*, a framework that allows *continuous* queries on sensor data. This way, they are able to capture the fact that the underlying processes that create the sensor data are also continuous. Examples are the temperatures and humidity readings of a region over time. *Pulse* converts the queries to simultaneous polynomial equations operating on continuous time models. In [Gao02], the authors address the problem of discovering relevant patterns in incoming streaming time series and reacting immediately. They specifically address the problem in which the patterns are time series stored in a database and the incoming stream is a continually appended time series. The proposed method uses the *FFT* to find cross correlations of time series and in this way identify the nearest neighbors of the incoming time series up to the current time position. To achieve fast response, the method uses prediction methods to predict future values and compute predicted distances and predicted errors. When the actual data arrive, the predicted distance and error is used to prune patterns that are not possible to be nearest neighbors. Experimental results confirm the performance gain attained by the algorithm.

In [Ath08], the authors propose a method, called *EBSM,* for *approximate* subsequence matching in large time series data sets using the *DTW* distance measure. The main idea is to map each database time series into a sequence of vectors, and in this way convert subsequence matching to *vector matching in an embedded space.* The embedding is performed by applying *DTW* between reference objects and the time series in the database. In the same way, at query time, the query object is mapped into the embedded space. The embedding process provides for significant computational gains, as vector matching is used to identify a small number of candidate matches. Another novel contribution of *EBSM* is that optimal matches do not have to have the same length as the query, and therefore, it utilizes *unconstrained DTW*. Experimental results show that *EBSM* increases the speed by over one order of magnitude in comparison with a brute force approach, while having less than 1% losses in retrieval accuracy. In addition, experimental results show that *EBSM* has the best performance in terms of accuracy versus efficiency in comparison with a state-of-the-art method utilizing *unconstrained DTW*.

Although a significant amount of work has been done for the discovery of temporal patterns in time series, relatively little work has been done for the mining of *time-interval-based* data. As described in [Vil00], in time-interval mining techniques, an event is considered to be *active* for a period of time, instead of being considered as an instantaneous occurrence. This can find applications in areas such as healthcare, where the patient history (presurgery preparation, surgery, recovery) can be viewed as a series of interval events. Note that it is possible for an interval event to be contained in another interval event. The authors in [Vil00] propose a technique to find such temporal containment relationships.

In [Höp02b], a temporal pattern is a set of intervals and their relationships, where the latter are described using *Allen's* temporal relations. The authors discuss how to find patterns that lead to the derivation of *informative rules*. The rules are ranked using the *J-measure*, discussed previously in this chapter. Another article that describes the discovery of temporal patterns for interval-based events is [Kam00]. The authors use a *SPADE*-like algorithm to discover patterns and *Allen's interval relations*.

In [Win07], Winarko and Roddick propose *ARMADA*, an algorithm for finding richer relative temporal association rules from interval-based data. The authors represent time series as *interval-based events*. The number of generated rules is reduced by introducing a maximum gap constraint. *ARMADA* extends the *MEMISP* algorithm [Lin02], which is described above, so it reads the database into memory and requires only one database scan. If the database is too large, as in the case of *MEMISP*, the database is divided into partitions and each partition is mined separately. Frequent patterns are placed in groups, where all patterns within one group have the same prefix. An index set is then constructed and long patterns are grown from shorter frequent ones, where only sequences indicated by the index set are considered. In [Kim96], a dynamic indexing structure, called TIP-index, is proposed for efficiency searching time series patterns.

Finally, in [Kum06], the authors address the problem of finding sequential patterns in a multi-database. They approach the problem by developing a method, called *ApproxMAP*, to summarize sequential pattern information in each local database. Then, only the summarized sequence information can be used as input to the global mining algorithm. Experimental results indicate that the algorithm is both scalable and efficient in finding long patterns in large sequence databases.

REFERENCES

[Abr01] Abramowicz, W. and J. Zurada, *Knowledge Discovery for Business Information Systems*, Springer, 2001.

[Abr99] Abraham, T. and J. F. Roddick, Incremental Meta-Mining from Large Temporal DataSets, *Proceedings of the 1st International Workshop on Data Warehousing and Data Mining*, pp. 41–54, Springer, 1999.

[Aga99] Agarwal, R., Aggarwal, C., and V.V.V. Prasad, Depth-First Generation of Large Itemsets for Association Rules, *IBM Technical Report RC 21538,* July 1999.

[Agr93] Agrawal, R, T. Imielinski, and A. Swami, *Mining Association Rules Between Sets of Items in Large Databases, Proc. ACM SIGMOD International Conference Management of Data*, vol. 22, pp. 207–216, 1993.

[Agr95] Agrawal, R. and R. Srikant, Mining Sequential Patterns, *Proceedings of the 11th Int. Conference on Data Engineering*, pp. 3–14, March 1995.

[Agr96] Agrawal, R. and R. Srikant, Mining Sequential Patterns: Generalizations and Performance Improvements, *Lecture Notes in Computer Science— Advances in Database Technology*, pp. 3–17, 1996.

[Ahm07] Ahmad, Y. and U. Cetintemel, Declarative Temporal Data Models for Sensory-Driven Query Processing, *Proceedings of the 4th workshop on Data Management for Sensor Networks: In Conjunction with 33rd International Conference on Very Large Data Bases, pp. 37–42, 2007.*

[Ahm07] Ahmed, T., B. Oreshkin, and M. Coates, Machine Learning Approaches to Network Anomaly Detection, http://www.usenix.org/event/sysml07/tech/full_papers/ahmed/ahmed.pdf, 2007.

[Ale00] Ale, M.J. and G.H.Rossi, An Approach to Discovering Temporal Association Rules, *Proceedings of the 2000 ACM Symposium on Applied Computing*, vol. 1, pp. 294–300, 2000.

[Ara06] Araujo, L., J.A. Cuesta, and J.J. Merelo, Genetic Algorithm for Burst Detection and Activity Tracking in Event Streams, *Lecture Notes in Computer Science*, vol. 4193, pp. 302–311, 2006.

[Are04] Aref, W.G., M.G. Elfeky, and A.K. Elmargamid, Incremental Online Merge Mining of Partial Periodic Patterns in Time-Series Databases, *IEEE Transactions on Knowledge and Data Engineering*, vol. 16, no. 3, pp. 332–342, 2004.

[Ath08] Athitsos, V. et al., Approximate Embedding-Based Subsequence Matching of Time Series, *Proceedings of the ACM SIGMOD International Conference on Management of Data*, pp. 365–378, 2008.

[Ayr02] Ayres, J. et al., Sequential Pattern Mining Using a Bitmap Representation, *Proceedings of Conference on Knowledge Discovery and Data Mining*, pp. 429–435, 2002.

[Aβf08] Aβflag, J. et al., T-Time: Threshold-Based Data Mining on Time Series, *Proceedings of the 24th International Conference on Data Engineering*, 2008.

[Bet98] Bettini, C. et al., Discovering Frequent Event Patterns with Multiple Granularities in Time Sequences, *IEEE Transactions Knowledge and Data Engineering,* vol. 10, no. 2, pp. 222–237, 1998.

[Blu83] Blum, R., Representation of Empirically Derived Causal Relationships, *Proceedings of International Joint Conference on Artificial Intelligence (IJCAI),* pp. 268–271, 1983.

[Bou07] Bouandas, K. and A. Osmani, Mining Association Rules in Temporal Sequences, *Proceedings of the 2007 IEEE Symposium on Computational Intelligence on Data Mining,* pp. 610–615, 2007.

[Bro02] Broersen, T.M. and S. de Waele, Frequency Selective Time Series Analysis, *IEEE Instrumentation and Measurement Technology Conference,* vol. 1, Anchorage, AK, pp. 775–780, May 2002.

[Bur01] Burdick, D., Calimlim, M., and J. Gehrke, MAFIA: A Maximal Frequent Itemset Algorithm for Transactional Databases, *Proceedings of the 2001 International Conference on Data Engineering,* pp. 443–452, April 2001.

[Cap02] Capelle, M., C. Masson, and J. -F. Boulicaut, Mining Frequent Sequential Patterns under a Similarity Constraint, http://liris.cnrs.fr/~jboulica/ideal02.pdf, 2002.

[Che98a] Chen, X. and I. Petrounias, Language Support for Temporal Data Mining, *Proceedings of 2nd European Symposium on Principles of Data Mining and Knowledge Discovery,* PKDD'98, pp. 282–290, Springer-Verlag, Berlin, 1998.

[Che98b] Chen, X., I. Petrounias, and H. Heathfield, Discovering Temporal Association Rules in Temporal Databases, *Proceedings International Workshop on Issues and Applications of Database Technology,* pp. 312–319, 1998.

[Che00] Chen, X. and I. Petrounias, Discovering Temporal Association Rules: Algorithms, Language, and System, *Proceedings of the 16th International Engineering Conference on Data Engineering,* p. 306, 2000.

[Che02] Chen, Y. et al., Multi-dimensional Regression Analysis of Time-Series Data Streams, *Proceedings of the 28th International Conference on Very Large Data Bases,* pp. 323–334, 2002.

[Che05] Chen, G., X. Wu, and X. Zhu, Sequential Pattern Mining in Multiple Streams, *Proceedings of the 5th IEEE International Conference on Data Mining,* pp. 585–588, 2005.

[Che07a] Chen, M., S.C. Chen, and M.L. Shyu, Hierarchical Temporal Association Rules Mining for Video Event Detection in Video Databases, *IEEE 23rd International Conference on Data Engineering,* pp. 137–145, 2007.

[Che07b] Chen, Y. et al., SpADe: On Shape-Based Pattern Detection in Streaming Time Series, *Proceedings of the IEEE 23rd International Conference on Data Engineering,* pp. 786–795, 2007.

[Chi03] Chiu, B., E. Keogh, and S. Lonardi, Probabilistic Discovery of Time Series Motifs, *Proceedings of the ACM SIGKDD Conference,* pp. 493–498, 2003.

[Chi05] Chin, S.C., A. Ray, V. Rajagopalan, Symbolic Time Series Analysis for Anomaly Detection: A Comparative Evaluation, *Signal Processing,* vol. 85, pp. 1859–1868, Elsevier, 2005.

[Cho08] Choudhary, R., S. Mehta, and A. Bagchi, On Quantifying Changes in Temporally Evolving Dataset, *Proceedings of the ACM Conference on Information and Knowledge Management (CIKM)*, pp. 1459–1460, Napa Valley, California, 2008.

[Cot03] Cotofrei, P. and K. Stoffel, Higher Order Temporal Rules, *Lecture Notes on Computer Science 2657*, pp. 323–332, Springer, 2003.

[Das96] Dasgupta, D. and S. Forrest, Novelty Detection in Time Series Data Using Ideas from Immunology, *5th International Conference on Intelligent Systems*, 1996.

[Das98] Das, G. et al., Rule Discovery from Time Series, *Proceedings of the ACM SIGKDD Conference*, pp. 16–22, 1998.

[Dhi03] Dhilon, I.S. and S. Sra, Modeling Data Using Directional Distributions, *Technical Report TR-06-03*, University of Texas, Dept. of Computer Science, February 2003.

[Die85] Dietterich, T.G. and R.S. Michalski, Discovering Patterns in Sequences of Events, *Artificial Intelligence*, vol. 25, pp. 187–232, 1985.

[Dom01] Domingos, P. and G. Hulten, Mining High Speed Data Streams, *Proceedings of KDD*, pp. 71–80, 2001.

[Dur98] Durbin, R. et al., *Biological Sequence Analysis: Probabilistic Models of Proteins and Nucleic Acids*, Cambridge University Press, 1998.

[Elf05] Elfeky, M.G., W.G. Aref, and A.K. Elmargarmid, Periodicity Detection in Time Series Databases, *IEEE Transactions on Knowledge and Data Engineering*, vol. 17, no. 7, pp. 875–887, 2005.

[Fle06] Fleischman, M., P. Decamp, and D. Roy, Mining Temporal Patterns of Movement for Video Content Classification, *Proceedings of ACM Conference on Multimedia Information Retrieval*, pp. 183–191, 2006.

[Fuj05] Fujimaki, R., T. Yairi, and K. Machida, An Approach to Spacecraft Anomaly Detection Problem Using Kernel Feature Space, *Proceedings of the ACM SIGKDD Conference*, pp. 401–410, 2005.

[Gal01] Galic, S. and S. Loncaric, Cardiac Image Segmentation Using Spatio-Temporal Clustering, *Proceedings of the SPIE*, pp. 1199–1206, 2001.

[Gan02] Ganti, V., J. Gehrke, and R. Ramakrishnan, Mining Data Streams under Block Evolution, *SIGKDD Explorations*, vol. 3, no. 2, pp. 1–10, 2002.

[Gao02] Gao, L. and X.S. Wang, Continually Evaluating Similarity-Based Pattern Queries on a Streaming Time Series, *Proceedings of the 2002 ACM SIGMOD International Conference on Management of Data*, pp. 370–381, 2002.

[Ge00] Ge, X. and P. Smyth, Deformable Markov Model Templates for Time-Series Pattern Matching, *Technical Report,* Dept. of Information and Computer Science, University of California, Irvine, 2000.

[Guh03] Guha, S., D. Gunopulos, and N. Koudas, Correlating Synchronous and Asynchronous Data Streams, *Proceedings of the ACM KDD Conference*, pp. 529–534, 2003.

[Gui05] Guil, F. et al., An Iterative Method for Mining Frequent Incremental Patterns, *Lecture Notes in Computer Science, Computer Aided Theory*, vol. 3643, pp. 189–198, 2005.

[Gwa03] Gwadera, R., M.J. Atallah, and W. Szpankowski, Reliable Detection of Episodes in Event Sequences, *Proceedings of 3rd IEEE International Conference on Data Mining*, pp. 67–74, 2003.

[Han98] Han, J., W. Gong, and Y. Yin, Mining Segment-wise Periodic Patterns in Time Related Databases, *Proceedings of Fourth International Conference on Knowledge Discovery and Data Mining*, pp. 214–218, AAAI Press, Menlo Park, 1998.

[Han99] Han, J., G. Dong, and Y. Yin, Efficient Mining of Partial Periodic Patterns in Time Series Database, *Proceedings of the 15th International Conference on Data Engineering*, pp. 106–115, 1999.

[Han00] Han, J. et al., FreeSpan: Frequent Pattern-Projected Sequential Pattern Mining, *Proceedings of the ACM SIGKDD Conference*, pp. 355–359, Boston, MA, 2000.

[Han04] Han, J. et al., Mining Frequent Patterns without Candidate Generation: A Frequent-Pattern Tree Approach, *Data Mining and Knowledge Discovery*, vol. 8, no. 1, pp. 53–87, Springer, 2004.

[Het02] Hetland, M.L. and P. Sætrom, Temporal Rule Discovery Using Genetic Programming and Specialized Hardware, *Proceedings of 4th International Conference on Recent Advances in Soft Computing*, 2002.

[Het05] Hetland, M.L. and P. Sætrom, Evolutionary Rule Mining in Time Series Databases, *Machine Learning*, pp. 107–125, 2005.

[Hil07] Hill, J.D., B.S. Minsker, and E. Amir, Real-Time Bayesian Anomaly Detection for Environmental Sensor Data, *American Geophysical Union*, Fall 2007.

[Hoc02] Hochheiser, H., Interactive Querying of Time Series Data, *Proceedings of CHI*, Minneapolis, Minnesota, 2002.

[Hon01] Honda, R. and O. Konishi, Temporal Rule Discovery for Time Series Satellite Images and Integration with RDB, *Proceedings of the 5th European Conference on Principles of Data Mining and Knowledge Discovery, Lecture Notes in Computer Science*, pp. 204–215, 2001.

[Hon02] Hong, P. and T. Huang, Automatic Temporal Pattern Extraction and Association, *Proceedings of the IEEE ICASSP Conference*, vol. 2, pp. 2005–2008, 2002.

[Höp02a], Höppner, F., Learning Dependencies in Multivariate Time Series, *Proceedings of the ECAI'02 Workshop on Knowledge Discovery in Spatio-Temporal Data*, pp. 25–31, 2002.

[Höp02b] Höppner, F. and F. Klawonn, Finding Informative Rules in Interval Sequences, *Intelligent Data Analysis*, vol. 6, no. 3, pp. 237–256, 2002.

[Hua07] Huang, J. and W. Wei, Efficient Algorithm for Mining Temporal Association Rule, *International Journal of Computer Science and Network Society*, vol. 7, no. 4, pp. 268–271, April 2007.

[Ima94] Imam, I.F., An Experimental Study of Discovery in Large Temporal Databases, *Proceedings of the 7th International Conference on Industrial and Engineering Applications of Artificial Intelligence and Expert Systems, IEA/AIE*, pp. 171–180, 1994.

[Jen06] Jensen, K.L. et al., A Generic Motif Discovery Algorithm for Sequential Data, *Bioinformatics*, vol. 22, no. 1, pp. 21–28, 2006.

[Jia04] Jia, S., Y. Qian, and G. Dai, A Hybrid Online Series Pattern Detection Algorithm, *Proceedings of the 5th World Congress on Intelligent Control and Automation,* pp. 1876–1879, 2004.

[Kam00] Kam, P.S. and A.W.C. Fu, Discovering Temporal Patterns for Interval Based Events, *Proceedings of DaWaK-00,* 2000.

[Keo02] Keogh, E., S. Lonardi, and B.Y. Chu, Finding Surprising Patterns in a Time Series Database in Linear Time and Space, *Proceedings of the ACM SIGKDD Conference,* pp. 550–556, 2002.

[Keo05] Keogh, E., Lin, J., and A. Fu, HOT SAX: Finding the Most Unusual Time Series Subsequence: Algorithms and Applications, *Proceedings of the 5th IEEE International Conference on Data Mining,* pp. 226–233, 2005.

[Kim96] Kim, Y., Y. Park, and J. Chun, A Dynamic Indexing Structure for Searching Time Series Patterns, *IEEE Computer Society,* pp. 270, 1996.

[KouURL] Koundourakis, G., M. Saraee, and B. Theodoulidis, Data Mining in Temporal Databases, *Citeseer.*

[Kum06] Kum, H.C., J.H. Chang, and W. Wang, Sequential Pattern Mining in Multi-Databases via Multiple Alignment, *Data Mining and Knowledge Discovery,* vol. 12, pp. 151–180, 2006.

[Lax06] Laxman, S. and P.S. Sastry, A Survey of Temporal Data Mining, *Sadhana,* vol. 31, part 2, pp. 173–198, April 2006.

[Lee04] Lee, W.J. and S.J. Lee, Discovery of Fuzzy Temporal Association Rules, *IEEE Transactions on Systems, Man, and Cybernetics,* vol. 34, no. 6, pp. 2330–2342, 2004.

[Ler04] Lerner, A. et al., Fast Algorithms for Time Series with Applications to Finance, Physics, Music, Biology, and Other Suspects, *Proceedings of SIGMOD,* pp. 965–968, 2004.

[Li04] Li, D., L. Jiang, and J.S. Deogun, Temporal Knowledge Discovery with Infrequent Episodes, *Proceedings of the 2004 National Conference on Digital Government Research,* pp. 1–2, 2004.

[Li08] Li, C. and J. Wang, Efficiently Mining Closed Subsequences with Gap Constraints, *Proceedings of the SIAM Conference,* pp. 313–322, April 2008.

[Lin02] Lin, M.Y. and S.Y. Lee, Fast Discovery of Sequential Patterns by Memory Indexing, *Data Warehousing and Knowledge Discovery, Lecture Notes in Computer Science,* vol. 2454, pp. 227–237, 2002.

[Lin04] Lin, J. et al., Visually Mining and Monitoring Massive Time Series, *Proceedings of KDD,* pp. 460–469, 2004.

[Lin05] Lin, J., E., Keogh and S. Lonardi, Visualizing and Discovering Non-trivial Patterns in Large Time Series Databases, *Information Visualization,* vol. 4, no. 2, pp. 61–82, 2005.

[Lin07] Lin, N.P. et al, Discover Sequential Patterns in Incremental Database, *International Journal of Computers,* vol. 1, no. 4, pp. 196–201, 2007.

[Low99] Lowe, A., R. W. Jones, and M.J. Harrison, Temporal Pattern Matching Using Fuzzy Templates, *Journal of Intelligent Information Systems,* vol. 13, pp. 27–45, 1999.

[Ma01] Ma, S. and J.L. Hellerstein, Mining Partially Periodic Event Patterns with Unknown Periods, *Proceedings of the 17th International Conference on Data Engineering,* pp. 205–214, 2001.

[Man97] Mannila, H., H. Toivonen, and A. Verkamo, Discovering of Frequent Episodes in Event Sequences, *Data Mining and Knowledge Discovery,* vol. 1, pp. 259–289, 1997.

[Mar05] Marre, M.S. et al., An Approach for Temporal Case-Based Reasoning: Episode-Based Reasoning, *Lecture Notes in Computer Science,* vol. 3620, pp. 465–476, Springer, 2005.

[Mas98] Masseglia, F., F. Cathala, and P. Poncelet, The PSP Approach for Mining Sequential Patterns, *Lecture Notes in Computer Science, Principles of Data Mining and Knowledge Discovery,* vol. 1510, pp. 176–184, 1998.

[Mas03] Masseglia, F., P. Poncelet, and M. Teisseire, Incremental Mining of Sequential Patterns in Large Databases, *Data and Knowledge Engineering,* vol. 46, no. 1, pp. 97–121, 2003.

[Mas05] Masseglia, F., M. Teisseire, and P. Poncelet, Sequential Pattern Mining: A Survey on Issues and Approaches, http://www-sop.inria.fr/axis/personnel/ Florent.Masseglia/International_Book_Encyclopedia_2005.pdf, 2005.

[Min06] Minnen, D. et al., Activity Discovery: Sparse Motifs from Multivariate Time Series, *The Learning Workshop,* Snowbird, UT, 2006.

[Min07] Minner, D. et al., Discovering Multivariate Motifs Using Subsequence Density Estimation and Greedy Mixture Learning, *AAAI,* pp. 615–620, 2007.

[Mör04] Mörchen, F., and A. Ultsch, Mining Hierarchical Temporal Patterns in Multivariate Time Series, S. Biundo, T.W. Frühwirth, G. Palm (Eds), *KI 2004: Advances in Artificial Intelligence, Proceedings 27th Annual German Conference in AI,* pp. 127–140, Ulm, Germany, Springer, Heidelberg, 2004.

[Mör06] Mörchen, F., Algorithms for Time Series Knowledge Mining, T. Eliassi-Rad et al., eds., In *Proceedings The Twelveth ACM SIGKDD International Conference on Knowledge Discovery and Data Mining,* Philadelphia, PA, pp. 668–673, 2006.

[Mör07a] Mörchen, F. A. Ultsch, Efficient Mining of Understandable Patterns from Multivariate Interval Time Series, *Data Mining and Knowledge Discovery,* vol. 15, no. 2, Springer, pp. 181–215, 2007.

[Mör07b] Mörchen, F., Unsupervised Pattern Mining from Symbolic Temporal Data, ACM *SIGKDD Explorations,* vol. 9, no. 1, pp. 41–55, 2007.

[Ng07] Ng, V. et al., Incremental Mining for Temporal Association Rules for Crime Pattern Discoveries, *Proceedings of the 18th Australian Database Conference,* 2007.

[Ngu08] Nguyen, P. and N. Shiri, Fast Correlation Analysis on Time Series DataSets, *Proceedings of the 17th ACM Conference on Information and Knowledge Management,* pp. 787–796, 2008.

[Oat02] Oates, T., PERUSE: An Unsupervised Algorithm for Finding Recurring Patterns in Time Series, *IEEE International Conference on Data Mining,* pp. 330–337, 2002.

[Pap04] Papadimitriou, S., A. Brockwell, and C. Faloutsos, Adaptive, Unsupervised Stream Mining, *The VLDB Journal*, vol. 13, pp. 222–239, 2004.

[Pap05] Papadimitriou, S., J. Sun, and C. Faloutsos, Streaming Pattern Discovery in Multiple Time Series, *Proceedings of the 31st VLDB Conference*, pp. 697–708, 2005.

[Pap06] Papadimitriou, S. and P.S. Yu, Optimal Multi-Scale Patterns in Time Series Streams, *Proceedings of the ACM SIGMOD International Conference on Management of Data*, pp. 647–658, 2006.

[Par99] Parthasarathy, S. et al., Incremental and Interactive Sequence Mining, *Proceedings of the 8th International on Information and Knowledge Management*, pp. 251–258, 1999.

[Pei00] Pei, J., J. Han, and R. Mao, CLOSET: An Efficient Algorithm for Mining Frequent Closed Itemsets, *Proceedings 2000 ACM-SIGMOD International Workshop Data Mining and Knowledge Discovery*, pp. 11–20, May 2000.

[Pei04] Pei, J. et al., Mining Sequential Patterns by Pattern Growth: The PrefixSpan Approach, *IEEE Transactions on Knowledge and Data Engineering*, vol. 16, no. 10, pp. 1424–1440, Nov. 2004.

[Pev00] Pevzner, P.A. and S.H. Sze, Combinatorial Approaches to Finding Subtle Signals in DNA Sequences, *Proceedings of the 8th International Conference on Intelligent Systems for Molecular Biology*, pp. 269–278, 2000.

[Pol00] Policker, S. and A.B. Geva, Non-stationary Time Series Analysis by Temporal Clustering, *IEEE Transactions on Systems, Man, Cybernetics*, vol. 30, no. 2, pp. 339–343, April 200.

[Pov03] Povinelli, J. and X. Feng, A New Temporal Pattern Identification Method for Characterization and Prediction of Complex Time Series Events, *IEEE Transactions on Knowledge and Data Engineering*, vol. 15, no. 2, pp. 339–352, March/April 2003.

[Rad04] Radhakrishnan, R., A. Divakaran, and Z. Xiong, A Time Series Clustering Based Framework for Multimedia Mining and Summarization Using Audio Features, *Proceedings of the 6th ACM SIGMM International Workshop on Multimedia Information Retrieval*, pp. 157–164, 2004.

[Rai00] Rainsford, C.P. and J.F. Roddick, Visualization of Temporal Association Rules, *Springer*, vol. 1983/2008, pp. 167–190, 2000.

[Rod99] Roddick, J.F. and M. Spyliopoulou, A Bibliography of Temporal, Spatial, and Spatio-Temporal Data Mining Research, *ACM SIGKDD, Explorations Newsletter*, 1999.

[Rod02] Roddick, J. and M. Spyliopoulou, A Survey of Temporal Knowledge Discovery Paradigms and Methods, *IEEE Transactions on Knowledge and Data Engineering*, vol. 14, no. 4, pp. 750–767, July/August, 2002.

[Sak05] Sakurai, Y., S. Papadimitriou, and C. Faloutsos, BRAID: Stream Mining through Group Lag Correlation, *Proceedings of ACM SIGMOD Conference*, pp. 599–610, 2005.

[Sch98] Scholkopf, B., A. Smola, and K.R. Muller, Nonlinear Component Analysis as a Kernel Eigenvalue Problem, *Neural Computation*, vol. 10, pp. 1299–1319, 1998.

[Sha04] Shasha, D. and Y. Zhu, *High Performance Discovery in Time Series*, Springer, 2004.

[She06] Sheng, C., W. Hsu, and M.L. Lee, Mining Dense Periodic Patterns in Time Series Data, *Proceedings of the 22nd International Conference on Data Engineering,* pp. 115–115, 2006.

[Smy92] Smyth, P. and R.M. Goodman, An Information Theoretic Approach to Rules Induction from Databases, *IEEE Transactions on Knowledge and Data Engineering,* vol. 4, no. 4, pp. 301–316, 1992.

[Tom01] Tompa, M. and J. Buhler, Finding Motifs Using Random Projections, *Proceedings of the 5th International Conference on Computational Molecular Biology,* pp. 67–74, 2001.

[Tse05] Tseng, V.C., P.C. Tseng, K.W.C. Lin, Mining Temporal and Spatial Object Relations in Multimedia Contents, *International Conference on Wireless Networks, Communications, and Mobile Computing,* vol. 2, pp. 1371–1376, 2005.

[Tum06] Tumasonis, R. and R. Rastenis, New Statistical Characteristics for Mining Frequent Sequences in Large Databases, *Computer Modeling and New Technologies,* vol. 10, no. 4, pp. 46–52, 2006.

[Ude04] Udechukwu, A., K. Barker, and R. Alhajj, Discovering All Frequent Trends in Time Series, *Proceedings of the Winter International Symposium on Information and Communication Technologies,* vol. 58, Cancun, Mexico, 2004.

[Vil00] Villafane, R. et al., Knowledge Discovery from Time Series of Interval Events, *Journal of Intelligent Information Systems,* vol. 15, no. 1, pp. 71–89, 2000.

[Vla05] Vlachos, M., P. Yu, V. Castelli, On Periodicity Detection and Structural Periodic Similarity, *Proceedings of SIAM International Conference on Data Mining,* Newport Beach, CA, 2005.

[Vla06] Vlachos, M. et al., Structural Periodic Measures for Time Series Data, *Data Mining and Knowledge Discovery,* vol. 12, no. 1, pp. 1–128, 2006.

[Wan96] Wang, K. and J. Tan, Incremental Discovery of Sequential Patterns, *Proceedings of ACM SIGMOD Workshop on Research Issues on Data Mining and Knowledge Discovery,* Montreal, Canada, pp. 96–102, 1996.

[Wei05a] Wei, L. et al., Assumption Free Anomaly Detection in Time Series, *Proceedings of the 17th International Conference on Scientific and Statistical Database Management,* pp. 237–240, 2005.

[Wei05b] Wei, L. et al., Atomic Wedgie: Efficient Query Filtering for Streaming Time Series, *Proceedings of the 5th IEEE International Conference on Data Mining,* pp. 490–497, 2005.

[Win05] Winarko, E. and J.F. Roddick, Discovering Richer Temporal Association Rules from Interval-Based Data, *Lecture Notes on Computer Science 3589,* Springer, pp. 315–325, 2005.

[Win07] Winarko, E. and J.F. Roddick, ARMADA – An Algorithm for Discovering Richer Relative Temporal Association Rules from Interval-Based Data, *Data & Knowledge Engineering,* vol. 63, no. 1, pp. 76–90, 2007.

[Win08] Winarko, E. and J.F. Roddick, A Signature Based Indexing Method for Efficient Content-Based Retrieval of Relative Temporal Patterns, *IEEE Transactions on Knowledge and Data Engineering,* vol. 20, no. 6, pp. 825–835, June 2008.

[Wu05] Wu, C.L., J.L. Koh, and P.Y. An, Improved Sequential Pattern Mining Using an Extended Bitmap Representation, *Lecture Notes in Computer Science*, vol. 3588, pp. 776–785, 2005.

[Wu07] Wu, S.Y. and Y.L. Chen, Mining Non-ambiguous Temporal Patterns for Interval Based Events, *IEEE Transactions on Knowledge and Data Engineering*, vol. 19, no. 6, pp. 742–758, June 2007.

[Yan03a] Yan, X., J. Han, and R. Afshar, CloSpan: Mining Closed Sequential Patterns in Large Datasets, *Proceedings of 2003 SIAM International Conference Data Mining*, 2003.

[Yan03b] Yang, J., W. Wang, and P.S. Yu, Mining Asynchronous Periodic Patterns in Time Series Data, *IEEE Transactions on Knowledge and Data Engineering*, vol. 15, no. 3, pp. 613–628, 2003.

[Yan07] Yankov, D., E. Keogh, and U., Rebbapragada, Disk Aware Discord Discovery: Finding Unusual Time Series in Terabyte Sized Datasets, *Proceedings of the 7th IEEE International Conference on Data Mining*, pp. 381–390, 2007.

[Yi00] Yi, B.K. et al., Online Data Mining for Co-evolving Time Sequences, *Proceedings of the 16th International Conference on Data Engineering*, p. 13, 2000.

[Yun08] Yun, U., A New Framework for Detecting Weighted Sequential Patterns in Large Sequence Databases, *Knowledge-Based Systems*, vol. 21, no. 2, pp. 110–122, March 2008.

[Zak98] Zaki, M., N. Lesh, and M. Ogihara, PlanMine: Sequence Mining for Plan Failures, *Proceedings of the 4th International Conference on Knowledge Discovery and Data Mining*, pp. 369–373, 1998.

[Zak01] Zaki, M., SPADE: An Efficient Algorithm for Mining Frequent Sequences, *Machine Learning*, vol. 42, no. 1/2, pp. 31–60, 2001.

[Zak02] Zaki, M.J. and C.J. Hsiao, CHARM: An Efficient Algorithm for Closed Itemset Mining, *Proceedings of 2002 SIAM International Conference Data Mining*, pp. 457–473, April 2002.

[Zha03] Zhao, Q. and S.S. Bhowmick, Sequential Pattern Mining: A Survey, Technical Report, *Center for Advanced Information Systems, School of Computer Engineering, Nanyang Technological University*, Singapore, 2003.

[Zha06] Zhang, X. and D. Shasha, Better Burst Detection, *Proceedings of the 22nd International Conference on Data Engineering*, p. 146, 2006.

[Zha07] Zhang, S. et al., Mining Follow-Up Correlation Patterns from Time-Related Databases, *Knowledge and Information Systems*, vol. 14, no. 1, pp. 81–100, 2007.

[Zhu02] Zhu, Y. and D. Shasha, StatStream: Statistical Monitoring of Thousands of Data Streams in Real Time, *Proceedings of the Conference on Very Large Databases*, pp. 358–369, 2002.

[Zhu03] Zhu, Y. and D. Shasha, Efficient Elastic Burst Detection in Data Streams, *Proceedings of the 9th ACM SIGKDD International Conference on Knowledge Discovery and Data Mining*, pp. 336–345, 2003.

CHAPTER 6

Temporal Data Mining in Medicine and Bioinformatics

In this chapter, we summarize important recent contributions in the field of temporal data mining with applications to medical data. Many of these techniques, however, can be applied to other types of data as well. Temporal data mining in medicine has many important applications, such as disease detection, patient progress evaluation, and workflow optimization.

In Section 6.1, we discuss algorithms for the detection of temporal patterns, their classification and clustering in a variety of medical data ranging from ECGs to Genomic Databases. In Section 6.2, we discuss the construction of temporal mediators for a variety of applications, such as the answering of abstract queries on medical data. Finally, in Section 6.3 we discuss clinical workflow optimization in a temporal context. Additional bibliography is provided in Section 6.4.

6.1 TEMPORAL PATTERN DISCOVERY, CLASSIFICATION, AND CLUSTERING

6.1.1 Temporal Mining in Clinical Databases

In [Alt06], the authors discuss the many challenges presented by clinical trial databases, such as the ones used in Phase III clinical trials. The data quite often present significant heterogeneities:

- Different patient qualification criteria. Patients are not recruited on the same day and the results of the trial could be affected by the season.

- Patient health status changes.

- Multiple study sites.

- Multiple lab sources.

- Differences in error distributions.

The authors present a method for dealing with heterogeneous time series data that has two steps:

1. Identify homogeneous subgroups of the data and then mine them to identify global patterns. As a result, time series that belong in the same group have the same length and the same interval.

2. Apply traditional time series distance metrics, such as the Euclidean distance, the correlation coefficient, and the slope-wise comparison method, discussed in Chapter 2.

In [Blu83], the *Rx project* is described whose goals are the following: (1) increase the validity of causal relationships derived from clinical databases, (2) develop methods for intelligently aiding hypothesis testing in a database, and (3) automate the process of discovering causal relationships. The framework contains the following components:

- Knowledge base

- Discovery module

- Study module

- Statistical package

- Clinical database

Causal relationships are described with the following features: intensity, distribution, direction, mathematical form, setting, validity, and evidence.

In [Raj06], the application of *multirelational* data mining in time-oriented biomedical databases is investigated. In this type of data mining, *association rule mining* is used to extract association rules from multiple input tables. This data mining is called *Chronos Miner* and it utilizes as input a hierarchical view of relations. In [Raj07], Chronos Miner is applied to the discovery of temporal associations between newly arising mutations in the *HIV* genome and past drug regimens.

In [Ram00], the development of a temporal pattern discovery system for *course-of-disease* data is described. The system is called *TEMPADIS* and it was applied on a database of *HIV* patients. The framework was used to do *hypothesis testing*. Specifically, it was used to determine whether people with the same chronic illness had similar experiences over time. Eighteen variables, such as white blood cell count and drug types were extracted from the database. An *event* is the occurrence of one of these variables along with its value. A decision tree approach (C4.5) was used to determine the health status of a patient. There were five health status categories ranging from *asymptomatic, no specific therapy* to *severe/life-threatening illness.* The *GSP* algorithm was used to discover patterns in sequences of events across patients in the database.

In [Sal99], the problem of dealing with the huge volumes of data generated at Intensive Care Units is addressed. The focus of the approach is to identify trends in the data that characterize them as *steady, increasing,* or *decreasing.* The algorithm has three steps: filtering (to smooth the data), temporal interpolation, and temporal inference. Each interval is described as follows: $I(x, t_{begin}, t_{end}, a_i, b_i, \lambda_i, \mu_i)$, where x is the variable under study (e.g., heart rate), t_{begin} is the start time of the interval, t_{end} is the end time of the interval, a_i is the minimum value in the interval, b_i is the maximum value in the interval, λ_i is the absolute value of the gradient of x in the interval, and μ_i is the mean of x over the interval.

In [Abe08], the *IREIIMS* project is discussed, whose purpose is to determine relationships between complete medical data and assessed health status. The database consisted of 1800 patients with 130 items recorded for each patient. The data were analyzed using the *C4.5 decision tree algorithm.* In [Ira90], artificial intelligence, statistical analysis, and neural networks are investigated to discover causal relationships in a large clinical trial database.

In [Sil08], the authors present a biometric-data-driven approach to identify *adverse events* to prevent organ failure. The study was conducted on a large patient database consisting of 25,215 daily records from 4,425 patients in 42 European ICUs. The adverse events were based on the following physiological variables:

- Systolic blood pressure
- Heart rate

- Pulse oximeter oxygen saturation

- Urine output

These adverse events, in combination with patient history characteristics, such as age and diagnosis, were the input to a classifier. The output was one of three organ status stages: (1) *normal*, (2) *dysfunction*, and (3) *failure*. Two types of classification schemes were used: *neural networks* and *multinomial logistic regression (MLR)*. Neural networks outperformed *MLR*, where the *area under the ROC score* and the *Brier score* were used as the *discrimination* and *calibration* measures. Note that a Brier score is a scoring function that measures the success of probabilistic forecasts. It is equal to the mean squared difference between the probability of an event's occurrence and the measured outcome. The outcome is a binary number (0 or 1). The ROC curve plots false positive rate versus true positive, and the area under the ROC curve indicates the accuracy of the test. If the area is 1, then the classification accuracy is perfect. An area of 0.5 indicates a failed classification test.

6.1.2 Various Physiological Signal Temporal Mining

Physionet (http://www.physionet.org) is a repository of physiological signals and open source algorithms. Physionet is the gateway to PhysioBank and PhysioToolkit. The latter contains open source algorithms for data visualization, signal and time series analysis, modeling, and so on. The signal and time series analysis is divided in the following categories:

- *Physiologic signal processing.* Algorithms included are QRS detectors and Principal Component Analysis.

- *General signal processing.*

- *Frequency domain analysis of time series.* Algorithms included are computation of power spectrum and computation of cross spectrum between 2 signals.

- *Nonlinear analysis of time series.* Algorithms included are multifractal analysis, entropy analysis, de-trended fluctuation analysis, similarity computation, and classification of time series.

- *Cardiac signal analysis.* Algorithms included are guessing the number of missing heartbeats and a tachometer.

In [Hau97], the authors performed fractal analysis of the gait of elderly subjects and subjects with Huntington's disease. Ten elderly subjects, seventeen subjects with Huntington's disease, and twenty-two young healthy adults were included in the study. *De-trended fluctuation analysis* was applied to the gait signals and fractal analysis was also performed. A time series was determined to be *self-similar* if the fluctuations at different observation windows were a function of the window size with a power law relationship. It was found that stride-interval measurements were more random in elderly subjects and subjects with Huntington's disease than in normal subjects.

[Dio03] examines the *multifractality* of human gait. As the authors note, physiological time series can be either *monofractal* or *multifractal.* Monofractal signals have the same scaling feature in the entire signal and can therefore be considered stationary from a local scaling point of view. On the other hand, multifractal signals are much more complex and inhomogeneous than monofractals. Their results show that the gait of ill people is more complex than that of healthy people, and this information could be useful in rehabilitation treatment and the addition of prosthetic devices.

In [Rut09], the foot temperatures of diabetic patients with diabetic polyneuropathy and diabetic patients without polyneuropathy were collected as a time series signal during sleep and wakefulness. Various measures were used on the time series to determine whether changes in the foot temperature could be quantifiably different in patients with polyneuropathy versus patients without the disease. It was found that wavelet analysis of the foot temperature during sleep could detect differences between the patients with the disease and those without it. Also differences during sleep were detected in the standard deviations of the foot temperatures, the rate of change and the maximal temperature.

In [Fig97], the wavelet transform is used to analyze physiological time signals. The types of signals analyzed are *heart rhythm, chest volume,* and *blood–oxygen saturation data.* The wavelet coefficients are used to calculate the energy distribution of the frequency spectrum. An entropy-like function, the *ICF,* is employed to compute the disorder in the energy distribution. The authors concluded that the wavelet transform is a powerful tool to analyze the changes in physiological signals.

In [Vas06], S. Vasudevan constructed a Bayesian classifier for the classification of the gait of normal subjects as well as patients with ALS, Parkinson's, and Huntington's disease. The wavelet transform was used to transform the data. The data features used as input to the Bayesian classifier include the mean and variance of both the original signals and the histogram as well as the skewness and kurtosis of the histograms of the wavelet coefficients. The Bayesian was successful in classifying the data sets of the testing set in the correct class.

In [Mon02], the authors propose a method for instantaneous parameter estimation in cardiovascular time series by harmonic and time frequency analysis. There are two well-known methods, *smoothed pseudo Wigner–Ville distribution (SPWVD)* and *complex demodulation (CDM)*, to be used as time-varying spectral estimators of physiological signals. These methods provide very similar results in real cardiovascular time series, but they also have complementary properties:

- Phase relationship between respiratory activity and high frequency oscillation: Measurable by *CDM,* but not measurable by *SPWVD.*

- Noise presence and interactions between spectral components: Measurable by *SPWVD,* but not measurable by *CDM.*

The authors in [Mon02] propose a combination of the above methods to overcome their limitations and combine their advantages.

In [Gui01], the authors describe a method to identify temporal patterns in time series extracted from the monitoring of patients with sleep-related breathing disorders. These patterns are converted in a linguistic knowledge representation with several abstraction levels. The temporal patterns are discovered via *self-organizing neural networks.* In [Doj98], the authors use temporal constraint networks to recognize typical medical scenarios, which is fundamental to patient monitoring. In [Dev08], the authors describe the application of a modified *k-means* clustering algorithm using sequence alignment to derive patterns from intensive care data.

In [Sac05a], a novel algorithm based on the *Apriori algorithm* is used to extract complex patterns from temporal data. The algorithm was applied on two different biomedical domains. The first domain is the analysis of time series coming from *hemodialysis* sessions and the other is *gene expression* data. In [Mör05], *muscle activation patterns* are discovered using a hierarchical temporal rule language. In [Lin03], wavelets are used to find pattern structures in elderly patient *diabetic test data.* Then, the wavelet-transformed data are statistically analyzed to identify global patterns.

In [Ang07], *multiscale entropy* of three physiological signals was used to differentiate between healthy subjects and subjects with chronic heart failure. The physiological signals were heart rate, blood pressure, and lung volume. Results showed that multiscale entropy produces statistically significant differences between healthy and patient populations, which *confirms the complexity-loss theory of aging and disease* as the authors note.

In [Bel99], blood glucose measurements from home monitoring of diabetic patients are analyzed using *Bayes-based structural filtering* and then post-processed using *temporal abstractions*, to derive meaningful knowledge that correctly abstracts the complex underlying process.

In [Hir02a], Hirano and Tsumoto describe a new method for the extraction of similar patterns. The method extracts similarities of medical time series by comparing them at different scales and clusters them using *rough clustering*. The advantage of rough clustering is that it can cluster objects, even if their similarity is only expressed in relative terms (which is the result of the multiscale matching). They applied their method to 488 sets of chronic hepatitis data. In multiscale matching, time series data are examined at different scales, and therefore, conclusions are derived about global and local similarity. To facilitate computations, the time series were divided into a set of convex/concave segments, at the ends of which are *inflection points*, which are matched across different scales. The different segments were described using shape parameters, such as curvature.

Rough clustering is based on rough set theory, where each object in the set is modeled as a row in a table, while each of the attributes of the object is a column of the table. Rough clustering uses the concept of *indiscernible objects*. Specifically two objects are indiscernible with respect to set B, if they have the same attributes with respect to set B. Let us consider the example shown in Table 6.1. We see that *pizza, hamburger*, and *ice cream* objects are indiscernible under the perspective of *calories content* and *taste index*. However, if we also introduce the concept *sweetness*, then the *ice cream* object becomes discernible.

In [Tsi00], a method for *event detection* in medical time series is described. The method was applied to the development of intelligent patient monitoring in ICU units, such that true alarms can be detected.

TABLE 6.1

Name	Taste Index	Calories Content	Sweetness
Pizza	High	High	None
Hamburger	High	High	None
Ice cream	High	High	High

The author makes the interesting note that as many as 86% of alarms in ICU environments are false. The method consists of four steps:

- *Event identification.* Examples of events are apnea and a heart attack.

- *Annotated data collection.* A large amount of data needs to be collected where event occurrences and nonoccurrences are annotated.

- *Annotated data preprocessing.* The substeps here are feature derivation and class labeling. An example of a feature is the mean. These features are measured over a specified time interval.

- *Model derivation.* Two models were explored: *decision trees* and *neural networks.* The latter used the backpropagation algorithm. Performance of the models was evaluated using the area under the *ROC.*

The method was applied to physiological signals including the following: (1) *ECG heart rate,* (2) *ECG respiratory rate,* (3) *ECG pulse oximeter oxygen saturation,* and (4) *systolic blood pressure.* Experimental results showed that both decision trees and neural networks performed well on the test data sets. Some limitations were noted by the author, such as that data were collected once every 5 seconds and the data annotations were recorded to the nearest minute.

6.1.3 ECG Analysis

In [Sta98], a novel technique is described for analyzing *ECG* signals to detect ischemia. An *ECG* signal was shown in Chapter 2 (Figure 2.6) and it consists of the following perceptually important points *(PIP): P, Q, R ,S, T.* There is one segment in the *ECG* signal that detects ischemia, the *ST* segment. The technique used for the detection of ischemia is *PCA (Principal Component Analysis). PCA* aims at dimensionality reduction by finding correlations between variables. In this article, a new version of *PCA* is used: *nonlinear. Nonlinear PCA* is implemented using a multilayered neural network.

An interesting feature of the classification algorithm is that the training set consists only of normal training samples. It was shown that with this approach the abnormal beat detection sensitivity can be at 90%, while the normal beat detection sensitivity is at 80%. The learning is implemented using a *radial-basis function*, which forms local clusters on the principal component space.

Another article that deals with the analysis of *ECG* characteristics, using wavelets and neural networks is [Kim07]. The wavelet transform is used to preprocess *ECG* signals and classification is performed using the backpropagation algorithm (see next section). In [Daj01], modeling of ventricular repolarization time series is achieved using multilayer perceptrons.

In [Chu07], a method that detects anomalies in *ECG* is described. The method is based on the idea of describing anomalies as *discords*, which is explained in Chapter 5. The authors proposed a modification of the discord idea, called the *Adaptive Window–Based Discord Discovery*. The algorithm was tested on *ECG* data sets from the *MIT* database, and it achieved an accuracy rate that varied from 70% to 90%. Also, in [Lin05], Lin et al. use time series discords to find unusual episodes in electrocardiograms. In [Jos05], a novel algorithm for arrhythmia detection is described that utilizes *support vector machines* and *wavelets.*

A method for the *automatic* creation of a fuzzy expert system for beat classification is described in [Exa07]. Given a training set, a decision tree is created which yields a set of crisp rules. The crisp rules are then *fuzzified* and the parameters are optimized through global optimization. The expert system was employed for ischemic beat classification, with 91% sensitivity and 92% specificity accordingly. It was also applied to arrhythmic beat classification with a 96% average sensitivity and 99% average specificity.

In [Asl08], a method for arrhythmia classification is proposed, which is based on the heart-rate variability (*HRV*) signal. The *HRV* signal is the *R–R interval signal*. As the author notes, the advantage of the *HRV* signal is that it can be extracted even from a noisy *ECG*, with little loss of accuracy. The proposed classification algorithm combines *SVM* with *generalized discriminant analysis.* The author notes that the algorithm was able to classify the six types of arrhythmia with an accuracy that exceeded other previously reported results.

6.1.4 Analysis and Classification of EEG Time Series

In [Par07], the fractal dimension is used to detect patterns in irregular time series. The method was applied to *EEGs* to analyze brain activity. As the authors discuss, the fractal dimension is appropriate for analyzing signals that are extremely complex, such as signals that map brain activity. The fractal dimension was measured using *MATLAB*® and the method was able to detect the chaotic patterns effectively. In [Cha06a], time series classification of *EEG* data is performed for the detection of epilepsy. The method described in the article utilizes statistical cross validation and *support vector machines.*

In [Gar03], *linear discriminant analysis (LDA), neural networks,* and *support vector machines* are compared for EEG signal classification. In *LDA,* the probability density function of each class is modeled as a normal distribution, and a new data point is assigned to the class whose probability density function value is largest. The *neural network* algorithm used is the *error backpropagation algorithm,* and as its name implies, it computes the error between the desired output and the actual output of the network. The "blame" for the error is propagated to neurons of the previous level giving greater responsibility to the neurons that have a greater error between their output and the desired output. *Support vector machines* were discussed in the context of classification in Chapter 3. The results showed that for the EEG data, the *SVM* method worked the best; however, linear discriminant analysis and the neural network algorithms were not that far behind.

In [Mel05], the authors present a method for the detection of Alzheimer's disease using *EEG* measurements that are preprocessed using independent component analysis. The method proved particularly useful for the classification of patients in an initial stage of the disease.

6.1.5 Analysis and Clustering of fMRI Data

fMRI, which stands for *Functional Magnetic Resonance Imaging,* is used to measure hemodynamic changes as a result of a stimulation. In [Esp05], *Independent Component Analysis (ICA)* is performed on *3-D functional MRI* brain data sets. The result is the transformation of the data sets in brain activity patterns revealing *modes* of signal variability. Two types of *ICA* are performed: *spatial ICA (sICA)* and *temporal (tICA).* The latter refers to the statistical distribution of source signals across the sampled time-points. In the article, similarity measures on *ICA* patterns are applied to create groups (clusters) in multisubject studies. The similarity measure used is the absolute value of the mutual correlation coefficients between the *ICA* components either in space or in time. A similarity measure that combines temporal and spatial similarity based on the Pearson correlation coefficient is

$$S(i, j) = \lambda C_s(i, j) + (1 - \lambda)C_t(i, j) \qquad (6.1)$$

where $C_s(i, j)$ is the spatial correlation coefficient of independent components Ci and Cj. $C_t(i, j)$ is the temporal correlation coefficient between the associated time courses of activation of Ci and Cj.

fMRI can be viewed as a 4-D data cube (3-D space + time), where each data cube is known as a *voxel.* [Mey08] deals with the detection of *activated* voxels, that is, voxels that are affected by the stimulus time series. Projections of the *fMRI* data are constructed on a low dimensional subspace, created using a wavelet transformation. Besides reducing the dimensionality of the problem, the described approach also creates clusters that separate the activated regions from the background.

Another article that deals with clustering of *fMRI* time series is [Fad00]. The main idea is to utilize the *fuzzy c-means clustering algorithm* (described in Chapter 3) to perform clustering. Another contribution of the article is a voxel reduction algorithm that addresses the fact that the number of nonactivated voxels is significantly larger than the number of activated voxels. In other words, it addresses the problem that classification of active voxels is an ill-balanced problem, because of the big difference in the populations of the two classes. As an example of voxel removal, white matter voxels are removed. The authors have also constructed a framework to test statistical meaningfulness of clusters that allows them to keep the most relevant clusters.

In [Gau04], the authors propose a novel method, the *delay vector variance (DVV) method,* to determine the nature of biomedical time series. This is an important problem as it can be an indicator of the health status of a patient. Their method is applied to two types of biomedical signals, *heart rate variability* and *fMRI time series,* and it can determine whether the time series is *deterministic* or *stochastic* and *linear* or *nonlinear.* Experimental results were consistent with nonlinearities known to be present in the time series. In [Shi06], a novel algorithm is described that combines a *genetic algorithm* and *k-means clustering* to detect *fMRI activation.* The results show that the combined algorithm detects *fMRI* activation better than simple *k-means clustering.*

Finally, [Fre02] addresses the issue of motion correction in *fMRI* series and specifically the similarity measure that should be used for this purpose. This motion correction is performed by estimating the subject's motion, which is modeled as 3-D transformation and is aligned to the reference volume, by maximizing a similarity measure. The authors in [Fre02] experimented with the *mutual information measure,* the *Geman–McClure ratio,* and the *correlation ratio,* which all performed better than the typically used difference of squares measure.

6.1.6 Fuzzy Temporal Data Mining and Reasoning

In [Pal06], a framework is proposed that provides a description of the temporal evolution of diseases and its causal links, thus proposing a way to explain the temporal progress of diseases. This is achieved by defining a temporal behavior model coupled with contextual information about the specific disease. The temporal behavior model consists of *fuzzy temporal constraint networks (FTCN)*, which are a way to model imprecise temporal information. The proposed framework can be used not only for modeling temporal evolution of diseases, but also for decision support purposes. It consists of two modules: the *Temporal/Domain Module* and the *Decision Support Module*.

The Temporal/Domain module consists of an ontology that allows the incorporation of contextual information and a temporal reasoner that ensures consistency of temporal information and allows complex and fuzzy queries. The Decision Support module provides a model-based diagnosis that utilizes the temporal database, the domain ontology, a knowledge base, and the temporal reasoner.

In [Gev98], a method for state identification and *event prediction* in biomedical signals is proposed that is based on *hierarchical fuzzy clustering*. The algorithm exploits the advantages of hierarchical clustering, while overcoming its disadvantages using fuzzy concepts. The algorithm consists of a feature extraction step, which utilizes the wavelet transform and an iterative process (the *HUFC* algorithm*)*. The latter consists of a feature reduction step based on *Principal Component Analysis* and a *weighted fuzzy clustering algorithm.*

Each time the iterative procedure is repeated, it is applied to the sub-fuzzy clusters that were found in the previous step, where the clusters consist of all the data points with nonzero membership. The weights applied to each data point at each recursion are based on its membership value found in the previous iteration. As the author notes, the hierarchical nature of the algorithm reveals the inner structure of the data, while the number of clusters at each iteration depends on the nature of the data.

In [Gor06], a recursive algorithm for the segmentation of biological time series is proposed, which is based on *fuzzy clustering*. In [Kis01], fuzzy logic is employed to model the temporal distributions of symptoms and diseases for use in computer-diagnosis systems.

6.1.7 Analysis of Gene Expression Profile Data

The phenomenal growth of computational biology-related databases that contain gene sequences has created the need for many new types of data extraction. The advent of *microarray technology* has made it possible to study thousands of genes simultaneously, based on *gene expression profiles,* which are measurements of the genes' activity. Gene expression profiles can be obtained either at *specific time points* or at *successive time intervals.* In the latter case they are called *gene expression time series.* Gene expression profiles are useful in understanding many biological phenomena, such as the following:

- Finding the genes that participate in a biological process over time. Specifically, genes with similar gene profiles are likely to belong to the same cellular process.

- Discovering *cause-and-effect* relationships among genes.

- Understanding gene sequence structure, gene function, and regulation.

- Finding gene and protein interactions.

6.1.7.1 Pattern Discovery in Gene Sequences

An important problem in medicine is the discovery of patterns that appear in *DNA sequences.* In [Chu02], the authors define three categories of pattern discovery in *DNA sequences*:

- *Pattern detection*: The model for the pattern we are looking for is known.

- *Supervised learning*: Here we know the locations of the patterns and a functional form of them, such as a Markov model.

- *Unsupervised learning*: Here the locations and parameters of the pattern are unknown.

The authors in [Chu02] describe the simplest form of pattern discovery: the *fixed length noisy pattern.* Here the pattern has a fixed length and each position has a *most likely symbol* to occur, while other *noise* symbols can also occur with some probability. Given the letters *A,B,C,D,* an example of a pattern is *ABADC*. A *noisy* version of this pattern is *ABCDC*, where the second occurrence of the letter *A* has been replaced by the letter *C*. The difficulty in discovering this pattern depends on several factors, such as the length of the pattern, the size of the alphabet, and the frequency of occurrence.

For the solution of the unsupervised learning problem, where the pattern to be discovered is the fixed length noisy pattern, a novel algorithm is described in [Chu02] that utilizes *hidden Markov models.* A hidden state is assumed to exist for each position in the pattern. The authors use the *Bayes error rate* to characterize the *learnability* of patterns. The *Bayes error rate* is defined as the minimum error rate that can be achieved. This will happen, if given the class feature vector, and the true model is used to make the optimal Bayes decision. Another article that proposes *HMMs* to model biological sequences is [Bal94]. The sequences that are modeled are globins, kinases, and immunoglobulins. The authors used *HMMs* for both motif detection and classification.

In [Bai95], the *MEME* algorithm is described, which is an extension of the *expectation minimization (EM) algorithm* that is applied in identifying motifs in unaligned biopolymer sequences. A *motif* here is defined as a family of nucleic or amino acid subsequences. The novelties of the *MEME* algorithm are the following:

- It requires no knowledge about the width, position, or letter frequencies of motifs.

- It expands the range of problems that can be addressed by the *EM* algorithm.

- It is resistant to the presence of noise in data.

- It allows multiple appearances of a *motif* in a sequence.

- It can find distinct *motifs* when they appear in a single sequence or in different sequences.

In [Pev00], the problem of finding subtle patterns in *DNA* sequences is addressed. Specifically, the pattern finding problem is stated as follows: "Given a sample of sequences and an unknown pattern that appears at different unknown positions in each sequence, can we find the unknown pattern?" The authors approach the problem from a combinatorial point of view, describing an algorithm called the *WINNOWER* algorithm, which is successful in finding subtle patterns but is computationally expensive.

In [Sze02], a comparison of various *motif-finding* algorithms is performed on several biological samples. The algorithms compared are *WINNOWER, MEME, SP-STAR, CONSENSUS,* and *GibbsDNA.* It was found that all algorithms perform well on noncorrupted samples. In experiments, where at most one pattern (site) per sequence is allowed and random sequences with no patterns are added, *MEME* outperformed

SP-STAR and *CONSENSUS*. When multiple sites of patterns are allowed, *SP-STAR* performed slightly better. The authors conclude that they cannot recommend one single program for all situations and it might be wise to try all available programs.

In [Sac07], the authors use *precedence temporal networks (PTNs)* to represent temporal relationships in gene expression data with the ultimate goal of reconstruction of gene regulatory networks. *PTNs* are networks where nodes represent temporal patterns and edges represent synchronization or precedence relationships between the nodes.

In [Jia03b], the authors address the problem of discovering *coherent* patterns in time series gene expression data and in this way discover important cellular processes. Specifically, an interactive tool is proposed in the article, called the *coherent pattern index graph*. This graph is created using an attraction *tree structure*, where an index list is used to group together in the list genes that share a coherent pattern. In [Bar02], a novel approach to analyze gene expression time series data is proposed, where each profile is modeled as a *cubic spline*. The spline coefficients of genes that belong to the same class are constrained to have similar expression patterns. Their method is shown to be particularly effective for nonuniformly sampled data.

The problem of finding coherent gene clusters is also addressed in [Jia04a]. Two efficient algorithms are proposed in the article, called *Sample-Gene* search and *Gene-Sample* search. The main idea of both algorithms is to use a *depth-first* approach to find maximal coherent genes and remove (prune) unsuccessful patterns. In the *Sample-Gene* search approach, all combinations of samples are enumerated systematically. Then, a search is performed to find the *maximal subsets of genes* that form coherent gene clusters on the samples. In the *Gene-Sample* approach, the enumeration of genes is performed first and then we look for *maximal subsets of samples*.

Approximate similarity search (otherwise known as *homology search*) in genomic databases is an important task, since it can reveal functional and evolutionary relationships among proteins. A very famous article in this area is [Alt00]. In this article, Altschul et al. describe *BLAST*, a widely used tool today to perform approximate local similarity searches for *DNA* and protein sequences. *BLAST* works by optimizing the *maximal segment pair* score and, as its authors note, it is a simple, yet robust, algorithm. Two other articles that address this issue are [Sac08] and [Cao04]. In the latter article, given a query sequence, only certain segments in the databases are accessed, which are called *piers*. As the authors note, the *piers* are selected randomly from a data sequence *A*, such that at least one pier is contained within any subsequence of A that satisfies a minimum length requirement. *Piers* are stored in a hash table to improve the search speed. Experimental results show that the algorithm can detect similar biological sequences with very high sensitivity.

In [Sac08], *landmarks* are used for sequence representation. Specifically, the sequences are embedded in a vector space given their distances to landmark sequences. Landmarks offer a concise way to perform efficient indexing and similarity search. Specifically, the authors demonstrate that the similarity search in the embedded space can be orders of magnitude faster than in the original space.

6.1.7.2 Clustering of Static Gene Expression Data

A thorough survey of gene expression clustering techniques is presented in [Jia04a]. We will briefly review the discussion of these techniques in [Jia04a] and then discuss in detail clustering techniques for *gene expression time series*.

The resulting data set from a microarray experiment is a real-valued matrix where the rows represent genes and the columns represent samples. Clustering of microarray data can be focused either on *genes* or on *samples*. In the first case, *gene-based* clustering produces clusters of co-expressed genes, that is, genes with similar expression profiles. This could be an indicator that the genes participate in similar cellular processes. *Sample-based* clustering results in homogenous groups of samples. Each group could correspond to an observable characteristic, such as a cancer type. Note that observable characteristics are known as *phenotypes*.

Gene-based clustering has some unique challenges, such as noise-corrupted data and intersecting clusters. Several clustering techniques for genes are discussed in [Jia04a]. Some of them are general clustering techniques (not tailored for gene expressions), such as *Distance-based clustering* (Euclidean Distance and Correlation Coefficient), *K-means, Self-Organizing Maps*, and *Hierarchical Clustering*. Others have been developed specifically for gene expressions, such as the following:

- *UPGMA (Unweighted Pair Group Method with Arithmetic Mean)*. This is a hierarchical clustering approach, of the agglomerative type [Eis98].

- *DAA (Deterministic-Annealing Algorithm)*. This is also a hierarchical clustering approach, of the divisive type [Alo99].

- *CLICK (CLuster Identification via Connectivity Kernels)*. This is a graph-based approach [Sha00].

- *CAST*. This is also a graph-based approach [Ben99].

- *DHC*. This is a density-based hierarchical approach [Jia03a].

6.1.7.3 Clustering of Gene Expression Time Series

Surveys of *gene expression time series* clustering can be found in [Mol03] and [Bar04]. In addition to the challenges mentioned above for gene clustering, such as presence of noise, clustering of gene expression *time series* has additional challenges [Mol03]:

- They are very short (even as short as four samples).

- They can be unevenly sampled.

- They should be able to handle shifting and scaling of time series.

- The similarity measure should not be based on the intensity of the expression profile, instead it should be based on the changes in the intensity; that is, it should be shape-based.

6.1.7.3.1 Spline-Based Methods   In [Bar03], the authors develop a clustering method that takes into account the fact that there are missing points in gene expression time series. They model the profiles using *cubic splines* and constrain the spline coefficients in the same class to have similar expression patterns. Experimental results show that the algorithm can reconstruct missing points with 10% to 15% less error in comparison with previous methods and its application on *yeast-knock-out-data* produces biologically meaningful results. Another clustering algorithm that uses spline smoothing is described in [Dej07]. The spline smoothing parameter was determined using statistical and biological considerations. Then the clustering was performed on the derivatives of the smoothed curves. The clustering methods employed were *k-means* and *hierarchical* clustering.

In [Lua03], a *mixed-effects model* combined with *B-splines* is proposed to cluster gene expression time-series data. Here, each time measurement of a gene is modeled as the sum of the following:

- A smooth population mean. This is a function dependent on time and gene cluster.

- A spline function with random coefficients to model individual gene effects.

- Gaussian noise resulting from the measurement process.

The number of clusters was determined using the *Bayesian inference criterion*. The feasibility of the method was determined by applying it to a data set with evenly sampled points and another one with unevenly sampled points.

6.1.7.3.2 Model-Based Methods   In [Ram02], a model-based clustering method is described. The method utilizes autoregressive equations to model gene-expression dynamics and it proposes a novel agglomerative method to find the number of clusters. A Bayesian criterion is used to measure the similarity of the time series. An autoregressive model of order one was found to have the best performance. However, as mentioned in [Mol03], a possible disadvantage of the method is that shifting and scaling issues are not considered.

Another model-based approach is described in [Giu05], where a *sum-of-exponentials model* is fitted to the nonuniformly sampled data. A grid search is used to initialize the parameters. The authors also introduce a *minimum description length (MDL) criterion* for deciding on the number of exponentials in the model and a fitting procedure that experimental results show to have an accuracy close to the *Cramér–Rao lower bound*. This bound sets a lower bound on the variance of an estimated value. The clustering algorithm uses a *classification–expectation–maximization (CEM) approach*, where the time constants of the genes are modeled using a Gaussian mixture model. A model-based approach is also used in [Hea05]. Specifically, the time series are clustered by optimizing a joint probability model, which is computed using an expectation-maximization approach. *Bayesian hierarchical clustering* is used to find locally optimal clusters.

In [Sch03], *hidden Markov models* are used to cluster gene expression time series data. The method starts with a collection of *HMMs* that are in the form *(up–down-regulated)*. Then an iterative procedure is used to maximize the joint-likelihood of clustering [Mol03]. As noted in [Bar04], *HMMs* have a distinct advantage over other clustering methods, such as *hierarchical clustering* and *k-means*. These latter have the problem that they treat their input as a vector of independent samples and do not take into account the fact that the time series values are correlated. The authors in [Ji03] also propose a *hidden Markov model-based approach* for clustering gene expression time series data. Experimental results showed the following:

- The quality of their clustering is similar to that of two prevalent algorithms (*k-means* and *SOM*). Note that *SOM* is described below in this section.

- The algorithm can distinguish between function-related and regulation genes.

- It can harvest biologically meaningful gene clusters.

As mentioned in [Mol03], a difference of the methods in [Ji03] and [Sch03] is that in [Ji03], the parameters of the *HMMs* are initialized randomly and adjusted iteratively in a way that the greatest influence (in parameter reestimation) is given to sequences that have a higher probability for a specific *HMM*.

[Yua07] proposes an unsupervised *conditional random field (CRF)* for gene expression time series clustering. *CRF* is a probabilistic model. CRFs have the advantage that they allow arbitrary dependencies between observations. An interesting difference between this algorithm and most clustering algorithms is that cluster centers do not need to be computed. Instead, each sample finds its own cluster iteratively by allowing it to interact with its own voting pool. *A voting pool* of a sample consists of the most similar samples based on the Euclidean distance, the most dissimilar samples, and some samples selected at random. Another interesting feature of this article is that it lists the requirements for developing clustering algorithms for gene expression data:

- The temporal relationship between time points should be taken into account. Also the fact that the samples might be obtained in a non-uniform way should be taken into account.

- Relationships between clusters are also of interest because causality is of interest in gene time series experiments.

- Problems, such as missing data and noise, should be accounted for.

- It is better to use an unsupervised algorithm, because our current knowledge about time series of gene expressions is very limited.

Another model-based approach is proposed in [Yeu01a]. Here, the gene expression data are assumed to come from *a mixture of Gaussian transformations*. The *Bayes information criterion (BIC)* was used to select the number of clusters and compare different models. As the authors note, an advantage of their model-based approach over heuristic methods is that the determination of the number of clusters and the choice of the best clustering method become statistical model choice problems. In contrast, in heuristic methods, there is no established method to solve either problem. The authors applied their method to both real and synthetic data sets. On real data sets, the proposed model-based algorithm had similar performance to a leading heuristic algorithm (*CAST*), while on synthetic data sets, the model-based algorithm had superior performance.

Finally, in [Kun05] a method is described for the clustering of genes into transcriptional modules, where *transcription* is the process of copying a strand of DNA into a strand of RNA. The authors make the important note that genes with *similar expression profiles* can be *unrelated in a regulatory fashion*. Therefore, to derive similarity based on similar regulatory mechanisms, it is important to combine time series expression profiles with additional information. In this paper, this additional information is *genome-wide motif data* regarding regulatory sequences. Then a probabilistic model is used to combine the two types of information and an expectation minimization algorithm is used for the learning process. The authors also use splines to model the time series expression data. The method is validated on yeast cell cycle data, for which there are a large number of motifs regarding regulatory elements.

6.1.7.3.3 Fuzzy Clustering–Based Methods   In [Mar06], *fuzzy clustering* of *DNA* microarray data is described. In the current state of the art, such as in the technique described in [Tav99], *crisp clustering*, such as *k-means* is used to cluster microarray data. The problem with this approach is that genes can play multiple roles and therefore belong to multiple clusters. However, crisp clustering assigns genes to a single cluster. On the other hand, fuzzy clustering allows a gene to belong to multiple clusters.

A fuzzy clustering method is described in [Gas02], which is applied to yeast genomic data. In addition to allowing a gene to belong to more than one cluster, the authors note another significant advantage of fuzzy clustering. It is not as sensitive as *k-means* to the number of clusters. The authors propose a heuristic modification of fuzzy clustering where they combine it with *Principal Component Analysis (PCA)* and *hierarchical clustering.* Specifically, *PCA* is used to perform initialization, instead of using random initialization, which is typically used in fuzzy *c-mean* clustering. Another differentiation of the algorithm in [Gas02] is that the clustering is performed in three cycles, where the second and third cycle are performed on subsets of the data. In experimental results, the method was able to successfully identify clusters of functionally related genes.

6.1.7.3.4 Template-Based Methods   In [Tsi08], the main idea is to cluster expression profiles by measuring their similarity using *template matching*, after *DTW* is employed to align the profiles. Then, for each cluster a standardized version of the cluster profile is created using an aggregation algorithm. The aggregation algorithm is recursive and it is a hybrid algorithm in the sense that it employs a combination of three aggregation operations. The operators are the *arithmetic*, the *geometric*, and *harmonic means*. The reason for using three different aggregation operators is that each one has its own advantages and disadvantages. For example, the arithmetic one hides details and is influenced by very large values or very small values. On the other hand, the geometric and to a more extreme degree the harmonic means are very much influenced by small values close to 0.

The authors in [Ern05] propose a model-based approach, where a set of predefined model profiles, which constitute the *ground truth*, are used to find the distinct patterns expected to emerge from the experimental results. The motivation for the authors to use a template is that many patterns are expected to arise randomly. Two other works that use a template-based approach are [Sac05b] and [Hvi03]. In both works, the time series data are abstracted to fall in one of three categories (*increasing, decreasing, steady*) and then clustering is performed using the abstracted data.

6.1.7.3.5 Simulated Annealing–Based Method   In [Luk01], a robust algorithm for gene expression time series is proposed, which is based on an optimization technique called *simulated annealing*. This algorithm achieves the following optimizations simultaneously:

- *Finding the optimal number of clusters.* As the authors note, this is an advantage compared to techniques, such as hierarchical clustering, that leave this decision to the observer.

- *Finding the optimal distribution of genes over the clusters.* Again, the authors note, this is an advantage compared to techniques that use a cost function where only a local optimum can be guaranteed.

The similarity measure used for the gene profiles is the Euclidean distance. The *simulated annealing algorithm* is used to optimize the sum of distances of genes within a cluster. The efficiency of the algorithm is proven by applying it to genes whose correct distribution is known *a priori*. The algorithm is shown to successfully place more than 90% of the genes in the correct cluster.

6.1.7.3.6 A Self-Organizing Maps-Based Method   Finally, an article that is very often cited in the gene expression analysis and clustering literature is [Tam99]. The authors of the article use *self-organizing maps* to cluster gene expression data and they note that *SOMs* have the following advantages over other techniques, such as *hierarchical* and *k-means clustering*:

- They allow partial clusters.

- They facilitate visualization and interpretation.

- They have ease of implementation and good scalability to large data sets.

SOMs summarize a data set by extracting the most prominent patterns and placing them in the *SOM* such that they are neighbors.

6.1.7.3.7 A Systematic Analysis Framework of Gene Expression Data Sets   The authors in [Hye99] propose a framework for the analysis of gene expression data sets that performs the following tasks:

- Preprocessing of the data.

- Computation of similarity. The *jackknife correlation* is proposed as a similarity measure, which has the advantages of capturing the shape of the expressed pattern and being insensitive to outliers.

- Clustering. A novel clustering algorithm is proposed that has a *quality guarantee*. The main idea of the algorithm is the following: The clusters that are formed have a diameter that does not exceed a specific threshold value d. Any two objects that belong to the same cluster have a jackknife correlation that is at least $1-d$. Because the jackknife correlation ranges from -1 to 1, the diameter of the clusters ranges from 0 to 2.

- Analysis of the clusters to discover information about gene function and regulation.

6.1.7.3.8 The STEM Framework The authors in [Ern06] present *STEM*, which is a framework for the analysis of short time series gene expression data. *STEM* has the following unique features:

- It is the first software package targeted for *short* time series gene expression data (3–8 points).

- It introduces a novel clustering algorithm that utilizes a set of data-independent model profiles. Specifically, each gene is assigned to the model profile, with which it has the highest correlation. A permutation test is used to determine which models have a statistically significant higher number of genes assigned to them. Clusters of significant profiles are formed by utilizing standard hypothesis testing.

- *STEM* is integrated with an *ontological layer* (GO), which allows a biological interpretation of statistically significant patterns.

6.1.7.4 Additional Temporal Data Mining–Related Work for Genomic Data

In [Aac01], time warping algorithms are applied to *RNA* and protein expression data. The application area is the analysis of *RNA* expression data collected at selected time points to study biological processes that develop over time. An issue that appears in this analysis is that biological processes unfold at different rates in different organisms. For this reason, an important issue in biological process analysis is *time series alignment*. The authors demonstrate that time warping is superior to clustering in regards to mapping time states.

In [Raf05], the authors present a method to discover temporal relationships between genotype and clinical (phenotype) data. They applied their method in data relating to *HIV* drug resistance. The core of their method is *CQL*, a temporal query language that is based on *TSQL2*, the temporal extension to *SQL*. It allows queries that mix absolute time stamps and intervals. For example, a query could be "Select all patients who have taken drug X for more than one month since October 2007." In addition to *CQL*, the authors also present *EMF*, a framework that allows temporal aggregation, such as finding the *minimum, maximum, sum,* and *count.*

In [Sin05], the authors address the problem of finding an appropriate sampling rate in time series gene expression data. As the authors point out, there is a high cost associated with sample collection, while undersampling bears the risk of having an inaccurate representation of the data set. In their approach the sampling points are iteratively chosen using as the objective function the uncertainty in the estimated smoothing splines combined with active learning to choose the next sampling point. Specifically, having j previously sampled time points, the authors describe the procedure for choosing the $j+1$ point, as follows:

- Generate a smoothing function for the previously sampled points, using splines and choosing all the spline parameters based on cross validation.

- At all the sampled points compute locally sensitive confidence intervals and use active learning to find the next sampling point, based on the confidence intervals.

The authors point out that their algorithm can be extended to other types of data beyond gene expression time series data.

In [Bor06], the authors address the problem of *classifying* gene expression profiles. Note that most research in bioinformatics is focused on gene expression *clustering*. However, gene expression time series *classification* can be very useful in certain fields, such as pharmacogenics, where the question is how patients respond to certain drugs. The time series are modeled in the article as dynamic systems that evolve over time. This is consistent with what happens in practice as gene expression profiles change over time because of interactions between proteins and gene transcription. The classification approach followed in the article is *support vector machines (SVM)*. The method was tested for the prediction of responses of *multiple sclerosis* patients to *recombinant human interferon beta*. The average prediction accuracy reported in the article was 87.05%.

In [Qia01], the authors infer causal relationships from gene expression time series profiles using local alignment algorithms of time-shifted and inverted gene expression profiles. In [Hol01], *singular value decomposition (SVD)* is used to construct a time translation matrix, which in turn is used to predict future gene expression levels given the expression levels at an initial time.

In [Yeu01b], the authors propose a framework that allows the systematic comparison of various gene expression clustering methods. The framework is based on the *jackknife* approach, and it works by applying the clustering algorithm to all data but one experimental condition. The experimental condition that is left out is used to determine the predictive power of the remaining clusters.

In [Mar07], the authors address the problem of retrieving proteins based on structural or sequence-based similarity. For this purpose, they propose two normalized protein representations that allow fast protein retrieval based on sequence or structure. For the structural representation, the authors transform the 3-D structures into normalized 2-D distance matrices and then use a 2-D wavelet transform to generate the representations. In [Nam08], the authors address the problem of finding out how the expression of a gene affects the expression of other genes. They express it as a data mining problem in the form of temporal association rule identification.

6.1.8 Temporal Patterns Extracted via Case-Based Reasoning

In [Nil05], a clinical decision support system is described for aiding clinicians in the detection of respiratory sinus arrhythmia. Patterns are extracted from time series data, which are then transformed into wavelet coefficients. These patterns are matched for similarity against preclassified cases stored in a database. This type of reasoning is known as *case-based reasoning (CBR),* which is described in Chapter 2. In [Nil05], not all extracted patterns are matched against the stored cases. Instead, they are reduced by clustering, using the *k-means clustering* method.

Matching of cases is performed by matching individual wavelet coefficients. Each frequency band is assigned a weight. Higher frequency bands are assigned higher weights. The weight of each frequency band is

$$W = \frac{1}{2^{\varphi}} \tag{6.2}$$

where, for example, for the highest frequency band $\varphi=0$. The lowest band has a weight of 0.25. Besides the wavelet transform, the Fourier transform was also tried. However, with the *DFT*, the correct retrieval rate was reduced by 20%.

In [Fri02], the authors address the problem of recognizing critical situations derived from laboratory results in cases where the definition of *normal* varies from patient to patient. They developed a *case-based reasoning* algorithm, where the case base consists of kidney transplant recipients. The medical problem is recognition of acute rejections based on changes in *serum creatinine*. The measure of similarity is *Dynamic Time Warping* to be able to compare series of infrequent measurements at irregular intervals. As expected, the accuracy of the algorithm increased with the number of cases in the case base. Specifically, with the largest case base, the accuracy of the case-based reasoning algorithm reached 78%±2%, which is higher than the accuracy of experienced physicians.

In [Sch01], a case-based approach for the classification of medical time series is proposed. The method is applied to the early identification of renal transplantation patients who are likely to experience a transplant rejection. Their creatinine level is monitored over time and then a distance measure based on linear regression is used to find similar case in a patient library. The measured sensitivity is significantly higher compared to the physicians accuracy.

6.1.9 Integrated Environments for the Extraction, Processing, and Visualization of Temporal Medical Information

In [Abe07], the authors present a framework for the pattern extraction, rule induction, and rule evaluation in medical time series. The framework was applied on chronic hepatitis data and it overall does the following operations:

- Data preprocessing

- Mining (rule induction)

- Rule evaluation

- Visualization

In [Wei05], a framework is described that allows users to navigate through a large number of time series. The framework can create a bitmap of features that exist in a time series whose color depends on the relative frequency of these features. This way, it becomes easy for the user to quickly discover clusters and irregularities in the data set. The framework was applied to a variety of medical sets.

In [Car01] and [Alo03], an algorithm is described that finds similar patterns in a series of *isokinetic tests*. *Isokinetic* machines are used in physiotherapy to assess muscle function. Each datum in an isokinetic machine is a measurement of strength at a particular time. The goal of the authors was to detect patterns that appear in exercises done by patients with injuries and do not appear in exercises done by healthy subjects. A pattern search tree was built and an *Apriori*-type algorithm was used to find patterns. Two different algorithms were used: one that considered only identical sequences and another one that considered the Euclidean distance between patterns. The latter made a significant improvement in the discovery of patterns.

In [Rod03], the authors address the issue of exploratory medical knowledge discovery which deals with discovering data that confirm or reject a medical null hypothesis. The data mining process includes space and time data, particularly temporal data about episodes. The particularly useful data mining processes are as follows:

- *Anomaly detection.* This discovers possible changes in the strength of rules.

- *Difference detection.* This discovers differences between rule sets that should be ideally similar. These differences should be statistically significant. An example is the process of following specific clinical guidelines in various hospitals.

- *Longitudinal X analysis.* This can discover changes in the cohesion of medical data clusters.

- *Temporal/Spatial X analysis.* This can be used to discover causal relationships.

In [Pos07a], a framework called *PROTEMPA (Process Oriented Temporal Analysis)* is described that allows for the discovery of temporal patterns in retrospective data. The goal was to use these patterns to represent disease and patient care processes so that patient populations with these patterns could be extracted from clinical repositories to support clinical research, outcomes studies, and quality assurance. The framework was evaluated against standard *SQL* queries with the goal of identifying patients from a large clinical laboratory with the *HELPP* syndrome. Then the patients were categorized according to disease severity based on the time sequences of their lab results.

In [Aig06], the authors describe a method for the integrated visualization of computerized protocols and patient temporal data. This way multiple simultaneous views of different aspects of the treatment plans and patient data can be obtained. In [Fan07], the authors propose and compare different techniques for the visualization and exploration of time-varying medical image sets. Their main idea is to view a time varying medical image set as a 3-D array consisting of *voxels*. Each voxel contains a time-activity curve, for which they define three different similarity measures. In [Chi06], the author discusses the visualization of patient data at different temporal granularities on mobile devices. This is a challenging problem because of the limited display and storage capabilities of these devices. In [Ple03], a method for the visualization of time series volumetric data is described using animated *electro-holography*. The medical data were *MRI* volumetric data of brain lesions due to the progression of multiple sclerosis. In [Wan08], an interactive visual tool is presented that allows the user to display patient histories aligned using sentinel events, such as a severe asthma attack. This type of visualization allows the user to discover cause-and-effect relationships, such as *precursor, co-occurring*, and *following* events.

6.2 TEMPORAL DATABASES/MEDIATORS

6.2.1 Medical Temporal Reasoning

Temporal reasoning is closely related to temporal data mining and it is often used in clinical decision support in medicine. An excellent review of the subject can be found in [Aug05]. Publications of the last two decades are reviewed according to which stage of the clinical decision process is used: prognosis, diagnosis, and therapy management. An important article in medical temporal reasoning is [Ker95]. It defines a *time object* as the "coupling between a property and a time existence." Time objects can be used to model patient data, disorders, and clinical actions. In this article, the requirements for a *medical temporal reasoner* are described as follows:

- Time objects can be expressed in either absolute or relative terms. An example of absolute expression is a time-stamped medical event (such as date/time of admission to a hospital). Relative expression can be achieved using *Allen's relations*, such as *before, after, meets*, and so on.

- Time objects can be vague (fuzzy).

- Time objects can be expressed at various granularities.

- Causality can have explicit temporal constraints.

- Temporal data abstraction must be able to be performed in an efficient way.

- We must be able to perform temporal deduction and induction.

Medical data are inherently temporal and time concepts can be used to query electronic medical records and reason about clinical data with temporal attributes. In [Adl06], the authors describe the challenges and research directions in temporal representation and reasoning. They divided the research directions into four categories: (1) fuzzy logic, time, and medicine; (2) temporal data mining and reasoning; (3) health information systems, business processes, and their temporal aspects; and (4) temporal clinical databases. In [Zhu07], the authors perform a review of research in temporal reasoning for medical data, with emphasis on *medical natural language processing*. An important note that authors make is that most medical language processing systems today do not have a formal representation of time, and therefore, their ability to support temporal reasoning is limited.

In [Sha99], Shahar makes the important distinction between *temporal reasoning* and *temporal maintenance*. Temporal reasoning is mainly addressed by the artificial intelligence community, while temporal maintenance is mainly supported by the database community. The author proposes an integration of the two tasks through the concept of a *temporal mediator*. Critical to both areas is *temporal modeling*. One framework of temporal modeling that is discussed in the paper is the *KBTA* (the *knowledge-based temporal abstraction*) framework. The *KBTA* decomposes temporal knowledge into four different areas: *functional, structural, temporal-semantic (logical)*, and *temporal-dynamic (probabilistic)*. The *KBTA* framework uses its own ontology, and it has been evaluated in various clinical domains, such as oncology and *AIDS*. The *KBTA* framework and the corresponding temporal mediator are discussed more below.

6.2.2 Knowledge-Based Temporal Abstraction in Clinical Domains

Temporal abstractions (TAs) convert quantitative data to an interval-based qualitative representation. Examples of temporal abstractions are *trend TAs* that capture increasing/decreasing values in time series and *state TAs* that correspond to low, high, or normal values in the time series. A survey of the use of temporal abstractions in medicine can be found in [Sta07]. The focus of the article is temporal abstractions used in the intelligent analysis of clinical data, particularly multidimensional real-time patient data streams.

In [Sha96], the *KBTA* temporal framework is presented, which is implemented in a computer program called *RÉSUMÉ*. The structure {*parameter, value, context, interval*} is at the core of the framework, and it denotes that a specific parameter has a specific value with a certain context interpreted over a certain time interval. The *temporal abstraction task* is divided into the following subtasks:

- Temporal context restriction

- Vertical temporal inference

- Horizontal temporal inference

- Temporal interpolation

- Temporal pattern matching

The *context-forming* mechanism *allows abstractions to be formed within a specific context*. The *temporal inference* mechanism *allows the inference of logical temporal conclusions*, using temporal properties. The following characteristic examples are presented by the authors: (1) two consecutive 9-month pregnancies cannot be considered as a single 18-month pregnancy, and (2) a week of oscillating blood pressure does not necessarily mean that the patient will have an oscillating blood pressure each and every day.

Therefore, as we can see, context is very important in the definition of temporal concepts in the framework. The temporal interpolation mechanism bridges gaps between temporal intervals. Finally, the temporal pattern matching algorithm matches predefined complex patterns such as "multiple heart attacks" with clinical data. The *KBTA* framework can be applied in many clinical domains. Critical for the accomplishment of this is the organization of the domain knowledge in parameter, event, and context ontologies.

In [Ho03], a temporal abstraction framework is discussed for generalizing states and trends in hepatitis patient data. The data were collected between 1982 and 2001, and they contain about a thousand examinations. The goal of the temporal abstraction framework is twofold:

- Abstracting time-stamped data into episodes; specifically finding states (*high, low, medium*) and trends (*increasing, decreasing*).

- Finding temporal relationships between episodes.

In [Had07], temporal abstraction was applied to time-stamped data from multiple monitoring devices in intensive care units. The goal was to model the patient's evolution in an unsupervised way. The data were from a group of patients diagnosed with *Pulmonary Emphysema,* and the most frequently measured parameters were *Heart Rate, Arterial Blood Pressure (Systolic* and *Diastolic), Respiratory Frequency,* and *Oxygen Saturation.* Five behaviors of time-stamped data were defined as *stationary, positive trend, negative trend, positive jumps,* and *negative jumps.* The data consisted of 1-minute intervals that had 6 dimensions corresponding to the 6 forementioned measured physiological parameters (heart rate, etc.). *K–Nearest Neighbor* classification was used to classify the data into one of the five behaviors mentioned above.

In [Ter05], a new model for temporal medical data is proposed as well as a new type of temporal semantics, namely, *interval-based semantics.* However, it is argued in the article that both point-based semantics and interval-based semantics are needed and the temporal query language needs to be extended to deal with switching from one type of semantics to the other.

In [Bel00] and [Bel97], temporal abstractions are used for preprocessing and interpreting time series that are derived from monitoring diabetes patients. Specifically, temporal abstractions are used to create a new representation of the time series, which is useful for deriving metabolic control information about the patient. In [Van99], the authors describe a patient case report language that formalizes temporal knowledge as found in case descriptions. Finally, in [Ngu03], trends and states are discovered in *irregular* medical data after applying a *temporal abstraction method,* which was developed specifically for irregular medical data.

6.2.3 Temporal Database Mediators and Architectures for Abstract Temporal Queries

In [Ngu97], the authors describe a temporal mediator known as *Tzolkin,* which can be used to answer temporal abstract queries on a database of clinical data. *Tzolkin* can be used as a standalone system or it can be embedded in a larger architecture. In fact, it was embedded in such a larger architecture known as *EON. Tzolkin* is based on previous systems, known as *Chronos* and *RÉSUMÉ.* Another relevant article is [Sha99b], which describes a graphical automated tool that allows experts to enter knowledge necessary to form in *RÉSUMÉ* high level concepts from raw clinical data with a temporal component. The query language used in *RÉSUMÉ* is known as *CAPSUL* [Cha04], and it allows a wide variety of temporal patterns to be expressed, including four types of temporal constraints.

The query language used by *Tzolkin* is known as *SQLA*, which is a powerful query language, because it allows the retrieval of not only data stored in the database as primitives, but also *data that are stored in the database as abstractions of these primitives. Tzolkin* also allows batch computation, which entails generating and storing all possible abstractions for a given data set. A type of query that can be answered using Tzolkin is "Find a patient that has experienced a second heart arrhythmia episode within 5 weeks who is being treated under protocol #AAA."

IDAN [IDA04] is another framework that allows access to a clinical information system to answer abstract temporal queries, during the execution of a guideline-based protocol. *IDAN* integrates temporal data, semantic domain-specific information, and computational services. This information can be used directly by a physician or a decision support tool.

The IDAN architecture contains the following components:

- A temporal abstraction server

- A data analysis and visualization tool

- Medical Vocabulary Search Engine

- Local Clinical Database Term Converter

In [Mar08], Martins et al. evaluate *KNAVE-II*, which is a knowledge-based framework for the visualization and exploration of clinical data, concepts, and patterns. *KNAVE-II* mediates queries to *IDAN*. A balanced crossover study was performed to evaluate the satisfaction of a group of clinicians when answering queries about clinical data with temporal attributes using *KNAVE-II* and the standard methods of paper charts and electronic spreadsheet. Clinicians gave a better score to *KNAVE-II* than paper charts and electronic spreadsheets.

Temporal granularity is defined in Chapter 1, and it refers to the existence of temporal objects at different levels, such as weeks, days, months. *Temporal indeterminacy,* on the other hand, deals with the fact that some of the temporal locations are vague. The focus of [Com97] is the introduction of the query language *GCH-OSQL*, which stands for *Granular Clinical History–Object-Structured Query Language,* and an object-oriented data model (*Granular Clinical History–Object-Oriented Clinical Data Model*). The focus of the data model and the query language is the following areas:

- Dealing with different levels of granularity and indeterminacy of the valid time.

- Managing uncertainty in temporal relationships because of indeterminacies and different granularities.

- Modeling complex temporal information.

The query and data model were used on data from patients who underwent coronary heart angioplasty.

In [Boa05], Boaz and Shahar describe a distributed system that integrates medical knowledge bases with a time-oriented clinical database. The goal of the framework is to answer temporal queries about raw data and their abstractions in a distributed fashion. In [Sha98], Shahar et al. describe the *Asgaard* project, which is a framework for the critiquing and application of *time-oriented clinical guidelines. Clinical guidelines* are schemes that describe clinical procedure knowledge and are instantiated and refined in a dynamic way. In [Mos05], Moskovitch and Shahar describe the use of temporal data mining based on temporal abstractions. Another article that deals with the implementation of temporal constraints in clinical guidelines is [Ans06]. Different kinds of constraints are considered in this article, such as repeated actions and composite actions.

In [Oco07], the authors address the problem of knowledge querying of temporal patterns in clinical research systems. Using the *Semantic Web ontology, OWL,* and the *rule language SWRL,* they achieved integration of low-level representation of relational data with high-level concepts used in the management of research data. In [Tus07], an open-source platform, *SPOT,* is described which is an extension of a knowledge-based temporal abstraction framework. *SPOT* supports the *R*-statistical package and it uses *Protégé-OWL.* In [Sha08], Shankar et al. use an ontology to express temporal constraints for use in clinical trial protocols. The authors use a representative set of temporal constraints found in immune tolerance studies. Subsequently, they use the constraints to temporally annotate activities in a clinical protocol.

In [Oco06], an ontology-driven mediator for querying time-oriented biomedical data is described. Using *OWL* and *SWRL,* a temporal model is constructed that educates the mediator of the temporal semantics used in data analysis. An earlier version of the mediator called *Chronus II* is described in [Oco02]. This mediator extends the standard relational model and *SQL* to perform temporal queries. In [Oco01], *RASTA*, a distributed temporal abstraction system, is described which is used for knowledge-driven monitoring of clinical databases. An earlier version of a temporal management framework for clinical decision support systems is described in [Das92]. The framework was based on an extension of *SQL* and was applied to protocol-based care of HIV patients.

In [Spo07], an active database architecture is described for continuously arriving temporal raw data. The architecture performs knowledge-based incremental interval-based abstraction of complex concepts. The system was applied to a database of 1,000 patients monitored after bone-marrow transplantation. In [Das01], Das and Musen address the important problem of data heterogeneity in clinical repositories. For this purpose, they developed a temporal model that utilizes a set of twelve operators that map heterogeneous temporal data into a uniform temporal scheme representation. In [Pos07b], the authors describe a query language that allows the specification of temporal abstractions and the retrieval of relevant clinical data sets. In [Pla08], the authors describe a novel query tool for searching for temporal patterns in patient histories in a large operational system, specifically *Microsoft Amalga*.

6.2.4 Temporality of Narrative Clinical Information and Clinical Discharge Documents

In [Com95], the authors present a framework (*TIME-NESIS*) that allows the management of temporal information existing in narrative clinical data at different levels of abstraction granularity. The authors define *abstraction granularity* as granularity not directly tied to the time axis; instead it is tied to the ability to express complex temporal relationships, such as "the patient had high blood pressure during a period of two months." *TIME-NESIS* uses the hierarchy *instant, duration, interval.*

According to its authors, *TIME-NESIS* is the only framework that allows identification at a higher abstraction level of a time interval's starting point, ending point, and duration. As an example of the type of inferences that the framework allows, let us consider the following two temporal assertions: "The patient suffered an asthma attack on May 12, 2005, at 2.30 p.m., lasting 30 minutes." "The patient experienced a significant increase in blood pressure on May 12, 2005, at 2.45 p.m." The framework can determine that the increase in blood pressure happened *DURING* the asthma attack.

In [Hri05], the authors address the problem of modeling temporal information in medical narrative reports as simply a temporal constraint satisfaction problem. They focus on medical discharge summaries because they constitute a complex clinical picture that includes the patient's history, treatment, and planned course of treatment. In the article, the authors define a temporal problem with the following temporal entities: (1) instants in times, called time points; (2) the origin, a time point anchored in absolute time; and (3) constraints, which represent relationships between temporal entities. Note that constraints are transitive. *Medical events* are modeled as time intervals and encoded with a descriptive phrase and a unique event identifier. An interesting characteristic of medical events is that they can be reported at different granularities. For example, lab tests may be known to the minute, while for doctor exams, we might know just the day they occurred on. Other challenging features of temporal events are vagueness, ambiguity, and uncertainty. An example of vagueness is that the patient's fever started approximately 3 days ago. Temporal vagueness was encoded using deterministic ranges.

In [Zhu08], *TimeText*, which is a temporal reasoning system to process clinical discharge summaries, is described. Specifically, the system has a threefold function: (1) representation, (2) extraction, and (3) reasoning about temporal information in clinical text. The system was tested using twenty randomly selected discharge summaries and the authors concluded that *TimeText* was able to identify the majority of temporal information identified by domain experts, and it was able to answer simple temporal questions. Specifically, the system was able to capture 79% of 307 temporal relations, determined to be clinically important by the subjects and the raters. Also, the system answered 84% of temporal queries correctly.

6.2.5 Temporality Incorporation and Temporal Data Mining in Electronic Health Records

Electronic Health Records (EHR) are becoming commonplace in most hospitals and medical practices, and there is a U.S. government initiative for universal EHR adoption by 2014. *EHRs* store the most recent information about a patient's medical condition, but also information about a patient's medical history. In [Mal06], a middle layer framework, called the data representation layer (*DLR*), is constructed that uses object versioning to manage different versions of a patient's medical data. In other words, the framework allows the representation and storage of different temporal objects and temporal relationships among them.

The following four issues are addressed in [Mal06]:

- Mapping between object versions

- How to represent fuzzy (inexact) mappings

- How to handle errors in mappings (impossible mappings)

- How to represent causes of error changes

Each object version has its own time stamp. Data objects are validated against schema objects, and a bi-directional mapping between two time stamps is represented by its own mapping object. The framework supports both object and schema versioning.

A different approach to incorporate temporality in electronic health records is considered in [Gal04]. Temporal components are incorporated in the architecture of electronic health records. Three different *EHR* architectures are investigated in terms of temporality: *CEN*, *HL7*, and *openEHR*. The authors conclude that all three *EHR* architectures can have temporality components incorporated in them, because their reference models are generic enough to allow temporal structures. Even complex components, such as time series, can be described in the aforementioned architectures. The temporality incorporation can be achieved using either generic components or predefined specific structures. In [Mow08], the authors develop a schema for the annotation of contextual and temporal elements of various clinical conditions discussed in clinical reports.

In [Hud08], the authors discuss performing trend analysis on individual electronic health records to detect change in trends, appearance of new symptoms, and improvement or deterioration of health based on symptom status. An example application is the production of automatic alerts based on the comparison of new findings in the existing health status of the patient.

In [Tho07], *EHRs* were investigated as a means to improve the detection performance of a syndromic surveillance system. Alarms generated from *EHRs* were compared with alarms generated from more traditional data sources and a significant correlation was found to exist between them. Therefore, incorporation of *EHR*-generated alarms can be of benefit in syndromic surveillance system. Despite the many benefits of *EHRs*, a systematic review of related literature in [Poi05] resulted into the conclusion that adoption of *EHRs* did not result in a decreased documentation time for nurses and physicians.

In [Ich05], the authors note the three idiosyncrasies of *EHRs* that make many conventional data mining methods insufficient to handle them: (1) values are missing sometimes because patients do not undergo all examinations, (2) data include time series attributes recorded at irregular intervals, and (3) number of records increase every time a patient visits a hospital. The authors propose a graph method approach that uses a graph structure to represent medical data and reduce the graph using specified rules.

6.2.6 The BioJournal Monitor

In [Mör08], the *BioJournal Monitor*, a decision support system that allows the anticipation of annotations and emerging trends in biomedical literature, is described. The system has the following novel features:

- It continuously integrates information from different data sources, such as *PubMed, MeSH ontology*, and *Gene Ontology*.

- It deploys several state-of-the art text mining algorithms, such as named entity detection, relation extraction, classification, and clustering.

- It allows automatic prediction of *MeSH* metadata for new biomedical abstracts. MeSH is the ontology used by the *National Library of Medicine* to categorize and annotate new documents.

- It allows the development of new *MeSH* terms and the prediction of emerging trends very early. Typically new terms are entered in *MeSH* manually after a consensus in the scientific community has been reached. In other words, the *BioJournal Monitor* system has the *unique feature of predicting new established knowledge.*

- It presents a user interface that allows the entering of keywords and the presentation of relevant document clusters.

6.3 TEMPORALITY IN CLINICAL WORKFLOWS

6.3.1 Clinical Workflow Management

Giving the right information at the right time to the right person is a critical part of an efficient clinical management and, subsequently, of its *workflow management.* In [Dad97], the authors discuss the functionality for such an environment. Other relevant references are [Kuh94a] and [Kuh94b]. The authors in [Dad97] discuss some unique issues and challenges in a clinical environment:

- Physicians and nurses spend their time not only on medical tasks, such as diagnosis and treatment, but also on administrative tasks. They are also responsible for many patients with varying numbers and types of medical problems.

- For the diagnosis and treatment of a patient, several separate units may be involved.

- Treatment of a patient can follow well-known protocols; however, it can also be dynamic based on the individual health condition of a patient. Unexpected events can also change the flow of treatment of a patient.

- Processes can be short, such as result reporting, or long, such as chemotherapy delivery.

- Missing information may lead to delays or repeated procedures.

Given the above information and with the goal of optimized *patient-centric care*, the authors concluded that an efficient clinical workflow system must fulfill the following requirements:

- It must be easy to use and have acceptable performance that supports a large number of concurrent users.

- It should be highly reliable and support data consistency. Ontologies could be of use in the latter part.

- It should be temporally aware. This means it should alert the clinician about approaching patient treatment actions; it should know the maximum and minimum distance between various steps and inform the user about potential time-related problems.

- It must support ad hoc deviations from the premodeled workflow and integrate these changes in a seamless way. Deviations must be recorded and must not be complicated for the user. This means that remapping of input/output parameters and deletions of steps should be integrated seamlessly in the system and be transparent to the user.

- It should support cyclic tasks, such as chemotherapy treatment.

The article [Dad97] describes the *ADEPT* project, which addresses many of the above requirements, such as dynamic changes and incorporation of temporal aspects.

6.3.2 Querying Clinical Workflows by Temporal Similarity

Clinical guidelines are a key part of *evidence-based* medicine, and they describe the sequence of actions in a clinical protocol. Adhering to the temporal constraints of the guidelines is an important part of evaluating healthcare quality. In [Com07], clinical guidelines are modeled as workflow processes and the problem of assessing the temporal similarity of workflow cases is addressed. The authors applied their method to comparing workflows of the Italian guidelines for *Stroke Management*. Their workflow consisted of temporal event sequences, where the events are time stamped clinical events, such as lab results for a patient during a week. The similarity of events, because they correspond to single time points, is measured within the context of a temporal window.

The rules that a workflow process has to follow can be described in a *workflow schema*. Just like database schemas, workflow schemas describe the structure of the cases and can contain temporal constraints. Individual workflow cases represent instances of the workflow schema and can be different in their structure and temporal properties, such as order and temporal length of activities. An important problem in the clinical domain is to evaluate the similarity of various workflow cases with an optimal workflow case.

A workflow is modeled as a set of *labeled task intervals*, where each interval has a timestamp for its beginning time and a timestamp for its ending time. The main steps of the method described in [Com07], which compares the temporal similarity of two workflows, are shown below:

- Express both cases using temporal interval constraint networks.

- Compute the distance between intervals representing corresponding tasks *(intratask distance)*.

- Compute the distance between the relations representing corresponding tasks *(intertask distance)*. A relation between two tasks is represented using a qualitative value, such as *Allen's relations* (*before, after*) and a quantitative value, such as the distance between the end of the first task and the beginning of the second task. The distance between two *Allen relations* can be measured using graphs.

- Compute the overall similarity by combining the intertask and intratask similarity.

6.3.3 Surgical Workflow Temporal Modeling

In [Man07], the authors use business process intelligence techniques to model *surgical workflow*. Cases of surgical workflow consist of various steps such as preoperational planning, systems involved in the procedure, and post-operational data. Surgical workflows are modeled using concepts from *Business Process Modeling (BPM)* and *Workflow Management (WfM)*. Both *BPM* and *WfM* model processes as consisting of activities and relationships between activities. Then, multidimensional data modeling is used to transform the data in data cubes, such that they can be imported in a relational OLAP system. Each cube corresponds to a fact, and dimensions, that describe the fact, are modeled as the axes of the cube. So, in the case of surgical workflow, the data cube called *SURGERY* has dimensions *Location, Patient, Surgeon,* and *Discipline.*

The main idea behind transforming a *BPM* process into a multidimensional view lies in finding a common abstraction level between the two. This is done by mapping the vertical decomposition of a process (that consists of temporal and or logical components) to facts. This is done at two granularity levels: the process and the work step level. Horizontal decomposition (function, organization, etc.) is mapped to dimensional characteristics of facts. As a result, one can build multidimensional views on the workflow and create data queries even with high complexity.

Finally, [Bar00] introduces *PATIENT SCHEDULER*, a prototype that demonstrates how computer technology can be used to implement temporal coordination in a surgical department.

6.4 ADDITIONAL BIBLIOGRAPHY

In [Ber07], sequential pattern matching is expanded to take into account the time that elapses between the observations of clinical parameters. This is achieved by introducing the concept of *time-annotated sequences (TAS)*, which was implemented in clinical data pertaining to a liver transplantation. In [Sar01], an iterative procedure is used that segments a medical time series in progressively finer linear segments. Originally, the number of linear segments is estimated using the number of cycles in the data set. Then a greedy approach that is based on a multiobjective function and utilizes local and global information is used to estimate the linear segments. In addition, breakpoints are estimated in the data set by measuring the ability of each point in the data set to form *bends*.

In [Sat07], a medical time series clustering method is proposed, which utilizes *Formal Concept Analysis (FCA). FCA* is a data representation method, that identifies conceptual structures in data sets and allows the derivation of an ontology consisting of a concept hierarchy. In [Sar03], Sarkar and Leong present a novel *fuzzy* method for the computation of the fractal dimension, which then they apply to characterize medical time series from ICU units.

In [Mez07], a statistical approach is used to describe statistical deviations in medical care. In [Buc08], simulated outbreak data are used to evaluate the ability of surveillance algorithms to detect different types of outbreaks. In [Moh07], the authors describe a mathematical framework for early detection of temporal anomalies that can accompany, for example, an outbreak. A formal language for creating an event-centered representation of outbreak histories is described in [Cha06b]. The language is called the *SpatioTemporal Extended Event Language (STEEL)*. An important feature of the language is that it allows the representation of event aggregations, based on their spatiotemporal location. In [Tor09], the authors experimented with a real-time (open-end) version of the *DTW* for the recognition of motor exercises, in post-stroke patients. It was concluded that this version is suitable for the classification of truncated quantitative time series.

In [Leo07], a case-based architecture is used to configure temporal abstractions for hemodialysis data. In [Mos07], the authors describe the application of the *time-series knowledge mining* method, which is a method that discovers partially ordered coinciding time intervals, in ICU data. The aforementioned knowledge mining method was first described in [Mör05], and it utilizes a novel time series knowledge representation technique. This knowledge representation technique is a hierarchical interval language that utilizes the temporal concepts of *duration, coincidence* and *partial order* in interval time series.

In [Mal05], a temporal object model is used for the representation of different schemas, ontologies, and health data. In [Ver07], the authors use temporal abstraction for feature extraction from monitoring data.

The conclusion from their work is that induction of numerical features from data is preferable to extraction of symbolic features derived from medical language concepts. In [Guy07], the authors construct knowledge from time series data using a human–computer collaborative system that involves specific annotations.

In [Wen02] the authors discuss temporal knowledge representation for scheduling tasks in *clinical trial protocols*. An important note that the authors make is that clinical trial protocols are much more *prescriptive* than clinical guidelines and therefore, temporal constraints are more useful in protocols than guidelines. A temporal ontology is described in the article that is used to encode clinical trial protocols and applied in a prototype tool used to perform patient specific scheduling.

In [Wu05], the authors propose a novel subsequence matching technique that uses the internal structure of the data directly in the matching phase. A novel idea they propose is that of *subsequence stability* to evaluate the representativeness of a subsequence. The technique is applied to tumor motion respiratory analysis. In addition, it can be used for prediction and clustering purposes. In [Wu07], the author presents a method *(GWKMA)* to cluster gene expression data that utilizes a combination of a genetic algorithm and *weighted k-means*. The genetic aspect of the algorithm was introduced to overcome two of the *k-means* shortcomings, which are sensitivity to initial partitions and resulting in local minima. The algorithm was applied to a synthetic and two real-life data sets and in all cases it outperformed the *k-means* algorithm.

Finally, in [Sri03], the authors discuss how to summarize textually time series data extracted from neonatal monitoring. Some of their key findings are the following: (1) Raw data contain artifacts, from baby handling, for example, that have to be separated from the data that will be included in the summary; (2) the summary should include data falls and rises; and (3) the summary should contain actual numerical values. An important part of the method was content selection, that is, deciding what part of the data to use for the summary. For this purpose a segmentation technique, known as *bottom-up approach* (and discussed in previous chapters) was used. As the authors note, an advantage of this technique is that the number of segments depends on a threshold specified by the user, who, in this way, can control the amount of detail with which each part of the time series will be rendered. Another article that addresses neonatal time series data is [Law05], where a comparison of textual and graphical presentations of time series data is presented.

REFERENCES

[Aac01] Aach, J. and G.M. Church, Aligning Gene Expression Time Series with Time Warping Algorithms, *Bioinformatics*, vol. 17, no. 6, pp. 495–508, 2001.

[Abe07] Abe, H. et al., Developing an Integrated Time Series Data Mining Environment for Medical Data Mining, *Seventh IEEE International Conference on Data Mining*, pp. 127–132, 2007.

[Abe08] Abe, A. et al., Data Mining of Multi-Categorized Data, *Lecture Notes in Computer Science*, vol. 4944, pp. 182–195, 2008.

[Adl06] Adlassnig, K.P. et al., Temporal Representation and Reasoning in Medicine: Research Directions and Challenges, *Artificial Intelligence in Medicine*, vol. 38, no. 2, pp. 101–113, October 2006.

[Aig06] Aigner, W. and S. Miksch, CareVis: Integrated Visualization of Computerized Protocols and Temporal Patient data, *Artificial Intelligence in Medicine*, vol. 37, no. 3, pp. 203–218, July 2006.

[Alo99] Alon, U. et al., Broad Patterns of Gene Expression Revealed by Clustering Analysis of Tumor and Normal Colon Tissues Probed by Oligonucleotide Array, *Proceedings of National Academy of Science*, vol. 96, no. 12, pp. 6745–6750, June 1999.

[Alo03] Alonso, F. et al., Discovering Similar Patterns for Characterizing Time Series in a Medical Domain, *Knowledge and Information Systems*, vol. 5, pp. 183–200, 2003.

[Alt00] Altschul, S.F. et al., Basic Local Alignment Search Tool, *Journal of Molecular Biology*, vol. 215, no. 3, pp. 403–410, 2000.

[Alt06] Altiparnak, F. et al., Information Mining over Heterogeneous and High-Dimensional Time Series Data in Clinical Trial Databases, *IEEE Transactions on Information Technology in Biomedicine*, vol. 10, no. 2, pp. 254–263, April 2006.

[Ang07] Angelini, L. et al., Multiscale Analysis of Short Term Heart Beat Interval, Arterial Blood Pressure, and Instantaneous Lung Volume Time Series, *Artificial Intelligence in Medicine*, vol. 41, no. 3, pp. 237–250, November 2007.

[Ans06] Anselma, L. et al., Towards a Comprehensive Treatment of Repetitions, Periodicity, and Temporal Constraints in Clinical Guidelines, *Artificial Intelligence in Medicine*, vol. 38, no. 2, pp. 171–195, 2006.

[Asl08] Asl, B.M., Support-Vector Machine-Based Arrhythmia Classification Using Reduced Features of Heart Rate Variability Signal, *Artificial Intelligence in Medicine*, vol. 44, no. 1, pp. 51–64, 2008.

[Aug05] Augusto, J.C., Temporal Reasoning for Decision Support in Medicine, *Artificial Intelligence in Medicine*, vol. 33, no. 1, pp. 1–24, January 2005.

[Bai95] Bailey, T. and C. Elkan, Unsupervised Learning of Multiple Motifs in Biopolymers Using Expectation Minimization, *Machine Learning*, vol. 21, no. 1–2, pp. 51–80, 1995.

[Bal94] Baldi, P. et al., Hidden Markov Models of Biological Primary Sequence Information, *Proceedings of the National Academic Science*, vol. 91, pp. 1059–1063, 1994.

[Bar00] Bardram, J.E., Temporal Coordination-On Time and Coordination of Collaborative Activities at a Surgical Department, *Computer Supported Cooperative Work,* vol. 9, no. 2, pp. 157–187, May 2000.

[Bar02] Bar-Joseph, Z. et al., A New Approach to Analyzing Gene Expression Time Series Data, *Proceedings of RECOMB'02,* Washington, DC, pp. 39–48, 2002.

[Bar03] Bar-Joseph, Z. et al., Continuous Representation of Time Series Gene Expression Data, *Journal of Computational Biology,* vol. 10, no. 3–4: 341–356, 2003.

[Bar04] Bar-Joseph, Z. et al., Analyzing Time Series Gene Expression Data, *Bioinformatics,* vol. 20, no. 16, pp. 2493–2503, Oxford University Press, 2004.

[Bel97] Bellazzi, R., C. Larizza, and A. Riva, Temporal Abstractions for Pre-Processing and Interpreting Diabetes Monitoring Time Series, *Proceedings JCAI-97 Workshop on Intelligent Data Analysis in Medicine and Pharmacology,* pp. 1–9, 1997.

[Bel99] Bellazzi, R. et al., Intelligent Analysis of Clinical Time Series by Combining Structural Filtering and Temporal Abstractions, *Artificial Intelligence in Medicine, Lecture Notes in Computer Science,* vol. 1620, pp. 261–270, 1999.

[Bel00] Bellazzi, R. et al., Intelligent Analysis of Clinical Time Series: An Application in the Diabetes Mellitus Domain, *Artificial Intelligence in Medicine,* vol. 20, no. 1, pp. 37–57, September 2000.

[Bel05] Bellazzi, R. et al., Temporal Data Mining for the Quality Assessment of Hemodialysis Services, *Artificial Intelligence in Medicine,* vol. 34, no. 1, pp. 25–39, 2005.

[Bel07] Bellazzi, R. and B. Zupan, Towards Knowledge-Based Gene-Expression Mining, *Journal of Biomedical Informatics,* vol. 40, pp. 787–802, 2007.

[Ben99] Ben-Dor, A., R. Shamir, and Z. Yakhini, Clustering Gene Expression Patterns, *Journal of Computational Biology,* vol. 6, no. 3–4, pp. 281–297, 1999.

[Ber07] Berlingerio, M. et al., Time Annotated Sequences for Medical Data Mining, *ICDM Workshops,* pp. 133–138, 2007.

[Blu83] Blum, R., Representation of Empirically Derived Causal Relationships, *Proceedings of the International Joint Conferences on Artificial Intelligence (IJCAI),* pp. 268–271, 1983.

[Boa05] Boaz, D. and Y. Shahar, A Framework for Distributed Mediation of Temporal-Abstraction Queries to Clinical Databases, *Artificial Intelligence in Medicine,* vol. 34, no. 1, pp. 3–24, May 2005.

[Bor06] Borgwardt, K., S.V.N., Vishwanathan, and H.P., Kriegel, Class Prediction from Time Series Gene Expression Profiles Using Dynamical Systems Kernels, *Pacific Symposium on Biocomputing,* vol. 11, pp. 547–558, 2006.

[Buc08] Buckeridge, D. et al., Predicting Outbreak Detection in Public Health Surveillance: Quantitative Analysis to Enable Evidence-Based Method Selection, *Proceedings of the AMIA Annual Symposium,* Washington, DC, pp. 76–80, 2008.

[Cao04] Cao, X. et al., Piers: An Efficient Model for Similarity Search in DNA Sequence Databases, *Proceedings of ACM SIGMOD,* vol. 33, no. 2, pp. 39–44, 2004.

[Car01] Caraca-Valente, J.P. and I. Lopez-Chavarrias, Discovering Similar Patterns in Time Series, *Proceedings of KDD Conference*, pp. 497–505, 2000.

[Cha04] Chakravarty, S. and Y. Shahar, CAPSUL: A Constraint-Based Specification of Repeating Patterns in Time-Oriented Data, *Annals of Mathematics and Artificial Intelligence*, vol. 30, no. 1–4, pp. 3–22, 2004.

[Cha06a] Chaovalitwongse, W.A., O. A. Prokopyev, and P. M. Pardalos, EEG Time Series Classification: Applications in Epilepsy, *Annals of Operations Research*, vol. 148, pp. 227–250, Springer, 2006.

[Cha06b] Chaudet, H., Extending the Event Calculus for Tracking Epidemic Spread, *Artificial Intelligence in Medicine*, vol. 38, no. 2, pp. 137–156, October 2006.

[Chi06] Chittaro, L., Visualization of Patient Data at Different Temporal Granularities on Mobile Devices, *Proceedings of Conference on Advanced Visual Interfaces (AVI)*, pp. 484–487, 2006.

[Chu02] Chudova, D. and P. Smyth, Pattern Discovery in Sequences under a Markov Assumption, *Proceedings of 8th ACM SIGKDD International on Knowledge Discovery and Data Mining*, pp. 153–162, 2002.

[Chu07] Chuah, M. and F. Fu, ECG Anomaly Detection via Time Series Analysis, *ISPA Workshops*, pp. 123–135, 2007.

[Com95] Combi, C., F. Pinciroli, and G. Pozzi, TIME-NESIS: A Data Model in Managing Time Granularity of Narrative Clinical Information, *Proceedings of the 5th International Conference on Artificial Intelligence in Medicine*, pp. 397–398, 1995.

[Com97] Combi, C., G. Cucchi, and F. Pinciroli, Applying Object-Oriented Technologies in Modeling and Querying Temporally Oriented Clinical Databases Dealing with Temporal Granularity and Indeterminacy, *IEEE Transactions on Information Technology in Biomedicine*, vol. 1, no. 20, pp. 100–127, June 1997.

[Com07] Combi, C. et al., Querying Clinical Workflows by Temporal Similarity, *Artificial Intelligence in Medicine*, vol. 4594, pp. 469–478, 2007.

[Dad97] Dadam, P., M. Reichert, and K. Kuhn, Clinical Workflows—The Killer Application for Process-Oriented Information Systems, *Ulmer-Informatik Berichte*, Nr. 97-16, Fakultät for Informatik, Universität Ulm, November 1997.

[Daj01] Dajani, R. et al., Modeling of Ventricular Repolarization Time Series by Multi-Layer Perceptrons, *8th Conference on Artificial Intelligence in Medicine (AIME)*, pp. 152–155, 2001.

[Das92] Das, A.K. et al., An Extended SQL for Temporal Data Management in Clinical Decision-Support Systems, *Sixteenth Annual Symposium on Computer Applications in Medical Care*, Baltimore, MD, pp. 128–132, 1992.

[Das01] Das, A.K., and M.A. Musen, A Formal Method to Resolve Temporal Mismatches in Clinical Databases, *Proceedings of the AMIA Annual Symposium*, pp. 130–134, 2001.

[Dej07] Déjean, S. et al., Clustering Time-Series Gene Expression Data Using Smoothing Spline Derivatives, *EURASIP Journal on BioInformatics and Systems Biology*, vol. 2007, 2007.

[Dev08] Devissher, M. et al., Pattern Discovery in Intensive Care Data through Sequence Alignment of Qualitative Trends Data: Proof of Concept on a Diuresis Data Set, *Proceedings of the ICML/UAI/COLT 2008 Workshop on Machine Learning for Health Care Professionals*, Helsinki, Finland, 2008.

[Dio03] Diosdado, A.M., J.L. del Rio Correa, and F. A. Brown, Multifractality in Time Series of Human Gait, *Proceedings of the 25th Annual IEEE EMBS Conference*, vol. 1, Cancun, Mexico, pp. 17–21, 2003.

[Doj98] Dojat, M., N. Ramaux, and D. Fontaine, Scenario Recognition for Temporal Reasoning in Medical Domains, *Artificial Intelligence in Medicine*, vol. 14, no. 1, pp. 139–155, September 1998.

[Eis98] Eisen, M. et al., Cluster Analysis and Display of Genome-Wide Expression Pattern, *Proceedings of National Academy of Sciences*, vol. 95, no. 25, pp. 14863–14868, 1998.

[Ern05] Ernst, J., G.J. Nau, and Z. Bar-Joseph, Clustering Short Time Series Gene Expression Data, *Bionformatics*, vol. 21, supp. 1, pp. i159–i168, 2005.

[Ern06] Ernst, J. and Z. Bar-Joseph, STEM: A Tool for the Analysis of Short Time Series Gene Expression Data, *Bioinformatics*, vol. 7, p. 191, 2006.

[Esp05] Esposito, F. et al., Independent Component of Analysis of fMRI Group Studies by Self-Organizing Clustering, *NeuroImage*, vol. 25, no. 1, pp. 193–205, 2005.

[Exa07] Exarchos, T.P. et al., A Methodology for the Automated Creation of Fuzzy Expert Systems for Ischemic and Arrhythmic Beat Classification Based on a Set of Rules Obtained by a Decision Tree, *Artificial Intelligence in Medicine*, vol. 40, no. 3, pp. 187–200, July 2007.

[Fad00] Fadili, M.J. et al., A Multi-step Unsupervised Fuzzy Clustering Analysis of fMRI Time Series, *Human Brain Mapping*, vol. 10, pp. 160–178, 2000.

[Fan07] Fang, Z. et al., Visualization and Exploration of Time Varying Medical Image Sets, *Graphics Interface Conference*, pp. 281–288, 2007.

[Fig97] Figliola, A. and E. Serrano, Analysis of Physiological Time Series Using Wavelet Transforms, *IEEE Engineering in Medicine and Biology*, vol. 16, no. 3, pp. 74–79, May/June 1997.

[Fre02] Freire, L., A. Roche, and J.F. Mangin, What Is the Best Similarity Measure for Motion Correction in fMRI Time Series, *IEEE Transactions on Medical Imaging*, vol. 21, no. 5, pp. 470–484, May 2002.

[Fri02] Fritsche, L. et al., Recognition of Critical Situations from Time Series of Laboratory Results by Case-Based Reasoning, *JAMIA*, vol. 9, pp. 520–528, 2002.

[Gal04] Gall, W., G. Duftschmid, and W. Dorda, Temporal Components in Architectures of Electronic Health Records, http://www.meduniwien.ac.at/msi/mias/papers/Gall2004a.pdf, 2004.

[Gar03] Garrett, D. et al., Comparison of Linear, Non-linear, and Feature Selection Methods for EEG Signal Classification, *IEEE Transactions on Neural Systems and Rehabilitation Engineering*, vol. 11, no. 2, pp. 141–144, June 2003.

[Gas02] Gasch, A.P. and M.B. Eisen, Exploring the Conditional Coregulation of Yeast Gene Expression through Fuzzy K-Means Clustering, vol. 3, no. 11, *Genome Biology*, 0059.1–0059.22, 2002.

[Gau04] Gautama, D., D.P. Mandic, and M.M. Van Hulle, A Novel Method for Determining the Nature of Time Series, *IEEE Transactions on Biomedical Engineering*, vol. 51, no. 5, pp. 728–736, May 2004.

[Gev98] Geva, A.B., Feature Extraction and State Identification in Biomedical Signals Using Hierarchical Fuzzy Clustering, *Medical and Biological Engineering and Computing*, vol. 36, pp. 608–614, 1998.

[Giu05] Giurcăneanu, C.D., I. Tăbus, and J. Astola, Clustering Time Series Gene Expression Data Based on Sum-of-Exponentials Fitting, *EURASIP Journal on Applied Signal Processing*, vol. 8, pp. 1159–1173, 2005.

[Gor06] Gorshkov, Y. et al., Robust Recursive Fuzzy Clustering-Based Segmentation of Biological Time Series, *International Symposium on Evolving Fuzzy Systems*, pp. 101–105, 2006.

[Gui01] Guimaraes, G. et al., A Method for Automated Temporal Knowledge Acquisition Applied to Sleep-Related Breathing Disorders, *Artificial Intelligence in Medicine*, vol. 23, no. 3, pp. 211–237, November 2001.

[Guy07] Guyet, T., C. Gatbay, and M. Dojat, Knowledge Construction From Time Series Data Using a Collaborative Exploration System, *Journal of Biomedical Informatics*, vol. 40, no. 6, pp. 672–687, Dec. 2007.

[Had07] Hadad, A.J. et al., Temporal Abstraction for the Analysis of Intensive Care Information, *16th Argentine Bioengineering Congress and the 5th Conference on Clinical Engineering*, 2007.

[Hau97] Hausdorff, J. M. et al., Altered Fractal Dynamics of Gait: Reduced Stride-Interval Correlations with Aging and Huntington's Disease, *Journal of Applied Physiology*, vol. 82, no. 1, pp. 262–269, January 1997.

[Hea05] Heard, N.A. et al., Bayesian Co-clustering of Anopheles Gene Expression Time Series: Study of Immune Defense Response to Multiple Experimental Challenges, *Proceedings of National Academy of Sciences*, vol. 102, no. 47, pp. 16939–16944, 2005.

[Hir02a] Hirano, S. and S. Tsumoto, Mining Interesting Patterns in Time-Series Medical Databases: A Hybrid Approach of Multiscale Matching and Rough Clustering. *AMIA Annual Symposium*, San Antonio, p. 1043, 2002.

[Hir02b] Hirano, S. and S. Tsumoto, Mining Similar Temporal Patterns in Long Time Series Data and Its Application to Medicine, *ICDM Proceedings of the 2002 IEEE Conference on Data Mining*, pp. 219, 2002.

[Ho03] Ho, T.B. et al., Mining Hepatitis Data with Temporal Abstraction, *Proceedings of the 9th ACM SIGKDD International Conference on Knowledge Discovery and Data Mining*, pp. 369–377, 2003.

[Hol01] Holter, N.S. et al., Dynamic Modeling of Gene Expression Data, *Proceedings of the National Academy of Sciences*, vol. 98, pp. 1693–1698, 2001.

[Hud08] Hudson, D. and E.M. Cohen, Temporal Trend Analysis in Personal Health Records, *30th Annual IEEE International Conference on Engineering in Medicine and Biology*, pp. 3811–3814, 2008.

[Hri05] Hripcsak, G. et al., Modeling Electronic Discharge Summaries as a Simple Temporal Constraint Satisfaction Problem, *JAMIA*, vol. 12, no. 1, pp. 55–63, Jan/Feb. 2005.

[Hvi03] Hvidsten, T.R., A. Laegreid, and J. Komorowski, Learning Rule-Based Models of Biological Process from Gene Expression Time Profiles Using Gene Ontology, *Bioinformatics,* vol. 19, pp. 1116–1123, 2003.

[Hye99] Hyer, L.J., S. Kruglyak, and S. Yooseph, Exploring Expression Data: Identification and Analysis of Coexpressed Genes, *Genome Research,* vol. 9, no. 11, pp. 1106–1115, 1999.

[Ich05] Ichise, R. and M. Numao, First-Order Rule Mining by Using Graphs Created from Temporal Medical Data, *Lecture Notes in Computer science,* vol. 34340, pp. 112–125, Springer, 2005.

[IDA04] IDAN: Development of a Distributed Temporal-Abstraction Mediator for Medical Database, http://www.openclinical.org/prj_idan.html, 2004.

[Ira90] Irani, E.A. et al., Formulating an Approach to Develop a System for the Temporal Analysis of Clinical Data, *Annals of Mathematics and Artificial Intelligence,* pp. 237–244, 1990.

[Ji03] Ji, X., J. Li-Ling, and Z. Sun, Mining Gene Expression Data Using a Novel Approach Based on Hidden Markov Models, *FEBS letters,* vol. 542, no. 1, pp. 125–131, 2003.

[Jia03a] Jiang, D., J. Pei, and A. Zhang, DHC: A Density-Based Hierarchical Clustering Method for Time-Series Gene Expression Data, *Proceedings of the 3rd IEEE International Symposium on Bioinformatics and Bioengineering,* 2003.

[Jia03b] Jiang, D., J. Pei, and A. Zhang, Interactive Exploration of Coherent Patterns in Time-Series Gene Expression Data, *Proceedings of SIGKDD,* pp. 565–570, 2003.

[Jia04a] Jiang, D., C. Tang, and A. Zhang, Cluster Analysis for Gene Expression Data: A Survey, *IEEE Transactions on Knowledge and Data Engineering,* vol. 16, no. 11, pp. 1370–1386, November 2004.

[Jia04b] Jiang, D. et al., Mining Coherent Gene Clusters from Gene-Sample-Time-Microarray Data, *Proceedings of KDD'04,* pp. 430–439, August 22–25, 2004.

[Jos05] Joshi, A. et al., Arrhythmia Classification Using Local Hölder Exponents and Support Vector Machine, *Lecture Notes in Computer Science,* pp. 242–247, Springer, 2005.

[Ker95] Keravnou, E.T., Modeling Medical Concepts as Time Objects, *Artificial Intelligence in Medicine,* vol. 934, pp. 65–78, 1995.

[Kim07] Kim, M.S. and H. Yang, A Study of ECG Characteristics by Using Wavelets and Neural Networks, Computer Systems and Applications, *IEEE International Conference on Computer Systems and Applications,* pp. 786–790, 2007.

[Kis01] Kiseliova, T. and C. Moraga, Modeling Temporal Distribution of Diseases and Symptoms with Fuzzy Logic, *IFSA World Congress,* vol. 3, pp. 1637–1641, 2001.

[Kuh94a] Kuhn, K. et al., A Conceptual Approach to an Open Hospital Information System, *Proceedings of the 12th International Congress on Medical Informatics,* (*MIE*), Lisbon, Portugal, pp. 374–378, 1994.

[Kuh94b] Kuhn, K. et al., An Infrastructure for Cooperation and Communication in an Advanced Clinical Information System, *Proceedings of the 18th Annual Symposium on Computer Applications in Medical Care,* Washington, DC, pp. 519–523, 1994.

[Kun05] Kundaje, A. et al., Combining Sequence and Time Series Expression Data to Learn Transcriptional Modules, *IEEE/ACM Transactions on Computational Biology and Bioinformatics,* vol. 2, no. 3, pp. 194–202, July–September 2005.

[Law05] Law, A.S. et al., A Comparison of Graphical and Textual Presentations of Time Series Data to Support Medical Decision Making in the Neonatal Intensive Care Unit, *Journal of Clinical Monitoring and Computing,* vol. 19, pp. 183–194, 2005.

[Leo07] Leonardi, G. et al., CBR for Temporal Abstractions Configuration in Haemodyalisis, www.cbr-biomed.org/workshops/ICCBR07/iccbr07wk-Leonardi.pdf, 2007.

[Lin03] Lin, W., M.A. Orgun, and G.J. Williams, Mining Temporal Patterns from Health Care Data, *Lecture Notes in Computer Science*, vol. 2454, pp. 329–332, Springer, 2002.

[Lin05] Lin, J. et al., Approximations to Magic: Finding Unusual Medical Time Series, *Proceedings of the 18th IEEE Symposium on Computer-Based Medical Systems*, pp. 329–334, 2005.

[Lua03] Luan, Y. and H. Li, Clustering of Time-Course Gene Expression Data Using a Mixed-Effects Model with B-Splines, *Bioinformatics*, vol. 19, no. 4, pp. 474–482, 2003.

[Luk01] Lukashin, A.V. and R. Fuchs, Analysis of Temporal Gene Expression Profiles: Clustering by Simulated Annealing and Determining the Optimal Number of Clusters, *Bioinformatics*, vol. 17, no. 5, pp. 405–414, 2001.

[Mal05] Mallaug, T. and K. Bratbergsensgen, Long-Term Temporal Data Representation of Personal Health Data, vol. 3631, pp. 379–391, *Proceedings of the Conference on Advances in Databases and Information Systems, Lecture Notes in Computer Science,* Springer, vol. 3631, 2005.

[Mal06] Mallaug, T. and K. Bratbergsensgen, Integrated Electronic Health Record Access by Object Versioning and Metadata, *Proceedings of the Conference on Advances in Databases and Information Systems (ADBIS), Research Communications*, 2006.

[Man07] Mansmann, S., T. Neumuth, and M.H. Scholl, Multidimensional Data Modeling for Business Process Analysis, *Proceedings of the 26th International Conference on Data Modeling,* 2007.

[Mar06] Maraziotis, I.A., A. Dragomir, and A. Berezianos, Semi-Supervised Fuzzy Clustering Networks for Constrained Analysis of Time-Series Gene Expression Data, *Lecture Notes on Computer Science,* vol. 4132, Springer, pp. 818–826, 2006.

[Mar07] Marsolo, K. and S. Parthasarathy, On the Use of Structure and Sequence-Based Features for Protein Classification and Retrieval, *Knowledge and Information Systems*, vol. 14, no. 1, pp. 59–80, January 2008.

[Mar08] Martins, S.B. et al., Evaluation of an Architecture for Intelligent Query and Exploration of Time-Oriented Clinical Data, *Artificial Intelligence in Medicine*, vol. 43, no. 1, pp. 17–34, 2008.

[Mel05] Melissant, C. et al., A Method for Detection of Alzheimer's Disease Using ICA-Enhanced EEG Measurements, *Artificial Intelligence in Medicine*, vol. 33, no. 3, pp. 209–222, March 2005.

[Mey08] Meyer, F.G. and X. Shen, Classification of fMRI Time Series in a Low Dimensional Subspace with a Spatial Prior, *IEEE Transactions on Medical Imaging*, vol. 27, no. 1, pp. 87–98, January 2008.

[Mez07] Mezger, J. et al., A Statistical Approach for Detecting Deviations from Usual Medical Care, *Proceedings of the AMIA Symposium*, p. 1051, 2007.

[Moh07] Mohtashemi, M., K. Yih, and K. Kleinman, Multi-Syndrome of Time Series: A New Concept for Outbreak Investigation, *Advances in Disease Surveillance*, vol. 2, p. 59, 2005.

[Mol03] Möller-Levet, C.S. et al., Clustering of Gene Expression Time Series Data, *Citeseer*, 2003.

[Mon02] Monti, A., C. Medigue, and L. Mangin, Instantaneous Parameter Estimation in CardioVascular Time Series by Harmonic and Time-Frequency Analysis, *IEEE Transactions on Biomedical Engineering*, vol. 49, no. 12, pp. 1547–1556, December 2002.

[Mör05] Mörchen, F., A. Ultsch, and O. Hoos, Discovering Interpretable Muscle Activation Patterns with the Temporal Data Mining Method, *Proceedings of the 8th European Conference on Principles and Practice of Knowledge Discovery, Lecture Notes in Computer Science*, vol. 3202, pp. 512–514, Springer, 2005.

[Mör08] Mörchen, F. et al., Anticipating Annotations and Emerging Trends in Biomedical Literature, *Proceedings of Thirteenth ACM SIGKDD International Conference on Knowledge Discovery and Data Mining*, pp. 954–962, 2008.

[Mos05] Moskovitch, R. and Y. Shahar, Temporal Data Mining Based on Temporal Abstractions, *IEEE ICDM-05 Workshop on Temporal Data Mining*, Houston, Texas, 2005.

[Mos07] Moskovitch, R. et al., Analysis of ICU Patients Using the Time Series Knowledge Mining Method, AIME 0'7, Conference *medinfo.ise.bgu.ac.il/med-Lab/MembersHome Pages/RobPapers/Moskovitch.ICU_TSKM.IDAMAP07. pdf*, 2007.

[Mow08] Mowery, D.L., H. Harkema, and W.W. Chapman, Temporal Annotation of Clinical Text, *BioNLP: 2008, Current Trends in Biomedical Natural Language Processing*, pp. 106–107, 2008.

[Nam08] Nam, H., K. Lee, and D. Lee, Identification of Temporal Association Rules from Time Series Microarray Data Set, *Proceedings of the ACM Workshop on Data and Text Mining Methods in Bioinformatics (DTMBIO)*, pp. 21–28, 2008.

[Ngu97] Nguyen, J.H. et al., A Temporal Database Mediator for Protocol-Based Decision Support, *Proceedings of the 1997 AMIA Annual Fall Symposium*, pp. 298–302, 1997.

[Ngu03] Nguyen, T.D., S. Kawasaki, and T.B. Ho, Discovery of Trends and States in Irregular Medical Temporal Data, *Discovery Science, Lecture Notes in Computer Science*, vol. 2843, pp. 410–417, 2003.

[Nil05] Nilsson, M., P. Funk, and N. Xiong, Clinical Decision Support by Time Series Classification Using Wavelets, *Proceedings of the IEEE Conference on Enterprise Information Systems (ICEIS)*, vol. 2, pp. 169–175, 2005.

[Oco01] O'Connor, M.J. et al., RASTA: A Distributed Temporal Abstraction System to Facilitate Knowledge-Driven Monitoring of Clinical Databases, *Proceedings of the 10th World Congress on Medical Informatics*, London, 2001.

[Oco02] O'Connor, M.J., S.W. Tu, and M.A. Musen, The Chronus II Temporal Database Mediator, *Proceedings of the AMIA Annual Symposium*, San Antonio, Texas, pp. 567–571, 2002.

[Oco06] O'Connor, M.J., R.D. Shankar, and A.K. Das, An Ontology-Driven Mediator for Querying Time-Oriented Biomedical Data, *19th IEEE International Symposium on Computer-Based Medical Systems*, Salt Lake City, Utah, pp. 264–269, 2006.

[Oco07] O'Connor, M.J. et al., Knowledge-Level Querying of Temporal Patterns in Clinical Research Systems, *12th World Congress on Medical Informatics*, Brisbane, Australia, 2007.

[Pal06] Palma, J. et al., Fuzzy Theory Approach for Temporal Model-Based Diagnosis: An Application to Medical Domain Domains, *Artificial Intelligence in Medicine*, vol. 38, no. 2, pp. 197–218, 2006.

[Par07] Paramanathan, P. and R. Uthayakumar, Detecting Patterns in Irregular Time Series with Fractal Dimension, *Proceedings of the International Conference on Computational Intelligence and Multimedia Applications*, pp. 323–327, 2007.

[Pev00] Pevzner, P.A. and S.H. Sze, Combinatorial Approaches to Finding Subtle Signals in DNA Sequences, *Proceedings of the International Conference on Intelligent Systems in Molecular Biology, AAAI Press*, pp. 269–278, 2000.

[Pla08] Plaisant, C. et al., Searching Electronic Health Records for Temporal Patterns in Patient Histories: A Case Study with Microsoft Amalga, *Proceedings of the American Medical Informatics Association (AMIA) Symposium*, pp. 601–605, 2008.

[Ple03] Plesniak, W. et al., Holographic Video Display of Time Series Medical Data, *Proceedings of the IEEE Visualization Conference*, p. 78, Seattle, WA, 2003.

[Poi05] Poissant, L. et al., The Impact of Electronic Health Records on Time Efficiency of Physians and Nurses: A Systematic Review, *JAMIA*, vol. 12, no. 5, pp. 505–516, 2005.

[Pos07a] Post, A.R. and J.H. Harrison, PROTEMPA: A Method for Specifying and Identifying Temporal Sequences in Retrospective Data for Patient Selection, *JAMIA*, vol. 14, pp. 674–683, 2007.

[Pos07b] Post, A.R., A.N. Sovarel, and J.H. Harrison, Abstraction-Based Temporal Data Retrieval for a Clinical Data Repository, *AMIA Annual Symposium*, pp. 603–607, 2007.

[Qia01] Qian, J. et al., Beyond Synexpression Relationships: Local Clustering of Time-Shifted and Inverted Gene Expression Profiles Identifies New, Biologically Relevant Interactions, *Journal of Molecular Biology*, vol. 314, no. 5, pp. 1053–1066, 2001.

[Raf05] Rafiq, M.I., M.J. O' Connor, and A.K. Das, Computational Method for Temporal Pattern Discovery in Biomedical Genomic Databases, *Proceedings of the IEEE Computational Systems Bioinformatics Conference*, pp. 362–365, 2005.

[Raj06] Raj, R., M.J. O'Connor, and A.K. Das, Multi-relational Data Mining of Time-Oriented Biomedical Databases, www-cs-students.stanford. edu/~rashmi/ publications/BCATS2006.pdf, 2006.

[Raj07] Raj, R., M.J. O'Connor, and A.K. Das, An Ontology-Driven Method for Hierarchical Mining of Temporal Patterns: Application to HIV Drug Resistance Research, *Proceedings of the AMIA Annual Symposium*, pp. 614–619, Chicago, IL, 2007.

[Ram00] Ramirez, J.C.G. et al., Temporal Pattern Discovery in Course-of-Disease Data, *IEEE Engineering in Medicine and Biology*, vol. 19, no. 4, pp. 63–71, 2000.

[Ram02] Ramoni, M.F., P. Sebastiani, and I.S. Kohane, From the Cover: Cluster Analysis of Gene Expression Dynamics, *Proc. National Academy of Sciences*, vol. 99, no. 14, pp. 9121–9126, 2002.

[Rod03] Roddick, J.F., P. Fule, and W.J. Graco, Exploratory Medical Knowledge Discovery: Experiences and Issues, *ACM SIGKDD Explorations Newsletter*, vol. 5, no. 1, pp. 94–99, July 2003.

[Rut09] Rutkove, S. et al., Impaired Distal Thermoregulation in Diabetes and Diabetic Polyneuropathy Foot Temperature in Diabetic Polyneuropathy, *Diabetes Care*, pp. 671–676, 2009.

[Sac05a] Sacchi, L. et al., Learning Rules with Complex Temporal Patterns in Biomedical Applications, *Lecture Notes in Artificial Intelligence*, pp. 23–32, 2005.

[Sac05b] Sacchi, L. et al., TA-Clustering: Cluster Analysis of Gene Expression Profiles through Temporal Abstractions, *International Journal of Medical Informatics*, vol. 74, pp. 505–517, 2005.

[Sac07] Sacchi, L. et al., Precedence Temporal Networks to Represent Temporal Relationships in Gene Expression Data, *Journal of Biomedical Informatics*, vol. 40, no. 6, pp. 761–774, 2007.

[Sac08] Sacan, A. and I.H. Toroslu, Approximate Similarity Search in Genomic Sequence Databases Using Landmark-Guided Embedding, *Proceedings of IEEE ICDE Workshop*, pp. 338–345, 2008.

[Sal99] Salatian, A. and J. Hunter, Deriving Trends in Historical and Real-Time Continuously Sampled Medical Data, *Journal of Intelligent Information Systems*, vol. 13, no. 1–2, pp. 47–71, 1999.

[Sar01] Sarkar, M. and T.Y. Leong, Top down Approaches to Abstract Medical Time Series Using Linear Segments, *Proceedings of the IEEE International Conference on Systems, Man, and Cybernetics*, vol. 2, pp. 765–770, 7–10 October 2001.

[Sar03] Sarkar, M. and T.Y. Leong, Characterization of Medical Time Series Using Similarity-Based Fractal Dimensions, *Artificial Intelligence in Medicine*, vol. 27, no. 2, pp. 201–222, 2003.

[Sat07] Sato, K. et al., Data Mining of Time-Series Medical Data by Formal Concept Analysis, *Knowledge-Based Intelligent Information and Engineering Systems, Lecture Notes in Computer Science*, vol. 4693, pp. 1214–1221, 2007.

[Sch01] Schlaefer, A., K. Schröter, and L. Fritsche, A Case-Based Approach for the Classification of Medical Time Series, *Medical Data Analysis, Lecture Notes in Computer Science*, vol. 2199, pp. 258–263, 2001.

[Sch03] Schliep, A., A. Schonhuth, and C. Steinhoff, Using Hidden Markov Models to Analyze Gene Expression Time Course Data, *Bioinformatics*, vol. 19, i264–i272, 2003.

[Sha96] Shahar, Y. and M.A. Musen, Knowledge-Based Temporal Abstraction in Clinical Domains, *Artificial Intelligence in Medicine*, vol. 8, no. 3, pp. 267–298, 1996.

[Sha98] Shahar, Y., S. Miksch, and P. Johnson, The Asgaard Project: A Task-Specific Framework for the Application and Critiquing of Time-Oriented Clinical Guidelines, *Artificial Intelligence in Medicine*, vol. 14, no. 1, pp. 29–51, 1998.

[Sha99a] Shahar, Y., Timing Is Everything: Temporal Reasoning and Temporal Data Maintenance, *Proceedings of Joint European Conference on Artificial Intelligence in Medicine and Medical Decision Making*, pp. 30–46, 1999.

[Sha99b] Shahar, Y. et al., Semi-automated Entry of Clinical Temporal Abstraction Knowledge, *Journal of the American Medical Informatics Association*, vol. 6, pp. 494–511, 1999.

[Sha00] Shamir, R. and R. Sharan, Click: A Clustering Algorithm for Gene Expression Analysis, *Proceedings of the International Conference on Intelligent Systems for Molecular Biology*, 2000.

[Sha08] Shankar, R.D. et al., An Ontological Approach to Representing and Reasoning with Temporal Constraints in Clinical Trial Protocols, *International Conference on Health Informatics*, vol. 1, pp. 87–93, 2008.

[Shi06] Shi, L., P.A. Heng, and T.T. Wong, Unifying Genetic Algorithm and Clustering Method for Recognizing Activated fMRI Time Series, *Lecture Notes in Computer Science*, vol. 3930, pp. 239–248, 2006.

[Sil08] Silva, A. et al., Rating Organ Failure via Adverse Events Using Data Mining in the Intensive Care Unit, *Artificial Intelligence in Medicine*, vol. 43, no. 3, pp. 179–193, 2008.

[Sin05] Singh, R. et al., Active Learning for Sampling in Time-Series Experiments with Application to Gene Expression Analysis, *Proceedings of the 22nd International Conference on Machine Learning*, pp. 832–839, 2005.

[Spo07] Spokoiny, A. and Y. Shahar, An Active Database Architecture for Knowledge-Based Incremental Abstraction of Complex Concepts from Continuously Arriving Time-Oriented Raw Data, *Journal of Intelligent Information Systems*, vol. 28, no. 3, pp. 199–231, 2007.

[Sri03] Sripada, S.G. et al., Summarizing Neonatal Time Series Data, *Proceedings of the 10th Conference on European Chapter of the Association for Computational Linguistics*, vol. 2, pp. 167–170, 2003.

[Sta98] Stamkopoulos, T., K. Diamantaras, and M. Strintzis, ECG Analysis Using Nonlinear PCA Neural Networks for Ischemia Detection, *IEEE Transactions on Signal Processing*, vol. 46, no. 11, pp. 436–447, Nov. 1998.

[Sta07] Stacey, M. and C. McGregor, Temporal Abstraction in Intelligent Clinical Data Analysis: A Survey, *Artificial Intelligence in Medicine*, vol. 39, no. 1, pp. 1–24, January 2007.

[Sze02] Sze, S.H., M.S. Gelfand, and P.A. Pevzner, Finding Weak Motifs in DNA Sequences, *Proceedings of Pacific Symposium on Biocomputing*, pp. 235–246, 2002.

[Tam99] Tamayo, P. et al., Interpreting Patterns of Gene Expression with Self-Organizing Maps: Methods and Application to Hematopoietic Differentiation, *Proceedings of the National Academy of Sciences,* vol. 96, pp. 2907–2912, 1999.

[Tav99] Tavazoie, S. et al., Systematic Determination of Genetic Network Architecture, *Nature Genetics,* vol. 22, no. 3, pp. 281–285, 1999.

[Ter05] Terenziani, P. et al, Extending Temporal Databases to Deal with Telic/Atelic Medical Data, *Proceedings of 10th Conference on Artificial Intelligence in Medicine, Lecture Notes in Computer Science,* Springer, vol. 3581, pp. 58–66, 2005.

[Tho07] Thompson, M.W., Correlation between Alerts Generated from Electronic Medical Record (EMR) Data Sources and Traditional Data Sources, *Advances in Disease Surveillance,* vol. 4, p. 268, 2007.

[Tor09] Tormene, P. et al., Matching Incomplete Time Series with Dynamic Time Warping: An Algorithm and An Application to Post-Stroke Rehabilitation, *Artificial Intelligence in Medicine,* vol. 45, no. 1, pp. 11–34, 2009.

[Tsi00] Tsien, C.L., Event Discovery in Medical Time-Series Data, *Proceedings of AMIA Symposium,* pp. 858–862, 2000.

[Tsi08] Tsiporkova, E. and V. Boeva, A Novel Gene-centric Clustering Algorithm for Standardization of Time Series Expression Data, *Proceedings of the 4th IEEE International Conference on Intelligent Systems,* pp. 12-8–12-13, 2008.

[Tus07] Tusch, G. et al., SPOT-Utilizing Temporal Data for Data Mining in Medicine, *Intelligent Data Analysis in Biomedicine and Pharmacology Workshop,* 2007.

[Van99] Van der Maas, A.A.F., A.H.M. terHofstede, and P.F. deVrie Robbe, Formal Description of Temporal Knowledge in Case Reports, *Artificial Intelligence in Medicine,* vol. 16, no. 3, pp. 251–282, July 1999.

[Vas06] Vasudevan, S., Classification of Diseases Based on Gait Dynamics, M.S. Project, UMASS-Dartmouth, Dept. of Electrical and Computer Engineering, 2006.

[Ver07] Verduijn, M. et al., Temporal Abstraction for Feature Extraction: A Comparative Case Study in Prediction from Intensive Care Data, *Artificial Intelligence in Medicine,* vol. 41, no. 1, pp. 1–12, September 2007.

[Wan94] Wang, J. et al., Combinatorial Pattern Discovery for Scientific Data: Some Preliminary Results, *Proceedings of the ACM SIGMOD International Conference on Management of Data,* pp. 115–125, 1994.

[Wan08] Wang, T.D. et al., Aligning Temporal Data by Sentinel Events: Discovering Patterns in Electronic Health Records, *Proceedings of ACM Computer Human Interaction Conference (CHI),* pp. 457–466, Florence, Italy, 2008.

[Wei05] Wei, L. et al., A Practical Tool for Visualizing and Data Mining Medical Time Series, *Proceedings of the 18th IEEE Symposium Computer-Based Medical Systems,* 2005, pp. 341–346, June 2005.

[Wen02] Weng, C., M. Kahn, and J. Gennari, Temporal Knowledge Representation for Scheduling Tasks in Clinical Trial Protocols, *Proceedings of the AMIA Symposium,* pp. 879–883, 2002.

[Wu05] Wu, H. et al., Subsequence Matching on Structured Time Series Data, *Proceedings of the SIGMOD Conference,* pp. 682–693, 2005.

[Wu07] Wu, F.X., A Genetic Weighted K-Means Algorithm for Clustering Gene Expression Data, *Second IEEE International Multi-symposium on Computer and Computational Sciences,* pp. 68–75, 2007.

[Yeu01a] Yeung, K.Y. et al., Model-Based Clustering and Data Transformations for Gene Expression Data, *Bioinformatics,* vol. 17, no. 10, pp. 977–987, 2001.

[Yeu01b] Yeung, K.Y., D.R. Haynor, and W.L. Ruzzo, Validating Clustering for Gene Expression Data, *Bioinformatics,* vol. 17, no. 4, pp. 309–318, 2001.

[Yua07] Yuan, Y. and C.T. Li, Unsupervised Clustering of Gene Expression Time Series with Conditional Random Fields, *Inaugural IEEE International Conference on Digital Ecosystems and Technologies,* pp. 571–576, 2007.

[Zhu07] Zhu, L. and G. Hripcsack, Methodological Review: Temporal Reasoning with Medical Data—A Review with Emphasis on Medical Natural Language Processing, *Journal of Biomedical Informatics,* vol. 40, no. 2, pp. 183–202, 2007.

[Zhu08] Zhu, L. et al., The Evaluation of a Temporal Reasoning System in Processing Clinical Discharge Summaries, *Journal of the American Medical Informatics Association,* vol. 15, no. 1, pp. 99–106, 2008.

CHAPTER 7

Temporal Data Mining and Forecasting in Business and Industrial Applications

In this chapter, we will examine the following topics:

- Temporal Data Mining applications in enhancement of business strategy and customer relationships, such as applications of temporal data mining in CRM, event-based marketing, and integration of temporal research in business strategy (Section 7.1).

- Workflow management and related topics, such as temporality in supply chain management, temporal data mining to measure operations performance, and real-time data mining using Web services (Section 7.2).

- Miscellaneous applications of temporal data mining in industry, such as temporal management of RFID data (Section 7.3).

- Summary of advances that have been made in the field of financial forecasting, such as demand forecasting, volatility forecasting, and outlier detection in financial time series (Section 7.4).

7.1 TEMPORAL DATA MINING APPLICATIONS IN ENHANCEMENT OF BUSINESS AND CUSTOMER RELATIONSHIPS

CRM stands for *customer relationship management*, and it is a business process whose goal is to track the transactions/behavior of customers, through marketing, sales, and services, to understand customer needs better and enhance the company-client relationships. However, *CRM* needs to be understood really well to deliver true value to the company. An interesting article that describes the four perils of *not* implementing *CRM* correctly is [Rig02].

Temporal data mining can play an important role in implementing *CRM*, by allowing the company to understand how the purchases/preferences of the customer change over time, such that customers can be retained and also be approached with the right product at the right time.

A type of *CRM* for which temporal data mining is particularly useful is *Event-Based Marketing*, which is explained in the next section.

7.1.1 Event-Based Marketing and Business Strategy

Event-Based Marketing (EBM) is an important tool today in business, particularly for continuing and enhancing relationships with existing customers. It can be implemented in the form of a database trigger that signals a change in the status of the customer. Such changes can be the following:

- Maturing of a timed deposit. This could mean that it is time to approach the customer with new financial products.

- Change of last name.

- A large deposit in a money market account.

- A large withdrawal or an overdraft.

- Unusual activity on the credit card of the customer. For example, very expensive purchases or purchases in various foreign locations.

- Anniversaries.

- Purchase of a home.

- Purchase of a specific product. This could mean that the customer could be interested in another similar or better kind of product.

However, one has to be careful in filtering the above information before approaching the customer. Detecting the events is the first and easier step. The more difficult step is to determine how to act upon this information. Contacting the customer for every single event detected in the database can simply be annoying to the customer. Also, it is very important to combine *event information* about the customer with *background information* about the customer. For example, if we know that the customer is not interested in stock products, there is no point in contacting him even if there is a recent large deposit in his account.

In addition, the channel of contact has to be carefully chosen (e-mail, letter, phone call) and ensure that the most effective channel is chosen. Another important decision is when to contact the customer. Sometimes it is important to act quickly, especially when dealing with customer retention issues. Sometimes it is important to pay attention to *combination of events*. For example, just having a few expensive purchases on a credit card could be normal. However, if the purchases are combined with an overdraft, then this is probably a reason for concern. Another example of combined events is the change of the last name of the customer along with a purchase of an expensive home. This probably indicates a lifestyle change for the customer, which might mean that she is interested in purchasing certain financial or other kind of products.

A very important point that needs to be made here is that all actions that are implemented as part of *EBM must be completely in agreement with the business strategy of the organization.* For example, a customer might fit the profile for a certain financial product, but if the company is looking into phasing out this product, there is no point in contacting the customer. This is the reason that *EBM* is most successful in a transparent and agile organization, where, once the event information about a customer becomes available, the right channels and people will be mobilized to approach the customer.

e-Commerce has dramatically changed the marketplace today. A question that often arises is how to deploy *CRM* to increase the profitability of a dot-com site. Of particular interest is the understanding of customer needs in *three temporal dimensions: Presales, During Sales, and Post Sales.* Temporal data mining can then be applied in customer databases to gain insights as to how the different site features are used by the customers.

[Ahm08] examines the customer's perception of value from airline e-ticketing Web sites, when these sites deploy *e-CRM*. The main goal of *e-CRM*, as in the case of traditional *CRM,* is customer attention and retention [Wan07], [Phe01]. As noted in [Phe01], one problem that dot-com companies face is customer retention. Customers may never complain about a problem, if the necessary channels to express their concerns do not exist, but they will not make another purchase and they will also tell others about their bad experience. Desirable features of dot-com sites that relate to enhancing presales, sales, and post-sales *CRM* are discussed in [Anc01], [Ahm08], [Ros05], [Fei02], and [Kha05].

7.1.2 Business Strategy Implementation via Temporal Data Mining

Business strategy is the formulation of actions and processes that will allow a business to achieve its mission and realize its vision. A key part of formulating the right business strategy is understanding the environment in which the business operates. The *five forces framework* published by M. Porter in 1979 in Harvard Business Review [Por79] is a widely used framework for this purpose. It concentrates on the following forces that shape the marketplace:

- Buyers
- Suppliers
- Rivals
- Substitutes
- Potential entrants

Data mining and particularly temporal data mining can be instrumental in tracking the changes in these forces over time. For example, a temporal data mining system can be put in place that will track the price of substitute products, prices offered by suppliers, and rival product prices and offers. In a nutshell, temporal information mining allows an organization to stay up to date about changes in its environment and *be agile*, such that it will make the right move at the right time.

Tracking business activities over time can yield important insights as to how to enhance business strategy. Applying temporal data mining techniques to business data can provide data for the following questions:

- Which of our products or services are the most profitable?
- Which of our products or services are the most distinctive?

- Which of our customers are the most satisfied?

- Which customers, channels, or purchase occasions are the most profitable?

- Which of the activities in our value chain are the most different and effective?

Answers to these questions will unveil whether the business strategy is successful or show whether what we are doing is aligned with strategy. It can also allow the business to increase its competitive advantage by lowering cost and enhancing product differentiation.

7.1.3 Temporality of Business Decision Making and Integration of Temporal Research in Business

An interesting question in temporal data mining that pertains to business strategy is how the speed of past decisions has influenced the success of the organization.

In [Per02], the relationship between speed and decision making is explored. Using causal loop diagrams, the problem of *speed trap* is discovered, where decisions are made too fast. This is a pathological condition in organizations, where, while the requirement for speed seems to come from external conditions, in reality it comes from the organization's past emphasis on speed.

The work of Perlow and Okhuysen shows the mutual causality between action and structure of the organization. Fast decision making changes the environment in which the business operates, such that fast decision making has to continue. The speed trap is connected with four feedback loops:

1. *Accelerating aspirations loop*: Increased urgency and elevated aspirations lead to commitments that have to be realized within a smaller window of time.

2. *Closing window of opportunity*: Increased commitments shorten the required time of response and further increase the need of urgency.

3. *Mounting threats:* The shortened time horizon and increased commitments lead to mounting threats.

4. *Slippery slope:* All of the above problems lead to a decreased attention to decision making and increase of problems.

In [Anc01], three categories were identified to integrate temporal research in business organizations:

1. *Conceptions of time:* This includes conventional clock time, which is linear and continuous, and lifecycle, such as human lifecycle and lifecycle of software development.

2. *Mapping activities to time:* Rate, duration, and allocation fall in this category. The five subcategories of this category are the following:

    a. Single activity mapping to the time continuum. Examples include rate, duration, and allocation.

    b. Repeated mapping of the same activity multiple times on the time continuum. Examples include cycle and frequency.

    c. Single activity transformation mapping. Examples include life cycles and deadlines.

    d. Multiple activity mapping of two or more activities on the continuum. Examples include synchronization and reallocation of activities.

    e. Comparison of multiple temporal maps.

3. *Actors relating to time:* Here there are two subcategories:

    a. Temporal perception variables. This captures how actors perceive the continuum.

    b. Temporal personality variables. This captures how actors act in regards to the continuum.

A number of articles deal with temporal modeling of venture creation ([Bir92], [Lic05], [Liao05]). Specifically, [Bir92] presents a model that deals with entrepreneurial goal-setting and the ability to control/predict durations of various stages of the new venture creation process. The authors in [Liao05] collect data from 668 nascent entrepreneurs and use complexity theory to explain their findings, which suggest that the venture creation process is a nonlinear process whose stages are not easily identifiable. Finally, the authors in [Lic05] address the issue of whether there exists a correlation between successful venture creations and temporal characteristics of start-up behaviors. They found that, in contrast to discontinued start-up efforts, successful new ventures exhibited a slower pace of activities over a longer period of time, with a peak of activity at the beginning or end of their efforts.

7.1.4 Intertemporal Economies of Scope

A popular topic in strategic management is how to add value by simultaneously sharing resources between businesses within the same firm. [Hel04] deals with the topic of *intertemporal economies of scope*, that is, how corporations can add value by sharing resources between businesses within the firm over time, such as resources freed by exiting a market. They conclude that intertemporal economies of scope can be aided by a decentralized and modular structure of the different business entities and also related product diversification. This is particularly useful in dynamic markets where demand and technologies change frequently.

Economies of scope for two products, X and Y, at a certain time t are defined as

$$C(X, Y) < C(X, t) + C(Y, t) \qquad (7.1)$$

This means that the cost of producing products X and Y together is less than that for producing each product separately. For example, [Kim96] shows that semiconductor firms build on prior knowledge when entering related markets.

In [Hel04], intertemporal economies of scope are defined as

$$C(X, t-1, Y, t) < C(X, t-1) + C(Y, t) \qquad (7.2)$$

This equation states that the cost of producing X at time $t - 1$ and product Y at time t is less than producing the products at their respective times, but in separate firms. The gain comes from the fact that when the firm stops production of product X at time $t - 1$, all resources can be transferred to the production of product Y. A specific business was examined in the article (Omni Corporation). Several insights were gained from this study, such as the combination of a modular, decentralized structure and business recombination strategies to help minimize adjustment costs. Introduction of two generations of a product can also be helpful in achieving both intratemporal and intertemporal economies of scope.

7.1.5 Time-Based Competition

In a landmark article in *Harvard Business Review,* George Stalk, Jr., introduces the notion of *time-based competition* [Sta88]. This is a type of strategy to achieve competitive advantage, where reducing the time it takes to go through the different phases of a lifecycle is the most important factor. As a model, Stalk discusses the idea of the *flexible factory* implemented by Japanese companies. For example, *Honda* achieved its competitive advantage by implementing structural changes that enabled faster operational processes. Temporal data mining can be used to extract information from company data to determine how temporally efficient are the different cycles of the lifecycle.

As discussed in [Tim09], there are two main forms of time-based competition: *fast-to-market* and *fast-to-product.* In the fast-to-market approach, what needs to be mined is information as to how temporally efficient the product development phase is. An example of fast-to-market company is *Sun MicroSystems.* As discussed in [Tim09], products that are introduced on time, but are fifty percent over budget, generate higher profits than projects that do not exceed their budget, but are introduced six months late. In the fast-to-product approach, what is reduced is the response time to customer orders. Therefore, the information that needs to be mined is (1) how temporally efficient the manufacturing phase is and (2) the product delivery speed. A well-known example of a fast-to-product company that manages to significantly reduce the inventory levels is *Walmart.*

The authors in [Vic95] identify 10 steps for a company to successfully implement time-based competition. I will mention here the steps that can be improved by using temporal data mining on company data:

- Analyze business processes.

- Develop a measurement system that focuses on time.

- Increase the speed of new product introduction.

- Examine all managerial decision possibilities in terms of time.

In [Sta98], Stalk introduced some time-related rules that if applied in a company, can help the company achieve significant financial gains. Temporal data mining can be used on company data to decide how closely these rules are followed. Below is a brief description of two of these rules:

- *The 0.5 to 5 rule.* This rule describes the poor productivity of many companies. Most products receive only 0.5% to 5% of the overall time, while the value delivery system receives considerably more time. For example, in a computer factory, that does not have an agile and flexible delivery system, it takes significantly less time to put a computer together than to gather all the required parts.

- *The 3/3 rule.* It is quite common that during 95% to 99.95% of the time, a product is waiting instead of receiving value. Waiting time is almost equally allocated into the following categories:

  1. Completion of the overall product batch.

  2. Completion of the batch ahead.

  3. Waiting for management to execute the decision to move the product batch to the next step of the process.

Another interesting observation made by Stalk is that companies that reduce by 25% the time it takes to produce a product achieve as much as a 20% reduction of costs and a doubling of productivity.

7.1.6 A Model for Customer Lifetime Value

In [Ros03], the authors discuss a model for the *Customer Lifetime Value* in the telecommunications industry, to develop a model of the effect of various marketing activities on this value. *Customer LifeTime Value (LTV) represents the total net income a company can expect from a customer.* Below, important points from the work in [Ros03] are summarized:

An LTV model consists of three components:

- Customer's value over time, *v(t)*.

- Length of service *(LOS)*. This is modeled using a hazard function:

$$h(t) = \frac{f(t)}{S(t)} \tag{7.3}$$

where *S(t)* is a survival function that describes the probability that the customer will be active at time *t*. The function *f(t)* is the customer's instantaneous probability that he will churn at time *t*:

$$f(t) = -\frac{dS(t)}{dt} \tag{7.4}$$

- A discounting factor, $D(t)$. This describes the current value of each $1 earned in the future.

Then, the LTV model can be expressed as

$$LTV = \int_{0}^{\infty} S(t)v(t)D(t)dt \tag{7.5}$$

The value of a customer can be estimated in a practical way from various factors, such as his usage, price plan, etc. Regarding the LOS, Rosset et al. propose a *market-segment-based* approach. A *market segment* is a population of customers, who for marketing purposes can be treated as a homogeneous group. Specifically, they propose the following nonparametric approach for the computation of the hazard function:

$$h(t) = \frac{\sum_i I(t_i = t)I(c_i = 1)}{\sum_i I(t_i = t)} \tag{7.6}$$

where $\sum_i I(t_i = t)$ is the number of customers whose current tenure is t months and $\sum_i I(t_i = t)I(c_i = 1)$ is the number of customers whose current tenure is t months and churned this current month.

An important contribution of [Ros03] is the calculation of the *effect of a company's marketing efforts on customer LTV.* Let us assume that a company wants to know the effect on a specific market segment of an incentive that has the following parameters associated with it:

- C, which is the cost of the effort to the company for the incentive

- G, which is the cost incurred if the customer accepts an incentive

- P, the probability that the client will accept the offer

Then, the very interesting and yet intuitive result is derived in the article, regarding the difference between the LTV after the marketing incentive (LTV_{new}), and the LTV before the marketing incentive (LTV_{old}):

$$LTV_{new} - LTV_{old} > 0 \Leftrightarrow vp > 2(PG+C)/(ph(h-1)) \tag{7.7}$$

As the equation shows, the marketing incentive *should only be suggested to customers with a high enough value weighted risk vp*, where p is the churn probability of the customer for the next month. For the derivation of the above equation, some assumptions were made, such as (1) customer value is constant over time and is not affected by the incentive, (2) the incentive includes a commitment for h months at least, and (3) the churn risk is constant for each customer in the segment and for any time horizon.

7.2 BUSINESS PROCESS APPLICATIONS

7.2.1 Business Process Workflow Management

An important aspect of every efficient business environment is to be able to deliver the right information at the right time to the right person and provide her with the right tools to take an action congruent with the business strategy and goals. *Workflow management technology* can offer a solution to this problem by taking a process-oriented approach, rather than function-oriented, which is the case with most application software. For the implementation of workflow management systems (*WFM*), important concepts are interprocess communication and run-time exchange of output parameters, which will become the input to another process. *Therefore, through the use of a WFM system, information provided from a temporal data mining framework can be seamlessly integrated with business processes, such as shipping and CRM.*

A detailed discussion on *WFM* can be found in [Dad97]. Examples of such systems are *FlowMark* [Ler97] and *WorkParty* [Rup97]. Most systems are designed for well-structured business processes and allow little modification in the task sequences [Ell95], [Nut96]. However, there are some advanced systems that allow the users to deviate from the predefined task sequences [Kar90], [Geo95], [Hsu96]. These systems allow the change of the control flow, but the software developers have to be responsible for the integrity of the data and their correct data flow. There are also advanced *WFM* systems that allow *late binding* [Han96], [Hag97] and adaptive *WFMs* that allow dynamic changes at runtime [Jab97]. Many of these latter systems are based on the object migration model concept.

7.2.2 Temporal Data Mining to Measure Operations Performance

Temporal mining of company operations data can be used to obtain temporal measures of its operations and allow the company to gain insights as to how its processes can be improved and even optimized. These processes can be the individual steps of an assembly line or an individual project. Some of these measures are as follows:

- *Task time*: The time it takes to complete an individual step.

- *Capacity*: Amount of work that can be processed in a unit of time. For example, number of computers assembled per day.

- *Bottleneck*: The step with the lowest capacity; in other words, the step with the most room for improvement.

- *Throughput time*: The time it takes for all steps to be completed, i.e., the time it takes to complete one unit of production.

- *Cycle time*: The time it takes for the completion of a process, for example, if there are 3 workers working on the assembly of a computer and it takes 3 hours for the assembly, the operator's cycle time is 1 hour.

7.2.3 Temporality in the Supply Chain Management

Supply chains today face the need to become more flexible and adaptable to meet the challenges of increasing competition and respond to events in a real-time manner. As discussed in [Ste00], even a small change in a supply chain can have very negative impact on the overall performance of the company.

In [Liu07], seven patterns of such events are described and then combined in a *Petri net* which can be used to perform simulation of various events. *Petri nets* are directed graphs consisting of two kinds of nodes called *places* (drawn as circles) and *transitions* (drawn as boxes). Petri nets can be used to model processes that involve coordination, perform cause-effect analysis, forecast the impact of events, monitor the supply, and intelligently react to events.

Unexpected events in a supply chain can be *inaccurate forecast, product out of stock, delayed shipment*, etc. These events can cause other events and there can be a chain reaction of events with very negative impacts for the business. A supply chain can be considered as a series of synchronous and asynchronous events, where these events can be either external, such as *shipment delayed*, or task-related, such as *product out of stock*. These events can be simple or aggregate consisting of multiple events.

In [Doz07], the role of information technology (IT) is discussed in optimizing the performance of the supply chain. IT generally helps a company to have a dynamic and agile supply chain that responds effectively to external forces. The novel idea in the article is that it proposes a solution to responding to external forces by the strategic placement of IT technology along the supply chain. The evolution of the product through the supply chain is considered similar to the flow of a fluid in a pipe. Value is added at each step of the supply chain by adding components, refining the product, etc.

This movement of a fluid through a pipe can be modeled with the physics of propagation of waves. As is shown in the article, the addition of information technology changes the *rate* of flow of the fluid through the pipe. Information technology can be positioned *strategically* to allow the chain to resonate with the propagating waves, such that the production time is optimized.

7.2.4 Temporal Data Mining for the Optimization of the Value Chain Management

Value chain in an organization is the combination of all business processes where value is being added. Figure 7.1 shows the value chain starting from procurement and ending with the purchase of a product.

An important question in value chain management that can be addressed by temporal mining of business operations data is which areas can be temporally optimized so as to increase business profitability and customer satisfaction. This is particularly important today that real-time management of the value chain is becoming prevalent, given the increasing global competition, as addressed in [Cha07b]. Below are given some of the areas that could benefit from temporal data mining:

- Speed of material movement.

- Temporal efficiency of sending orders to suppliers, receiving orders from suppliers.

- Efficiency of real-time inventory management to find the best order quantity.

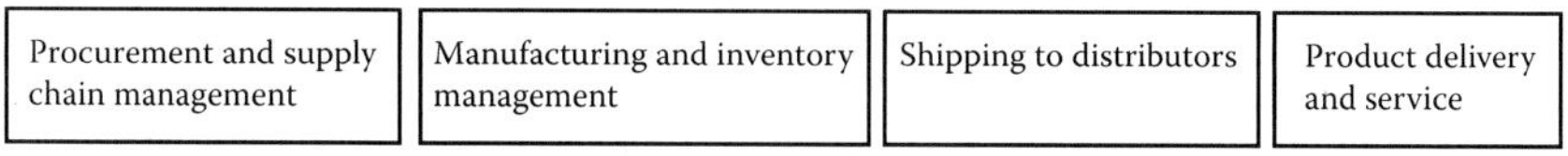

Procurement and supply chain management	Manufacturing and inventory management	Shipping to distributors	Product delivery and service

FIGURE 7.1 Value chain management.

- Speed of manufacturing process.

- Management of RFID data.

- Effectiveness of real-time quality management.

- Real-time tracking of customer movements using Web usage mining for marketing and advertising purposes. This is addressed in Chapter 8.

- Effectiveness of real-time monitoring at the point of sale.

- Temporal effectiveness of incorporating business intelligence (BI) methods in the value chain. Examples include incorporation of BI in supplier selection, capacity planning, route optimization.

This way temporal data mining can help reach larger business goals, such as price optimization and workforce allocation optimization.

7.2.5 Resource Demand Forecasting Using Sequence Clustering

In [Dat08], a novel algorithm for the clustering of projects that use similar resources over the project life cycle is described. The results of the algorithm are then combined with domain expertise to build a meaningful project taxonomy. The taxonomy is linked with project resource requirements and used for resource forecasting.

As discussed by the authors, there are three novel contributions of the article:

- Formulation of the project taxonomy building problem as a sequence clustering problem.

- A clustering algorithm that includes improvements of existing algorithms. The basis of the algorithm is a first-order *Hidden Markov model* to model the transitions of a project through the different phases during its lifecycle. The duration of each state is modeled using a gamma distribution. Because a model is used for the state duration, a more accurate name for the resulting model is *Hidden semi-Markov models (HsMMs)*.

- Description of the usage of the project taxonomy to forecast resource demand. The effectiveness of the approach is demonstrated on labor claim data from IBM.

7.2.6 A Temporal Model to Measure the Performance of an IT Project

[Gem07] proposes a temporal model for the measurement of information technology project performance, called *TMPP*. Project performance can be measured using two types of project outcomes: (1) *process outcomes*, which in turn can be measured by schedule variance and budget variance; and (2) *product outcomes*, which can be measured by percentage of expected quality delivered and percentage of expected benefits delivered.

As noted in the article, the novel contribution of this work is the modeling of the following:

- Temporal differences in risks. Specifically, risks are separated into earlier (*a priori*) risks and later (*emergent*) risks. *A priori* risks are further categorized into (1) *structural*, such as size, budget, and technical complexity, and (2) *knowledge-related*, such as competence of the manager and technical skills of team members. *Emergent* risks are further categorized into (1) *organizational support* risks, such as lack of sponsor, and (2) *volatility* risks, such as changing project targets.

- Influence of *a priori* risks on *emergent* risks and influence of the latter on performance.

- Interaction between project management practices and different risks.

Using data from 194 project managers, a *partial least square analysis* was performed, which resulted in several key findings, some of which are shown below:

- Project size and technical complexity have a statistically significant relationship with volatility risk.

- There is a statistically significant *inverse* relationship between volatility risk and organizational support.

- There is a statistically significant *inverse* relationship between project management practices and volatility risks.

The authors also note some limitations of the model, including the need for a more sophisticated temporal model, such as one that considers the point of time that a specific manager intervention occurs (closer to the project's deadline or at the beginning of the project).

7.2.7 Real-Time Business Analytics

In [Kel07], Jeff Kelly asked the opinions of experts about the trends in business intelligence (BI) for 2008. Demand for real-time business analytics was in the list, which is a result of the move that many companies make to an *agile business model*. Let us see the list of trends, where all trends are intertwined with the need for real time business analytics:

- Operational Business Intelligence will continue to grow. As a result, the importance of temporal mining of operational data will continue to grow as well. As also discussed in [Koh02], while in the past, success stories ended with a novel statistical analysis or analytical results, today there is a need to make BI results a *starting* point for business decisions.

- Business Intelligence will become pervasive. In other words, BI will become *verticalized* and permeate all levels of the value chain, as operations performance is directly linked to financial performance.

- Event driven analytics will play an increasingly important role in business. In other words, there will be increasing usage of tools that capture real-time business information, filter it, apply rules, and trigger, for example, alerts to either personnel or database updates. An example of real-time business information is data coming from performance management systems.

- Business Intelligence will become truly real time via the use of complex event processing (*CEP*), which involves the visualization and correlation of complex event streams.

7.2.8 Choreographing Web Services for Real-Time Data Mining

One of the goals of temporal data mining is to help the user find the right information at the right time. When the information to be mined is available over the Internet, *Web services* can be deployed for this purpose. *Web services* use the *service-oriented architecture (SOA) paradigm* and are a standardized way for machine–machine interoperability using standards, such as *XML, SOAP, and WSDL*.

For example in [Mit05], a Web services framework is described that allows dynamic integration of data mining Web services in real time for medical image processing. Which Web service gets invoked at the next step depends on the information that was mined in the previous step. This is achievable through Web services choreographing, which is possible through such standards as *BPEL (Business Process Execution Language)*. This language allows the definition of partners and roles involved in the execution of the process. It also allows the definition of activities involved in the process and which Web service is to be invoked depending on the output of the previous Web service.

7.2.9 Temporal Business Rules to Synthesize Composition of Web Services

Similarly to the work above, in [Yu07], Web services are composed on the fly using *BPEL*. The novel feature of this work is the definition of a pattern-based specification language called *PROPOLS*, which is used for verifying *BPEL* programs. The reason is that when services are dynamically composed, we must ensure that requirements are met. Verification by developing a model for the behavioral properties of the application and then applying the model can ensure this.

PROPOLS describes the sequence of the business rules that occur in the business domain. Each rule is translated into an automated finite state machine *(FSA)*. A set of business rules can be modeled using a composite *FSA*. Every business rule is mapped to one *FSA*.

PROPOLS has two types of patterns: *basic patterns* and *composite patterns*. The basic patterns are as follows:

- X precedes Y.

- X leads Y.

- X is absent.

- X is universal.

- X exists at most n times.

- X exists at least n times.

7.3 MISCELLANEOUS INDUSTRIAL APPLICATIONS

7.3.1 Temporal Management of RFID Data

RFID technology uses radio frequencies to automatically identify and track movable objects, and it has become very prevalent in warehouses for inventory management and product tracking. A *RFID* reading consists of the following values:

- The Electronic Product Code (*EPC*) of the reader

- The *EPC* of the product object

- The timestamp of the observation

A seminal article in the area is [Wan05]. Below we describe some key concepts from this article. *RFID* entities fall into the following categories:

- *Objects*: These include all objects tagged with an *EPC*, such as products, boxes, trucks.

- *Sensors/readers*: Each reader has unique *EPC* and uses RF signals to read the *EPC* codes of tagged objects.

- *Locations*: Locations can be symbolic, such as a warehouse, or actual, such as a geographic location.

- *Transactions*: An example is a checkout operation that involves the reading of the *EPC* codes of all products being purchased.

RFID data have the following characteristics:

- *Temporal*: Sensor observations of product objects are time-stamped.

- *Dynamic*: There are three things that change with time for the product objects: location (reader or object location change), containment relationships, and *EPC* transactions. Containment is modeled using a hierarchical relationship, for example, a toolbox containing a set of tools.

- *Sometimes inaccurate*: Although *RFID* technology is constantly improving, there are still inaccuracies in data, such as duplicate readings or missing readings. These need to be semantically filtered.

- *Large volume and streaming*: There is a tremendous volume of *RFID* generated in an automated way.

- *Integration*: *RFID* systems need to be integrated with product monitoring applications and generate business logic data.

In [Wan05], a framework is described for the management of *RFID* data, which has the following key characteristics:

- *Dynamic Relationship ER model*: This model maintains history of events and state changes.

- *Query Support*: The system supports *RFID* object tracking and monitoring.

- *Automatic Data Acquisition* and *Transformation*: The system implements automatic data configuration and filtering.

- *Adaptability* and *Portability*.

The database developed in [Wan05] has the following types of tables:

- *Static* tables: Sensor, object, location, transaction.

- *Dynamic relation* tables: Observation, containment, object location, containment location, transaction item.

An example of a query for tracking an object is

```
SELECT * FROM OBJECTLOCATION WHERE epc = 'X'.
```

7.3.2 Time Correlations of Data Streams and Their Effects on Business Impact Analysis

In [Say04], a novel method for extracting time correlations among multiple data streams is described. The method can be used in *Business Impact Analysis*, since it can be used to tell us the impact of a change in one numeric variable on a number of other numeric variables. For example, these time correlations can describe the following relationships:

- Web server performance and Web application performance.

- Web application performance and business activity execution time, such as total time for the completion and shipping of an order.

- Business activity execution time and daily revenue.

The method can be summarized as follows:

- The data are summarized at different granularities (weeks, days, etc.).

- Change points are detected using *CUSUM* [Pag54], which uses cumulative summing of points and comparison to a threshold to detect positive or negative changes.

- Generate correlation rules by merging and comparing time data streams.

Each time correlation rule includes information about the following:

- *Direction:* The direction is positive if increase in one variable results in increase in another variable. For example, increase in number of orders results in increase in revenue.

- *Sensitivity:* This measures the magnitude of change in one variable because of change in another variable. For example, increase in the number of orders has very little effect on the average time spent to prepare an order.

- *Time Delay:* This measures how long a change in one variable takes to affect another variable. For example, decreased server performance has an immediate effect on orders placed.

- *Confidence:* This shows how confident we are about the relationship between two variables.

7.3.3 Temporal Data Mining in a Large Utility Company

In [Heb01], applications of temporal data mining in *EDF*, the national French electric power company, are discussed. A time series, also called a *load curve*, is used to represent the electric power consumption of a customer. A load curve is not recorded for every customer, just for large customers, such as companies. It is typical for a load curve to record data every 10 minutes. Analysis of load curves can be used either to predict consumption or to make pricing decisions.

A typical data mining task is to cluster load curves that have similar shape characteristics. The method used is a *Self-Organizing Map* that produces clusters organized in a spatial way. The author also describes work in progress where, before clustering, the load curves are transformed into a symbolic form.

7.3.4 The Partition Decoupling Method for Time-Dependent Complex Data

In [Lei08], a novel computational tool is described that can reveal the underlying structure of complex time-dependent data, such as stock market data and the flow of oxygen in the brain.

To capture interdependencies between different components of a complex system, the tool looks for patterns in the "flow" between the different components. For example, in the case of the equities market, the authors created a map of different sectors and industries, and the level of capital flow between companies, industries, sectors, etc. The *Partition Decoupling* method finds regions where the flow is more than expected at random. This way, scale-dependent descriptions of the network structure of complex systems are created.

7.4 FINANCIAL DATA FORECASTING

7.4.1 A Model for Multirelational Data Mining on Demand Forecasting

Demand forecasting is critical for the success and growth of every business. However, it is a difficult task, since there are many parameters that can influence demand, such as economic conditions, competitor environment, and government regulations. An excellent treatise of the subject can be found in [Mou05]. Traditionally demand forecasting methods include the following:

- Simulation

- Asking expert advice

- Market surveys

- Decision support systems, rule-based systems that operate in an *if-then* mode

- Extrapolation: Using historical data to extrapolate the future

- Causal models, such as regression

- Neural networks

- Data mining

Two data mining algorithms are proposed in [Din04]: a pure classification model and a hybrid clustering classification model. Both of the models handle data in many dimensions. The data that were used consisted of the following dimensions: stores and their facts, products and their attributes, and shoppers and their demographic characteristics.

The pure classification model uses a *k–Nearest Neighbor* classification algorithm and operates on a data set that is created after joining the multiple relations of the store. In the hybrid clustering/classification algorithm, the data set is divided by the clustering algorithm into small groups based on their similarity. The clustering algorithm used is the *K–Mean Mode Clustering Technique* and the classifier used is the *k–Nearest Neighbor* algorithm. The article results show that the hybrid classification algorithm slightly outperforms the pure classification algorithm in terms of accuracy, but it is also significantly faster.

7.4.2 Simultaneous Prediction of Multiple Financial Time Series Using Supervised Learning and Chaos Theory

In Chapter 4, for illustration purposes, we utilized various linear techniques, such as regression and exponential smoothing, to predict values in the *S&P500* time series. The problem with applying these techniques to a financial series is that financial series are *nonstationary*, and therefore, linear techniques are in practice inadequate to perform reliable prediction. Nonlinear techniques are much more suitable for performing modeling and prediction of financial time series.

A popular way to do financial time series forecasting is to embed the time series in *a state space* (*chaos theory*) and combine it with supervised learning to do forecasting. The goal of chaos theory is to find the optimal embedding parameters for a time series. The process of embedding a time series in a state space can be described as follows: The time series is converted to a series of vectors: $x_d, x_{d+p}, x_{2d+p}, \ldots, x_{nd+p}$, where n is the embedding dimension and d is the separation.

Such an algorithm is described in [Edm94]. To perform the embedding, first the separation is estimated and then the embedding dimension is calculated using initially the *Auto Mutual information*, where the latter measures the amount of information common in two time series. Originally, the time series is embedded using a small dimension, for example, 2.

Then a search is performed for each embedded point to find its nearest neighbor and the Euclidean distance is measured. The embedding dimension is increased by 1, and then the new distance is calculated. If the new distance is significantly greater than the previous distance, the points are marked as *False Nearest Neighbors*. The process is repeated with ever increasing dimensions. When the number of *False Nearest Neighbors* gets close to zero, the process is terminated. At termination, the current dimension is returned as result.

7.4.3 Financial Forecasting through Evolutionary Algorithms and Neural Networks

In [Pav06], a time series forecasting method is presented that is used to forecast two daily foreign exchange spot rate time series. The proposed methodology utilizes principles from neural networks, clustering, and chaotic analysis and evolutionary computation. Clustering is initially used to identify neighborhoods in the time series and then neural networks are used to model the dynamics of each neighborhood.

In general, the steps of the algorithm are as follows:

- Divide the data into a training set and a testing set.

- Determine the minimum dimension for phase space reconstruction.

- Identify the clusters in the training set. This is achieved by using the *k-windows* clustering algorithm. This algorithm places a *d*-dimensional window that contains all patterns that belong to a single cluster. Originally, the windows are moved without altering their size and their centers are placed at the mean of the patterns that are currently included in the window. In a second stage, the windows are enlarged to allow inclusion of as many patterns as possible that belong to the cluster.

- For each cluster in the training set, train a different feedforward neural network (*FNN*), using only the patterns in that cluster. *FNNs* are parallel computational models that mimic the human brain and gain knowledge from their environment. Each *FNN* is trained with the help of a *differential evolution* algorithm and the *particle swarm optimization* method.

- Assign the patterns of the test set to the created clusters.

- Use the *FNNs* to obtain the forecasts.

The developed algorithm was applied to two daily foreign exchange time series: that of the Japanese yen against the United States dollar and that of the United States dollar against the British pound.

In [Li05], an *ARMA* model is combined with a *GRNN neural network*. The *GRNN* can be considered as an alternative to the back propagation training algorithm for a feedforward neural network. First, *ARMA* modeling is used to construct a model of the financial time series. Then this model is used as part of the training data for the neural network.

In [Ma07], genetic programming (*GP*) is used to forecast the *S&P500* time series. The results show that *GP* performs well in forecasting a financial time series. Below are some characteristic results, where the Mean Absolute Error was used to estimate forecasting accuracy. For 25 *GP* generations, the minimum forecasting accuracy was 71.23%, while the maximum forecasting accuracy was 74.4%. For 50 GP generations, the minimum forecasting accuracy was 73.4%, while the maximum forecasting accuracy was 76.7%.

In [Pan97], the authors discuss a method for the prediction of financial time series using neural and fuzzy tools. They applied their work to the *S&P500* series with promising results. They point to the work of Lapedes and Farber [Lap87] as an example that neural networks can successfully be used to predict financial time series. In the article, an important point is raised that quite often we are not interested in predicting just a single point in the series, instead we are interested in characterizing the future behavior of the series in terms of three concepts: the *trend, horizon*, and *confidence*. The *confidence* is a way to express the predictability of the series. *Horizon* and *confidence* are inversely related, since the longer we go out in time, in other words the greater the horizon, the lesser the confidence of the prediction. The training set for the neural networks is partitioned into three subgroups, relating to the volatility (*volatility up, down, same*), and this way three neural networks are trained. The training of the neural networks was done for 15,000 epochs of the training data.

As discussed in [Han97], state tax revenue forecasts are important for the state's budgeting process. Neural networks and traditional time series forecasting methods, such as exponential smoothing and *ARIMA* are used in Utah's budgeting process. This paper describes the use of a neural network to forecast revenues for the Office of the Legislative Analyst in the State of Utah. The type of neural network used is a *Time Delay Neural Network (TDNN)*. The *TDNN* is a multilayer feedforward network with output and hidden neurons replicated in the model. This is how the *TDNN* is constructed: Each connection is set to an interval back in time. The first connection is set at the current time, the second connection is set to one period past, etc. In conclusion, the authors note that neural networks can produce significant insights in business cycles.

In [Sit00], a neural network that models the long- and short-term behavior of the *S&P 500* time series is presented. The authors describe the fluctuations in stock market data as the result of three types of variations:

- Trend

- Periodic variations

- Day-to-day variations

They also describe how the *random walk model* can be suitable for explaining the fluctuations in the stock market data. Because stock market data have an exponential form and neural nets are not that suitable for modeling exponential data, the time series is considered as the sum of two components: a long-term f component and a residual-term r component. The long-term f is modeled using a *least-square mean fitted* function, while a *time delay neural network* is used to model the residual as a random walk. The *S&P* time series data for 22 years was chosen to test the model.

As the authors in [Cro05] note, every decision in business has a utility in either *increasing profit* or *decreasing cost*. Therefore, the utility of a forecast must be evaluated according to its ability to enhance a business decision. The authors discuss the concept of *cost-sensitive learning*, where there is a cost assigned to every false and true prediction. These costs are used to drive the selection and parameterization of a classifier for a particular application, such as the prediction of demand. The paper discusses symmetric and asymmetric error functions as ways to measure the performance of neural network training and the authors conclude that asymmetric cost functions may allow robust minimization of relevant costs.

In [Gil01], the authors convert the time series in a symbolic representation with a *self-organizing map* and then use grammatical inference with *recurrent neural networks (RNNs)*. The authors chose to use *RNNs*, because they are biased toward learning patterns that have a temporal order and are less likely to learn random events. However, *RNNs* are sensitive to noise, with a tendency to ignore long-term dependencies and favor short-term dependencies. Financial times tend to be nonstationary with a significant amount of noise. The problem is addressed by converting the time series into a sequence of symbols using a *self-organizing map*. This quantization leads to a reduction of noise.

7.4.4 Independent Component Analysis for Financial Time Series

In [Oja98], the authors discuss independent component analysis (*ICA*) for financial time series for two different purposes: (1) decomposition of parallel financial time series of weekly sales into basic factors and (2) time series prediction by linear combinations of *ICA* component predictions. The starting point is to convert a time series into a signal vector $x(t) = (x_1(t), x_2(t), \ldots x_m(t))^T$. Then the *ICA* problem can be approached as

$$Y = Wx \qquad (7.8)$$

where the goal is to find the matrix W that makes the elements of Y statistically independent. The ith row of W, W_i, can be found by maximizing the kurtosis of Y_i. Another way of describing the same problem is

$$x_i(t) = \sum a_{i,j} s_j(t) \qquad (7.9)$$

where $x_i(t)$ is a time series, $a_{i,j}$ are known as the mixing coefficients, and $s_j(t)$ are the underlying independent components. Then, the prediction is performed in the *ICA* space. The FastICA algorithm was used that simultaneously estimates the mixing coefficients and the underlying independent components. Applying the algorithm on 10 foreign exchange rate time series showed that *ICA* prediction outperformed direct prediction.

7.4.5 Subsequence Matching of Financial Streams

[Wu04] discusses a novel method for subsequence matching over massive financial data streams. As discussed in the article, the movement of the stock market can be approximated using a piecewise linear representation that has the *zigzag* shape. The contributions of [Wu04] are the following:

- A novel segmentation and pruning algorithm that takes as input raw financial data and produces a piecewise linear representation, where the end points are in a zigzag shape. The segmentation is not performed directly over the price data; instead it is performed over the *Bollinger Band Percent*. *Bollinger Bands* are financial indicators which are represented by curves drawn above and below a moving average of the financial time series. The three *Bollinger Bands* are defined as follows:

$$middle\_band = p\text{-}period\_moving\_average \qquad (7.10)$$

$$upper\_band = middle\_band + 2p\text{-}period\_standard\_deviation \ (7.11)$$

$$lower\_band = middle\_band - 2p\text{-}period\_standard\_deviation \ (7.12)$$

The actual metric used in the article is the *Bollinger Band percent*, which is defined as

$$\%b = \frac{close\_price - lower\_band}{upper\_band - lower\_band} \qquad (7.13)$$

- A novel similarity measure is introduced that utilizes a metric distance function based on the permutation of the subsequence. The purpose of the permutation is to ensure that two subsequences have the same end points.

- Event-driven subsequence matching is performed, where the end points of the piecewise linear representation drive the subsequence matching. An event is a new end point.

- A new trend definition is used that takes four values: UP, DOWN, NOTREND, UNDEFINED.

7.4.6 Detection of Outliers in Financial Data

In [Wan06], a framework is presented to identify outliers in financial time series data. The outliers are identified in high dimensional space. The ultimate goal of the framework is to enable portfolio diversification by identifying outliers. *Outlier trading* is quite often used in financial markets and it is a way to maximize financial return with the tradeoff of higher risk. Traditionally, the correlation coefficient between two financial time series can be used to determine their similarity.

The proposed framework consists of a novel horizontal and vertical dimensionality reduction algorithm. In the vertical dimension, an attribute selection algorithm is introduced that reduces the vertical dimension by identifying the most significant attributes. In the horizontal dimension, a piecewise linear representation based on segmentation and pruning is used to reduce the horizontal points of the selected attributes. Finally, customized hierarchical agglomerative clustering is used to cluster time series based on multidimensional information.

7.4.7 Stock Portfolio Diversification Using the Fractal Dimension

In [Kan04], a method for stock portfolio diversification is proposed, which is based on the measurement of the *fractal dimension*. The *fractal dimension* is a measure of how a *fractal* (a self-similar structure) fills the space in which it is embedded. The most common way to measure the fractal dimension is the *box counting* method. In this method, an evenly spaced grid is superimposed on the fractal structure, and the number of cells that are encountered by the structure is counted. Then, a plot is created where the x-axis is the increasing size of the grid cells and the y-axis is the number of cells encountered by the fractal structure for each increasing cell size. The fractal dimension is then computed from the slope of the plot.

The fractal dimension computed in [Kan04], using box-counting, is the *correlation fractal dimension*. This fractal dimension is given by this formula:

$$D = \frac{log \sum_i p_i^2}{log\, r} \tag{7.14}$$

where r is the cell size and p_i is the number of cells encountered by the structure. Reduction of the fractal dimension makes the algorithm eliminate stocks that are highly correlated and adds new uncorrelated stocks to the portfolio.

7.5 ADDITIONAL BIBLIOGRAPHY

There are quite a few books devoted to the subject of financial forecasting. Two of them dedicated to the subject of financial *time series* forecasting are [Mak97] and [Fra98]. A book dedicated to the *analysis* of financial time series is [Tsa05]. In [Tak98], multiple time series analysis is used to analyze competitive marketing behavior. In [Cha07a], a brief overview of time series analysis techniques is presented and in [Mil06] nonlinear time series analysis of business cycles is discussed in detail.

In [Kim03], the authors use support vector machines *(SVMs)* for financial time series forecasting. The authors make some very interesting comparisons of *SVMs* with neural networks: (1) *SVMs* may provide a global optimum solution and they are unlikely to have a problem with over-fitting. (2) Neural networks seek to minimize the misclassification error, and they may be stuck in a local solution.

In [Nay07], the authors use *temporal pattern matching* for the prediction of stock prices. Their technique forms a cluster around the query sequence. The cluster is formed using the *best fit error margin* idea. In the results section, they show that of the 800 predictions, 400 had an error rate of 1.5% or less. In [Chu02], the authors present a segmentation method of stock data utilizing evolutionary concepts. This method can be combined with a pattern matching algorithm that utilizes perceptually important points.

In [Bou08], an *ensemble of classifiers* is used to perform time series prediction. Using an ensemble of classifiers usually achieves better results than a single classifier. The base learner is a radial basis function neural network (RBFN). RBFNs are known for their predictive power; however, their performance can vary and for this reason it is a good idea to combine them with other learners.

Li and Han in [Li07] propose an algorithm to find the top k subspace anomalies in multidimensional time series data which can be used for market analysis. Market segment data consist of many dimensions such as age, gender, and income. Two examples of anomalies can be found in the trend and magnitude of the data. They applied their algorithm to both real and synthetic data, which demonstrated the effectiveness of their solution. In [Kov08], the authors discuss two different approaches to predict customer offer prices that can potentially result in orders in the supply chain management domain. The proposed algorithms utilize genetic algorithms and neural network techniques.

The authors in [Chi07] address the problem of discovering linkages between financial time series. In other words, the problem is finding a relationship, most likely nonlinear, between some predictor time series, such as interest rates and currency exchange rates, and a target series, such as the stock price of a company. They propose an extension of an existing genetic algorithm inspired by organizational theory, *DSMDGA (Dependency Structure Matrix Driven Genetic Algorithm)* [Yu04], to learn linkage relationships from the observed real values.

In [Mil08], an *OWL*-based semantic language is described, called *TOWL*, which allows the representation of time, change, and state transitions. The base class of *TOWL* is *TemporalEntity* with two subclasses: *Interval* and *Instant*. The language is then used to create a Web-based application that contains financial data and stock recommendations. The stock recommendations are in the form of *(buy/hold/sell)*.

[Dec07] discusses progress in a European *R&D* project, *MUSING*, that deals with the development of semantic resources, including a *temporal ontology*, for Business Intelligence applications. *MUSING* provides tools for semantic *BI* in three domains:

1. *Financial Risk Management*: This provides a framework to build a profile of a company from public and private data that include scores, indices, and so on.

2. *Internationalization:* This deals with the creation of a platform that allows a company to support foreign market access.

3. *IT Operational Risk*: This deals with risk assessment originating in IT enterprise systems.

[The02] discusses the *TEMPORA* project, which consists of three models that capture the following information in the business domain:

1. The *Entity Relationship Model* deals with the structural aspects. It contains the concepts of entity, relationship, and attribute.

2. The *Process Interaction Diagram* deals with the behavioral aspects.

3. The *Conceptual Rule Language (CRL)* applies to the modeling of the business rules, which deal with either the structural or behavioral aspects.

REFERENCES

[Ahm08] Ahmad, I. and A.R. Chowdhury, Electronic Customer Relationship Management (eCRM), Lulea University of Technology, Sweden, Master Thesis, Continuation Courses, Department of Business Administration and Social Sciences, 2008.

[Anc01] Ancona, D.G., G.A. Okhuysen, and L.A. Perlow, Taking Time to Integrate Temporal Research, *Academy of Management Review*, vol. 26, pp. 512–529, October 2001.

[Bir92] Bird, J.B., The Operations of Intentions in Time: The Emergence of the New Venture, *Entrepreneurship: Theory and Practice*, vol. 17, 1992.

[Bou08] Bouchachia, A. and S. Bouchachia, Ensemble Learning for Time Series Prediction, *The 1st International Workshop on Nonlinear Dynamic Systems and Synchronization*, 2008.

[Cha07a] Chakraborti, A., M. Patriarca, and M.S. Santhanam, Financial Time Series: A Brief Overview, *Proceedings of the International Workshop Econophys-Kolkata*, 2007.

[Cha07b] Chan, J.O., Real-Time Value Chain Management, *Communications of the IIMA*, August 2007.

[Chi07] Chiotis, T. and C.D. Clack, Non-linearity Linkage Detection for Financial Time Series Analysis, *Proceedings of GECCO Conference*, pp. 1179–1186, 2007.

[Chu02] Chung, F.L. et al., Evolutionary Time Series Segmentation for Stock Data Mining, *Proceedings of the IEEE Conference on Data Mining*, p. 83, 2002.

[Cro05] Crone, S., S. Lessmann, and R. Stahlbock, Utility Based Data Mining for Time Series Analysis, Cost Sensitive Learning for Neural Network Predictors, *Conference on Knowledge Discovery in Data*, pp. 59–68, 2005.

[Dad97] Dadam, P., M. Reichert, and K. Kuhn, Clinical Workflows—The Killer Application for Process-Oriented Information Systems, *Ulmer-Informatik Berichte, Nr. 97-16, Fakultät for Informatik, Universität Ulm*, November 1997.

[Dat08] Datta, R., J. Hu, and B. Ray, Sequence Mining for Business Analytics: Building Project Taxonomies for Resource Demand Forecasting, *Applications of Data Mining in E-Business and Finance*, IOS Press, vol. 177, pp. 133–141, 2008.

[Dec07] Declerck, T. et al., Integration of Semantic Resources and Tools for Business Intelligence, In International Workshop on Semantic-Based Software Development, Montreal, Canada, 2007. Also in: http://www.coli.uni=saarland.de/publications/show.php?author=Thierry_Declerck.

[Din04] Ding Bhavin, Q., Q. Ding, and B. Parikh, A Model for Multi-relational Data Mining on Demand Forecasting, http://citeseerx.ist.psu.edu/viewdoc/summary?doi=10.1.1.11.1705, 2004. http://cs.hbg.psu.edu/~ding/publications/249-R.pdf

[Doz07] Dozier, K. and D. Chang, The Impact of Information Technology on the Temporal Optimization of Supply Chain Performance, *Proceedings of the 40th Hawaii International Conference on System Sciences*, pp. 57–57, 2007.

[Edm94] Edmonds, A., D. Burkhardt, and O. Adjei, Simultaneous Prediction of Multiple Financial Time Series Using Supervised Learning and Clustering, *IEEE World Congress on Computational Intelligence*, vol. 5, pp. 3158–3163, 1994.

[Ell95] Ellis, C. A., Keddara, K., and Rozenberg, G. , Dynamic Change within Workflow Systems, *Proceedings of COOCS*, Milipitas, CA, pp. 10–21, 1995.

[Fei02] Feinberg, R.A. et al., The State of Electronic Customer Relationship Management in Retailing, *International Journal of Retail and Distribution Management*, vol. 30, no. 10, pp. 470–481, 2002.

[Fra98] Franses, P.H., *Time Series Models for Business and Economic Forecasting*, Cambridge University Press, 1998.

[Gem07] Gemino, A., B. Reich, and C. Sauer, A Temporal Model of Information Technology Project Performance, *Journal of Management Information Systems*, vol. 24, no. 3, pp. 9–44, 2007.

[Geo95] Georgakapoulos, D., M. Hornick, and A. Sheth, An Overview of Workflow Management, *Distributed and Parallel Databases*, vol. 3, pp. 119–153, 1995.

[Gil01] Giles, C.L., S. Lawrence, and A.C. Tsoi, Noisy Time Series Prediction Using a Recurrent Neural Network and Grammatical Inference, Machine Learning, vol. 44, no. 1–2, pp. 161–183, 2001.

[Hag97] Hagemeyer, J. et al., Flexibility in Workflow Management Systems, in Likowski, R. et al. (eds), *Usability Engineering*, Teubner, pp. 179–190, 1997.

[Han96] Han, Y. et al., Management of Workflow Resources to Support Runtime Adaptability and System Evolution, *Proceedings of the Int. Conf. on Practical Aspects of Knowledge Management, Workshop on Adaptive Workflows*, Basel, Switzerland, 1996.

[Han97] Hansen, J.V. and R.D. Nelson, Neural Networks and Traditional Time Series Methods: A Synergistic Combination in State Economic Forecasts, *IEEE Transactions on Neural Networks*, vol. 8, no. 4, pp. 863–873, July 1997.

[Heb01] Hebrail, G., Practical Data Mining in a Large Utility Company, *Qüestiió*, vol. 25, no. 3, pp. 509–520, 2001.

[Hel04] Helfat, C.E. and K.M. Eisenhardt, Inter-temporal Economies of Scope, Organizational Modularity and the Dynamics of Diversification, *Strategic Management of Journal*, vol. 25, pp. 1217–1232, 2004.

[Hsu96] Hsu, M. and C. Kleissner, Object Flow: Towards a Process Management Infrastructure, *Distributed and Parallel Databases*, vol. 4, pp. 169–194, 1996.

[Jab97] Jablonski, S., K. Stein, and M. Teschke, Experiences in Workflow Management for Scientific Computing, *Proceedings of the 8th International Workshop on Database and Expert Systems*, Toulouse, pp. 56–61, 1997.

[Kan04] Kantardzic, M., P. Sadeghian, and C. Shen, The Time Diversification Monitoring of a Stock Portfolio: An Approach Based on the Fractal Dimension, *ACM Symposium on Applied Computing*, pp. 637–641, 2004.

[Kar90] Karbe, B., N. Ramsperger, and P. Weiss, Support of Cooperative Work by Electronic Circulation Folders, *SIGOIS Bulletin*, vol. 11, no. 2/3, pp. 109–117, 1990.

[Kel07] Kelly, J., Business Intelligence Market Trends and Expert Forecasts for 2008, http://searchdatamanagement.techtarget.com/news/article/ 0,289142, sid91_gci1286652,00.html, December 2007.

[Kha05] Khalifa, M. and N. Shen, Effects of Electronic Customer Relationship Management on Customer Satisfaction: A Temporal Model, *Proceedings of the 38th Hawaii International Conference on System Sciences*, 2005.

[Kim03] Kim, K.J., Financial Time Series Forecasting Using Support Vector Machines, *Neurocomputing*, vol. 55, pp. 307–319, 2003.

[Kim96] Kim, D. and B. Kogut, Technological Platforms and Diversification, *Organization Science*, vol. 7, no. 3, pp. 282–301, 1996.

[Koh02] Kohavi, R., N. Rothleder, and E. Simoudis, Emerging Trends in Business Analytics, *Communications of the ACM*, vol. 45, no. 8, pp. 45–48, 2002.

[Kov08] Kovalchuk, Y. and M. Fasli, Adaptive Strategies for Predicting Bidding Prices in Supply Chain Management, *Proceedings of 10th International Conference on Electronic Commerce*, 2008.

[Lap87] Lapedes, A. and R. Farber, Nonlinear Signal Processing Using Neural Networks: Prediction and System Modeling, Los Alamos National Laboratory, Technical Report, LA-UR-87-2662, June 1987.

[Lei08] Leibon, G. et al., Topological Structures in the Equities Market Network, *Proceedings of the National Academy of Sciences*, vol. 105, no. 52, 2008.

[Ler97] Leymann, F. and D. Roller, Workflow-Based Applications, *IBM Systems Journal*, vol. 36, no. 1, pp. 102–123, 1997.

[Li05] Li, W.M., J.W. Liu, and J.J. Le, The Financial Time Series Forecasting Based on Proposed ARMA-GRNN Model, *Proceedings of the 4th International Conference on Machine Learning and Cybernetics, Guangzhou,* 18–21 August 2005.

[Li07] Li, X. and J. Han, Mining Approximate Top-K Subspace Anomalies in Multi-dimensional Time Series Data, *Proceedings of the 33rd International Conference on Very Large Databases,* pp. 447–458, 2007.

[Lia05] Liao, J., H. Welsch, and W.L. Tan, Venture Gestation Paths of Nascent Entrepreneurs: Exploring the Temporal Patterns, *The Journal of High Technology Management Research,* vol. 16, no. 1, pp. 1–22, 2005.

[Lic05] Lichtenstein, B.B., et al., Exploring the Temporal Dynamics of Organizational Emergence, Babson Conference, 2005.

[Li07] Li, X. and J. Han, Mining Approximate Top-K Subspace Anomalies in Multi-dimensional Time Series Data, *Proceedings of the 33rd International Conference on Very Large Databases,* pp. 447–458, 2007.

[Liu07] Liu, E., A. Khumar, and W. van der Aalst, A Formal Modeling Approach for Supply Chain Management, *Decision Support Systems,* vol. 43, no. 3, pp. 761–778, 2007.

[Ma07] Ma, I., T. Wong, and T. Sankar, Volatility Forecasting Using Time Series Data Mining and Evolutionary Computation Techniques, *Proceedings of the 9th Annual Conference on Genetic and Evolutionary Computation,* London, England, pp. 22–62, 2007.

[Mak97] Makridakis, S.G., S.C. Wheelwright, and R. J. Hyndman, *Forecasting: Methods and Applications,* Wiley, 1997.

[Mil06] Milas, C., P. Rothman, and D. van Dick, *Nonlinear Time Series Analysis of Business Cycles,* vol. 276, Elsevier Science, 2006.

[Mil08] Milea, V., F. Fransicar, and U. Kaymak, Knowledge Engineering in a Temporal Semantic Web Context, *8th International Conference on Web Engineering,* pp. 67–74, 2008.

[Mit05] Mitsa, T. and P. Joshi, An Evolvable, Composable Framework for Rapid Application and Development and Dynamic Integration of Medical Image Processing Web Services, *Proceedings of the IASTED Intl. Conference on Web Technologies, Applications, and Services,* (CD), pp. 7–12, Calgary, Canada, 2005.

[Mou05] Armstrong, J. S. and K.C. Green, Demand Forecasting: Evidence Based Methods, http://marketing.wharton.upenn.edu/ideas/pdf/Armstrong/Demand Forcasting.pdf, 2005.

[Nay07] Nayak, R. and P. te Braak, Pattern Matching for the Prediction of Stock Prices, *Proceedings of 2nd International Workshop on Integrating Artificial Intelligence and Data Mining,* pp. 95–103, 2007.

[Nut96] Nutt, G., The Evolution towards Flexible Workflow Systems, *Distributed Systems Engineering,* vol. 3, no. 4, pp. 276–294, 1996.

[Oja98] Oja, E., K. Kiviluoto, and S. Malaroiu, Independent Component Analysis for Financial Time Series, *Proceedings of ICONIP,* Japan, Oct. 1998.

[Pag54] Page, E.S., Continuous Inspection Schemes, Biometrika, 41, pp. 100–114, 1954.

[Pan97] Pantazopoulos, K.N., L.H. Tsoukalas, and E.N. Houstis, Neurofuzzy Characterization of Financial Time Series in an Anticipatory Framework, *Computational Intelligence for Financial Engineering*, pp. 50–56, 23–25 March 1997.

[Pav06] Pavlidis, N.G., D.K. Tasoulis, and M.N. Vrahatis, Financial Forecasting Through Unsupervised Clustering and Evolutionary Trained Neural Networks, *Proceedings of the Congress on Evolutionary Computation*, 2006.

[Per02] Perlow, L. and G. Okhuysen, The Speed Trap: Exploring the Relationship between Decision Making and Temporal Context, *The Academy of Management Journal*, vol. 45, no. 5, pp. 931–955, 2002.

[Phe01] Phelps, R.G., *Customer Relationship Management*, Thorogood Publishing Ltd., 2001.

[Por79] Porter, M.E., How Competitive Forces Shape Strategy, *Harvard Business Review*, March/April 1979.

[Rig02] Rigby, D. K., F. F. Reichheld, and P. Schefter, Avoid the Four Perils of CRM, *Harvard Business Review*, February 2002.

[Ros03] Rosset, S. et al., Customer Lifetime Value Models for Decision Support, *Data Mining and Knowledge Discovery*, vol. 7, pp. 321–339, 2003.

[Ros05] Ross, D.F., E-CRM from a Supply Chain Management Perspective, *Journal of Information Systems Management*, vol. 22, no. 1, pp. 37–44, 2005.

[Rup97] Rupietta, W., Business Processes and Workflow Management, in Bernus, P., K. Mertins, and G. Schmidt, (eds), *Handbook on Architectures of Informations Systems*, 1997.

[Say04] Sayal, M., Detecting Time Correlations in Time Series Data Streams, Hewlett-Packard Company, Citeseer, 2004.

[Sit00] Sitte, R. and J. Sitte, Analysis of the Predictive Ability of Time Delay Neural Networks Applied to the *S&P500* Time Series, *IEEE Transactions on Systems, Man, and Cybernetics, Part C: Applications and Reviews*, vol. 30, no. 4, pp. 568–572, November 2000.

[Sta88] Stalk, G., Jr., Time—The Next Source of Competitive Advantage, *Harvard Business Review*, pp. 41–51, July–August, 1988.

[Sta98] Stalk, G., Jr., The Time-Paradigm, Forbes, vol. 162, no. 12, pp. 213–214, November 1998.

[Ste00] Sterman, J. D., *Business Dynamics: Systems Thinking and Modeling for a Complex World*, Irwin McGraw-Hill, 2000.

[Tak98] Takada, H. and F.M. Bass, Multiple Time Series Analysis of Competitive Marketing Behavior, *Journal of Business Research*, vol. 43, no. 2, pp. 97–107, Elsevier Science, 1998.

[The02] Theodoulidis, B., P. Alexakis, and P. Loucopoulos, Verification and Validation of Temporal Business Rules, *DAISD*, pp. 1–15, 1992.

[Tim09] Time-Based Competition. *Reference for Business: Encyclopedia of Business*, 2nd ed., http://www.referenceforbusiness.com/management/Str-Ti/Time-Based-Competition.html, 2009.

[Tsa05] Tsay, R. S., *Analysis of Financial Time Series*, Wiley-InterScience, 2005.

[Vic95] Vickery, S. et al., Time-Based Competition in the Furniture Industry, *Production and Inventory Management Journal*, 4th quarter, pp. 14–21, 1995.

[Wan05] Wang, F. and P. Liu, Temporal Management of RFID Data, *VLDB Conference*, pp. 1128–1139, 2005.

[Wan06] Wang, D. et al., Hierarchical Agglomerative Clustering Based T-Outlier Detection, *Proceedings of the 6th IEEE International Conference on Data Mining*, pp. 731–738, 2006.

[Wan07] Wang, F. and M. Head, How Can the Web Help Build Customer Relationships? An Empirical Study on E-Tailing, *Information and Management*, vol. 44, no. 2, pp. 115–129, March 2007.

[Wu04] Wu, H., B. Salzberg, and D. Zhang, Online Event-Driven Subsequence Matching over Financial Data Streams, *Proceedings of the SIGMOD Conference*, pp. 23–34, June 13–18, 2004.

[Yu04] Yu, T.L. et al., A Genetic Algorithm Design Inspired by Organizational Theory: Pilot Study of a Dependency Structure Matrix Driven Genetic Algorithm, *Proceedings of Genetic and Evolutionary Computation Conference*, LNCS 2724:1620–1621, Springer Verlag 2004.

[Yu07] Yu, J. et al., Using Temporal Business Rules to Synthesize Service Composition Models, *Proceedings of the 1st International Workshop on Architectures, Concepts, and Technologies for Service-oriented Computing*, pp. 85–94, 2007.

CHAPTER 8

Web Usage Mining

8.1 GENERAL CONCEPTS

Temporal data mining, in the form of *Web usage mining*, is an important part of Web application development, in the sense that it allows Web developers to understand how users navigate and use the site. Web usage mining refers to the pages accessed by a user, time and duration of visits, as well as the user profile.

This way the Web developers can do the following:

- Improve the organization of the site.

- Develop adaptive and personalized sites. An example is Amazon™, which suggests books that match the user's interests based on his/her previous selections.

- Give to marketing personnel information that can allow the development of intelligent marketing techniques and improvement of customer relationships.

- Monitor network traffic.

Where do Web usage mining data come from? They come from Web server logs, proxy servers, and browsers. There are many ways to track site visits. Some are the following:

1. *Packet sniffing.* In this method, data are extracted directly from TCP/IP packets. When someone visits a Web site, he leaves his IP

address, time and date of access, the referring page, the agent, and so on.

2. *Cookies.* Here, user identification data are embedded directly in individual browsers. This is a more difficult type of Web usage mining because it requires permission by the users to allow cookies to track their usage.

Web usage mining consists of the following phases:

- Data preprocessing

- Pattern discovery

- Pattern analysis

Let us examine each one of these phases.

8.1.1 Preprocessing

Usage preprocessing is a difficult task because of the many stages data pass through to reach the Web server and the large amount of data needed to track users accurately. A list of problems encountered in preprocessing is described in [Sri00] and in [Ber02]. Here are some of these problems:

- Multiple IP address/single user. Tracking of a visitor becomes difficult when he/she accesses a site from different machines. The problem is that the user will have different IP addresses from session to session.

- Multiple agent/single user. A user who uses more than one browser even on the same computer will appear as different users.

- Overall, difficulty to obtain reliable data mainly happens because of proxy servers and anonymizers, dynamic IP addresses, and missing references due to caching. The problem of missing references due to caching can be dealt with by using temporal information and referrer logs. Proxy servers can lead to the mistake of having several users being identified as one. Techniques to overcome this problem include temporal information, IP address, browser agent, and so on.

Generally, as described in [Ber02], preprocessing consists of the following steps:

- *Data cleaning*, for example, removing irrelevant fields.

- *Data integration.* This is a more difficult task and it consists of many steps such as data synchronization from different Web server logs, integration of meta-data, and registration information about the user.

- *Data transformation.* This includes user identification, sessionalization, and page view identification, where the last is a set of objects and pages that contribute to a single view on the browser. *Sessionalization* refers to identifying individual sessions. A user session consists of the set of pages seen by the user during his visit to the site. One can use some heuristics to identify sessions. These are total session duration (time out intervals of 15 and 30 minutes are usually used) and page stay times, which must not exceed a maximum. In addition, a page must be reached from another page in the same session.

8.1.2 Pattern Discovery and Analysis in Web Usage

Discovering patterns in Web usage has many similarities with traditional association rule discovery and sequence mining, with the ultimate goal of discovering frequent itemsets. It is important to note that *order matters.* This means that the Web developer is not interested just in which pages the user accessed, but also the order in which he/she accessed them. There are different ways to analyze Web usage:

- *Statistical analysis*: Such an analysis can yield the following information: Most frequently accessed Web page; number of accessed Web pages; histogram of access to the different pages of a site for a specific time period, such as a week; average viewing time of a page; median viewing time of a page; maximum viewing time of a page; average length of a path to a site; and origin of access to the different pages of a site.

- *Path analysis*: Graph theory is quite often used as a tool in path analysis. Web pages can be modeled as nodes and hypertext links as edges. Path analysis yields the most frequent paths visited by the users [Liu04].

- *Association rule discovery*: This type of analysis discovers the pages that are accessed together in a user session and the support exceeds a certain threshold.

- *Sequential pattern discovery:* This type of analysis discovers patterns that appear in a sequence of site visits by a user. These can be useful in marketing and advertising applications.

- *Classification:* Classification groups users based on their profiles in certain categories, where each category is characterized by pre-defined attributes.

- *Clustering:* Clustering is used when the categories are not known. In this case, users are grouped based on the similarity of their browsing patterns.

8.1.3 Business Applications of Web Usage Mining

Web usage mining aims at the development of association rules and discovery of clusters for three different purposes:

1. Identifying groups of users with similar browsing patterns and sequentially accessed pages for marketing purposes. For example, assume there are 10 hotels on an island. However, only the visitors of the sites of two of these hotels often visit the site of a scuba diving school on that island. Therefore, it makes sense for the marketing person of the scuba diving school to put an advertisement on the sites of only those two hotels.

2. Identifying associated Web pages with the purpose of reorganizing the Web site. For example, the Web site administrator of a college has found that 80% of the students that visit the business school pages of the college also visit the athletic facilities pages in the same page. Therefore, he decides to put a hyperlink to the athletic facilities page in the business school page to attract more business school applicants. On the other hand, the prospective engineering students, besides the engineering department's site, also visit the library site. Therefore, he decides to put a direct hyperlink in the engineering department's site to the library site.

3. Identifying dependent Web sites for the purpose of better organizing business functions performed on the Internet. For example, what

are the pages that one will view and in what order for the purpose of buying a laptop online?

8.2 WEB USAGE MINING ALGORITHMS

8.2.1 Mining Web Usage Patterns

Mining Web usage patterns is becoming a key piece of business intelligence today. An important paper in the area is [Abr03], which describes an intelligent technique for Web usage mining, known as *i-Miner*. This algorithm uses the *c-means fuzzy clustering algorithm* [Bez81] to generate the clusters of usage. In this algorithm, the data are divided in clusters and a cluster center is computed, using a cost function of dissimilarity.

To improve the algorithm and optimally cluster usage interests, the authors use an evolutionary process to decide the number of clusters and also the center of the clusters. The algorithm was applied on data generated in a Web log during a week. Not all log data were analyzed. The data that were analyzed were the following:

- Domain byte requests

- Hourly page requests

- Daily page requests

These data were used as the focus of a cluster analysis. Other techniques were also applied on the data such as *Self Organizing Maps* (*SOMs*). In addition, *Artificial Neural Networks* and *Linear Genetic Programming* (*LGP*) were applied. However, *i-Miner* gave the best results overall.

In [Cad03], the authors present a method for model-based clustering and visualization of navigation patterns on a Web site. The site users are separated into clusters using first-order *Markov models*. A Web traffic visualization tool, *WebCANVAS*, is also implemented. The method was applied to Web traffic data from *msnbc.com* and yielded easy-to-understand insights. Another advantage of the method is that model-based clustering scales linearly with scale, while distance-based methods, such as *K-NN*, scale quadratically with data size.

In [Büc99], a method called *MiDAS*, is proposed for the discovery of navigational patterns from Web log files. In this method, the structure and content of Web log files is described using navigation templates that model domain knowledge, topology networks, and concept hierarchies.

The core of the method is a pattern tree which is a directed acyclic graph. In [Spi99], the authors propose *WUM*, a Web usage miner that focuses on user groups and also models navigational patterns as directed acyclic graphs.

In [Qam06], the authors propose a method for mining *stories* contained in *blogs* using community content and temporal clustering. Their method is called the *Content-Community-Time* model. Their experimental results showed the effectiveness of their method on several real-world data.

8.2.2 Automatic Personalization of a Web Site

Web site personalization deals with the modeling and grouping of Web pages and users, such that the appropriate users are directed to the right pages. As the authors in [Mul00] note, there are three aspects of a Web site that influence a user's satisfaction:

- Web site content

- Layout of individual pages

- Structure of the entire Web site

The authors in [Per98] address the issue of creating *adaptive Web sites* (i.e., sites that learn from visitors' access patterns to improve their structure and presentation). This is achieved by harvesting the information in Web server logs and automatically generating index pages for the Web site.

In [Mob00], a technique is described that utilizes mining of Web site usage to achieve automatic personalization. A user session consists of the following information:

- Who accessed the Web site

- What pages were accessed and the order in which the viewing happened

- How long each page was viewed

User transactions are modeled by preprocessing Web logs and grouping URL references. Three different data mining techniques are used on these patterns:

- Transaction clustering

- Usage clustering

- Association rule discovery

A transaction is defined as a set of visited URLs. The association rule discovery algorithm, *Apriori*, is used to discover URLs frequently appearing together in many transactions. Frequent URLs are chosen as those whose appearance support exceeds a certain support threshold. Another way to find URL clusters is by clustering transactions using similarity measures. Transactions are mapped into a multidimensional space as URL reference vectors. A *multivariate k-means* algorithm is used to achieve transaction clustering.

A recommendation engine finds a set of matching clusters using a matching score for the current active session and then it recommends a set of URLs within the cluster by using both the matching score and active physical link distance from the current active session. A recommendation score is computed and the pages whose recommendation score exceeds a certain threshold are added as a set of links to the latest page requested by the user.

[Lin07a] and [Lin07b] use graph theory to restructure the hyperlinks in a Web site to personalize to a user's needs. A graph consists of a set of vertices V and a set of edges E. For a Web site, let V be a set of Web pages $V = \{page1, page2, page3, ...\}$. While in graph theory, a neighbor of a vertex is an adjacent vertex, in a Web site the neighbors of a page are the set of pages that can be reached with a single click. In [Lin07b], an ideal hyperlink structure is one that satisfies the following criteria:

- Limited number of outgoing links from a page.

- Adequate number of incoming links.

- Most transfers between pages happen with two to three clicks.

Then in [Lin07a], some quantifiable measures of navigational ease are defined such as the following:

- Ease of navigation

- Minimum, maximum, average reach

In [Mor06], sequential pattern mining is used to develop a system that recommends links in an adaptive hypermedia educational system. The ultimate purpose of the system is to allow the teacher to personalize hypermedia courses for students, using information about the most used path by the students. This way the system can recommend potentially useful links to new students as they browse the course. The user of the system (the teacher) can select from three available sequential pattern mining algorithms: *AprioriAll*, *GSP*, and *PrefixScan* (all discussed in Chapter 5).

In [Oku09], the authors propose a recommendation method that is based not only on usage context, but also on the temporal information contained in usage context. They categorize the user contexts in the following categories: (1) contexts that occurred in the past, (2) contexts that occur when the user is just receiving a recommendation, and (3) contexts that may occur in the future.

In [Wex96], Wexelblat presents a method that facilitates site browsing by using assistive learning and personalized agent technologies to enable the user to operate in a *native information space*, instead of the site space. A similar approach is followed in [Joa97], where the path that is followed by many users for a certain search is the path suggested to inexperienced users.

8.2.3 Measuring and Improving the Success of Web Sites

In [Ho97], the author proposes four measures of *value creation* from the site visitor's point of view: *timely, custom, logistic, sensational*. These measures can be applied to three different business functions: *provision, processing*, and *promotion*. Here are some examples:

- *Custom value in provision*: database search capabilities, customized news report.

- *Logistic value in processing*: live customer service, job status tracking.

- *Sensational value in promotion*: outstanding Web site design, sweepstakes.

In [Spi01], the authors make an important note as to how the success of a Web site should be measured: *success should be measured according to the objective goals of the site's owner.* This type of thinking was initiated in [Ber96], where two measures of success are proposed:

- *Contact efficiency*: This is equal to the fraction of the users who spent at least a minimum amount of time exploring the site.

- *Conversion efficiency*: This is equal to the fraction of the users who after visiting the site, also bought something.

The authors in [Spi01] go one step further, where the goal of their research was not just to measure the success of a site, but to utilize the harvested information to *improve* the site. They propose a method to improve the success of a Web site by utilizing information from the site users' navigational patterns. The navigational pattern information is extracted using *WUM*, the Web usage miner discussed in Section 8.2.1.

In [Spi01], the following requirements are imposed for a method that improves the success of a Web site: (1) all the visitors of the site are taken into account, (2) it can be applied on a regular basis, and (3) it leads to a solid understanding of what are the shortcomings of the site and most importantly how the site can be improved. The authors deviate from the work in [Ber96], by focusing on *sessions* rather than on *users*. They make the following definitions:

- *Action page*: A page that when invoked, we know that the user is pursuing the site's goal.

- *Target page*: A page that its invocation signals the achievement of the site's goal.

- *Active session*: A session in which the user has at least one activity toward achieving the site's goal.

- *Customer session*: A session in which the user has reached the site's goal.

Using these definitions, the authors in [Spi01] redefine the notions of *contact efficiency* and *conversion efficiency* proposed in [Ber96] and instead they propose methods that measure the impact of each *page* on the success of the site. The authors applied their method on a real Web page containing an online catalog of products. Redesign of the site, based on information provided by the method in [Spi01], resulted in improvement of both the conversion efficiency and contact efficiency.

In [Hee01], the authors address the problem of how the Web site developer can identify the goals of their site's visitors, so that they can design it in a more efficient way. They introduce a novel method of clustering user groups called *Multi-Modal Clustering*, where they exploit many sources of information to cluster the users. They also introduce an interface to quickly explore these groups. The goal of *Multi-Modal Clustering* is to reveal patterns of site visits and identify categories of user needs.

The main idea *of Multi-Modal Clustering* is this: The features of Web pages, such as the words that appear in a page, are described using a multidimensional feature space. Then the profiles of users are modeled as multidimensional vectors that depend on the pages the users have visited. Finally, the user profiles are clustered to identify the most important user categories.

Regarding the creation of the user profile, the authors use the *longest repeating subsequence* (*LRS*) method, which was first described in [Pit99]. This method has the advantage that, while it reduces the complexity of the surfing path representation extracted from a set of raw site surfing data, it accurately represents the path. Let us see how the authors of [Hee01] define each one of the terms in *LRS*:

- *Longest*: This means that although the particular subsequence appears inside another repeated subsequence, there is at least one appearance of this subsequence where this is the longest repeating.

- *Repeating*: This means that the subsequence repeats more than a certain number of times, where this number can be defined using a threshold, T.

- *Subsequence*: A subsequence is a set of consecutive items.

Another novel contribution described in the article is the *Cluster Viewer*. This is a graphical interface that allows the user to view the cluster data and also refine the clusters. Some of the most interesting features of this cluster viewer are the following:

- For each cluster, the viewer can compute the most closely related Web pages.

- The cluster can be refined using merging and reclustering operators. Therefore, the user clusters can be examined at different granularities.

- The user can view the most highly valued dimensions (words in Web pages, for example) for each cluster.

- The user can check the accuracy of each cluster by drilling down to individual user paths.

The method was applied on three sites: (1) http://www.inxight.com, which is a provider of Web site related products; (2) the Web site of the Computer Science Department at the University of Minnesota; and (3) the Xerox Corporation home page. The sites provided a diverse mixture of traffic and user types and *Multi-Modal Clustering* was successful in identifying the major user types in all case studies.

8.2.4 Identification of Online Communities

In [Wan08], the authors address a problem of increasing importance: *identifying online communities*. They define the problem as that of grouping the event space according to the hidden "true" connections, and they propose a dynamic, generative model that is based on temporal events generated from online interactions. They model the system using *Hidden Markov Models,* specifically the *Mixed-Memory Aggregate Markov Chain Model (MAMC)*. The *Markov model* is estimated using an efficient iterative algorithm, based on maximum likelihood. Experimental results on real and synthetic data show that the algorithm can reveal novel insights about online interactions.

In [Chi07], the authors discuss how to perform structural and temporal analysis of the blogosphere using *community factorization*. In their work, a community consists of a set of blogs that communicate with each other based on a trigger event. *Community factorization* refers to the extraction of such communities from the blogosphere. The authors note the difference of identifying communities in a blog versus the Web. Blog entries are short-lived and, therefore, characterized by significant temporal locality. On the other hand, Web pages have a significantly longer lifetime in general and a new Web page can have a link to a very old one. In the approach described in the paper, the longer term graph structure that identifies the community (i.e., a Russian immigrant community, or a group interested in boating) is created from a series of short-term entries that are constructed from the threads of discussions in the blogs.

8.2.5 Web Usage Mining in Real Time

[Mas02] addresses the problem of how to perform Web usage mining in real time. Given that the behavior of site visitors can change frequently and unexpectedly, this is an interesting as well as a challenging problem. The algorithm is heuristically based and it has several novel features, such as finding frequent itemsets and exploiting the power of the machine of the user. The basis of the algorithm is a *genetic algorithm* approach. Specifically, at a first scan of the access log, a set of frequent itemsets is discovered.

Then a set of genetically inspired operators is applied, and a more refined set of frequent itemsets is generated. An example of such an operator is *crossover*, which chooses two frequent randomly selected sequences and proposes two new candidates coming from the joining. For example, given the sequences <page1, page5, page6> <page10, page11, page15>, the sequences <page1, page5, page10, page11> and <page5, page6, page11, page15> are proposed.

Another novel feature of the algorithm is that it exploits the computing power of the user's machine. This is achieved by sending a candidate set of sequences to each connected machine, which the connected machine compares with its navigation sequence. The purpose of this comparison is to return a percentage that indicates the similarity of the navigation sequence and the candidate sequence. Also the algorithm rewards long sequences if they are included in the navigation sequence.

8.2.6 Mining Evolving User Profiles

In [Nas08], a study on the Web site of a nonprofit organization is performed with the purpose of *mining evolving user profiles*. This information can be very useful in customer relationship management (CRM).

The authors propose an unsupervised learning approach, where the first step is to mine the Web logs to extract *user sessions*, which are then clustered into similar groups. User sessions are clustered using a hierarchical version of an unsupervised clustering technique (*H-UNC*), that is based on a genetic algorithm that evolves to a solution through selective reproduction. The similarity measure, used in *H-UNC*, between an input session s and a profile p is computed using the cosine similarity as shown in Equation 8.1:

$$Sim = \frac{\sum_{j=1}^{N} p_j s_j}{\sqrt{\sum_{j=1}^{N} p_j \sum_{j=1}^{N} s_j}} \qquad (8.1)$$

where N is the number of URLs. The authors also propose an expanded version of the above formula that takes into account an 'is-a' ontology information about the URLs and also external ontological information about different content items on the Web site.

Following user session clustering, relevant URLs are used to construct *user profiles* for each one of the clusters. An 'is-a' URL ontology is used to influence the computation of the similarity of URLs by taking into account the closeness of URLs in the hierarchy. The resulting profiles are enriched using additional information such as the following:

- Search queries: These are search queries performed by the user before visiting the particular Web site.

- Company information: This is information about both the companies that are being inquired and the companies doing the inquiring.

Finally, existing profiles are compared against already discovered profiles and their evolution is categorized. For each discovered profile, a measure of scale is defined that corresponds to the amount of variance of the user sessions around the cluster representative (to which this user profile corresponds). This way, a *boundary* is defined for each user profile. Two user profiles are *compatible* (and therefore could be related through *an evolution event*) if they have overlapping boundaries. The authors [Nas08] define broad categorization of user profile *evolution events*: *birth*, *death*, *persistence*, and *atavism*. The last one refers to a user profile that disappears and then reappears.

8.2.7 Identifying Similarities, Periodicities, and Bursts in Online Search Queries

In [Vla04], the authors address many of the research issues regarding online search queries, with the ultimate goal of identifying semantically similar queries. The novel contributions of the article are as follows:

- Development of a new compressed representation and indexing scheme for a time series. The *Fourier* transform is used to transform the series and then the k best coefficients are retained. The best coefficients are the ones with the most energy. As is shown in [Vla04], the reconstruction error, especially for time series with strong periodicities, is smaller when the coefficients with the most energy are used.

- Important period identification. The approach of the authors is based on what constitutes a nonperiodic time series, which is a sequence of points that are drawn independently from a Gaussian distribution. Therefore, they identify significant periods by identifying outliers according to an exponential distribution.

- Interactive burst discovery. First, the bursts are detected, and then they are compacted, such that one can pose queries about bursts. The compact representation of the bursts consists of a *start-date, end-date*, and *average value*.

Then the authors define the "*burst similarity*" of two time series X and Y. We will denote burst similarity as *Burst_sim*. Let us assume that BX and BY denote the burst features of time series X and Y:

$$BX = \{BX_1, BX_2,....BX_k\} \tag{8.2}$$

and

$$BY = \{BY_1, BY_2,....BY_m\} \tag{8.3}$$

Then,

$$Burst\_sim = \sum_{i=1}^{k} \sum_{j=1}^{m} intersect(BX_i, BY_j) \times similarity(BX_i, BY_j) \tag{8.4}$$

where

$$similarity(BX_i, BY_j) = \frac{1}{1 + avgValue(BX_i) - avgValue(BY_j)} \tag{8.5}$$

and

$$intersect = \frac{1}{2} \times \left(\frac{overlap(BX_i, BY_j)}{BX_i} + \frac{overlap(BX_i, BY_j)}{BY_j} \right) \tag{8.6}$$

where *overlap* calculates the amount of time intersection between two bursts.

In [Zha06], the authors address the problem of measuring similarities of Web search queries by exploring the click-through data logged by Web search engines. The main idea of the algorithm is to investigate the temporal characteristics, the explicit similarity, and the implicit semantics of the click-through data.

8.2.8 Event Detection from Web-Click-Through Data

In [Che08], a novel algorithm for event detection from *Web-click-through* data is described that is based on subspace analysis. *Web-click-through* data provide the following information about events: (1) event semantics, such as the pages visited by the user, and (2) time of events, that is, when the user accessed the pages. Given that a query session is defined by a query and the corresponding pages, the algorithm described in [Che08] consists of the following steps:

- Polar transformation. Each query session is mapped to a 2-D polar space point (r, θ), where the angle θ reflects the semantics of the query session and the radius r corresponds to the time of the query. Given a set of query sessions $\{S_1, S_2,..., S_N\}$, the radius that corresponds to session S_i is

$$r_i = \frac{T(S_i) - \min_j(T(S_j))}{\max_j(T(S_j)) - \min_j(T(S_j))} \qquad (8.7)$$

where $T(S_i)$ is the occurring time of session Si. The computation of the angle is more involved, which is expected since it reflects the semantic characteristics of the query. The main idea behind the computation of the angle is to compute a user session semantic similarity matrix and then perform *PCA* on it. Given two sessions:

$$S_1 = (Q_1, P_1) \qquad (8.8)$$
and
$$S_2 = (Q_2, P_2)$$

where Q_1, Q_2 represent the keywords and P_1, P_2 represent the visited pages, the similarity of the sessions can be defined as

$$Similarity(S_1, S_2) = \frac{a|Q_1 \cap Q_2|}{\max\{|Q_1|,|Q_2|\}} + \frac{(1-a)|P_1 \cap P_2|}{\max\{|P_1|,|P_2|\}} \qquad (8.9)$$

Having defined the similarity for two sessions, we can define a semantic similarity matrix and perform *PCA* on it. Let $\{f_1, f_2,..., f_n\}$ represent the first principal components of the sessions $\{S_1, S_2..., S_n\}$. Then the angle θ_i of session S_i can be defined as

$$\theta_i = \frac{f_i - \min_j(f_j)}{\max_j(f_j) - \min_j(f_j)} \times \frac{\pi}{2} \qquad (8.10)$$

- Subspace estimation. The main idea here is that semantically similar queries should map to points of similar angles and lie on a 1-D subspace. The algorithm utilized is the *Generalized Principal Component Analysis*, which assigns data to subspaces. The algorithm, however, can be sensitive to noise and the presence of outliers. An improvement to the algorithm is also presented in the article.

- Subspace pruning. Not every subspace contains events, and therefore, uninteresting subspaces need to be pruned. The criterion for subspace pruning is its entropy.

- Cluster generation.

8.3 ADDITIONAL BIBLIOGRAPHY

A thorough review of the subject of Web data mining can be found in [Liu06a]. An important paper in the area of Web usage mining is [Sri00]. The authors in [Coo97] describe a summary of the techniques in the field, while the authors in [Mob96] describe the first system for automated Web usage mining, *Web miner*. Two new languages related to Web usage mining are introduced in [Pun01]. One of these languages, *XGMML*, is for generating a Web graph. The other language, *LOGML*, is a Web log report language. First, a Web graph in *XGMML* format is generated from a Web site. Then, Web log reports are created in *LOGML* format using the Web graph and the Web log files.

8.3.1 Pattern Discovery

Significant work has been done in the discovery of association rules from Web usage statistics. References in this area include [Kit02], [Mas99], and [Pal02]. [Hog03] describes an algorithm for the extraction of temporal Web usage patterns using the adapted *Kohonen Self-Organized Map* based on rough set properties. In [Sun07], the authors study causal relations of queries in search engine query logs. In [Wu98], a method is proposed for finding the groups of most frequently visited pages.

In [Toy05], the authors describe *WebRelievo*, a system for the visualization and analysis of the evolution of the Web using time series graphs. In these graphs, nodes are Web pages and edges are relationships between pages. The authors applied *WebRelievo* to tracking trends in P2P systems and search engines for mobile phones.

8.3.2 Web Usage Mining for Business Applications

Significant work has been done in using information extracted from Web logs to reorganize the site and improve the business functions of the company, with applications in e-commerce, e-education, and so on. References in this area include [Pir96], [Büc98], [Che00], [Che97], and [Jes02], where algorithms are described that allow the most logical organization of the pages of a site to attract new customers and retain old customers. For example, in [Büc98], the authors describe data mining techniques for Internet data that can be combined with marketing data to discover actionable marketing information. These Internet data include server logs, query logs, and Web meta-data. After addressing resolution and data heterogeneity issues, the authors propose the construction of a data warehouse and present schematic designs for analytical and predictive activities.

In [Hal06], the role of temporal information is explored for the prediction of user navigation patterns on the Web, with respect to the URLs selected during navigation. An analysis is performed on *WAP* logs. In [Lin04], the authors address the problem of finding changes of clusters related to Web usage. Specifically, two types of cluster changes are addressed: changes in *cluster compositions* over time and changes in *cluster memberships* of Web users. This latter type of change can reveal valuable information as to the changing loyalty of the Web users of a site. The article proposes the idea of temporal cluster migration matrices (*TCMM*), which can be used as a visualization tool to analyze the change in a Web site usage and the loyalty of its specific users. Repetitive clustering is used to construct the *TCMM*, which stores the time series for cluster memberships. In other words, the *TCMM* gives a complete view of the user behavior over different periods.

In [Oya08], the authors present a method for determining whether Web data are for the same object or different objects. The main idea of their algorithm is to compute the probability that the data were generated for the same object with its attribute values going through changes over time. They also compute the probability that the data are from different objects, and they define similarities between observed data. The similarity is computed using agglomerative clustering. In [Liu06b], the authors explore whether there is a relationship between temporal and semantic similarity in Web query logs. Semantically related queries can be used in query suggestion and expansion. As the authors conclude, correlation among query log time series can find semantically coherent clusters, but about 27% of the clusters were not coherent.

Finally, in [Yan07], the authors propose a method to improve the performance of link *analysis* based on a time series of Web graphs. Specifically, a crucial component of link analysis in many algorithms, such as *PageRank*™, is page importance. However, the authors note that *PageRank* can have a problem of linking spam. To overcome this problem, the authors propose a more sophisticated method of calculating page importance called *TemporalRank*, where both the current Web graph and historical importance from previous Web graphs are combined to yield page importance. The reason is that spam pages have links that change in an irregular fashion and, therefore, demonstrate a different temporal link behavior than high quality pages. Experimental results show that incorporating historical information is an effective way of performing link analysis.

REFERENCES

[Abr03] Abraham, A., Business Intelligence from Web Usage Mining, *Journal of Information and Knowledge Management*, vol. 2, no. 4, pp. 375–390, 2003.

[Ber96] Berthon, P., L.F. Pitt, and R. T. Watson, The World Wide Web as an Advertising Medium, *Journal of Advertising Research*, vol. 36, no.1, pp. 43–54, 1996.

[Ber02] Berendt, B., B. Mobasher, and M. Spiliopoulou, Web Usage Mining for e-Business Applications, http://ecmlpkdd.cs.helsinki.fi/pdf/berendt-2.pdf, 2002.

[Bez81] Bezdec, J.C. *Pattern Recognition with Fuzzy Objective Function Algorithms*, Plenum Press, 1981.

[Büc98] Büchner, A.G. and M.D. Mulvenna, Discovering Internet Marketing Intelligence through Online Analytical Web Usage Mining, *Proceedings of ACM SIGMOD*, pp. 54–61, 1998.

[Büc99] Büchner, A.G et al., Navigation Pattern Discovery from Internet Data, *ACM Workshop on Web Usage Analysis and User Profiling (WEBKDD)*, pp. 25–30, 1999.

[Cad03] Cadez, I.V. et al., Model-Based Clustering and Visualization of Navigation Patterns on a Web Site, *Data Mining and Knowledge Discovery*, vol. 7, no. 4, pp. 399–424, 2003.

[Che97] Cheung, D.W., B. Kao, and J. Lee, Discovering User Access Patterns on the World Wide Web, *Knowledge-Based Systems*, vol. 10, pp. 463–470, 1997.

[Che00] Chen, P.M. and F.C. Kuo, An Information Retrieval Based on a User Profile, *The Journal of Systems and Software*, vol. 54, pp. 3–8, 2000.

[Che07] Chen, J. and T. Cook, Mining Contiguous Sequential Patterns from Web Logs, *Proceedings of the WWW Conference*, pp. 1177–1178, 2007.

[Che08] Chen, L., Y. Hu, and W. Nejdl, Using Subspace Analysis for Event Detection from Web-Click-Through Data, *Proceedings of the 17th International Conference on WWW*, pp.1067–1068, April 21–25, 2008.

[Chi07] Chi, Y. et al., Structural and Temporal Analysis of the Blogosphere Through Community Factorization, *Proceedings of the KDD Conference*, pp. 163–172, 2007.

[Coo97] Cooley, R., B. Mobasher, and J. Srinistava, Web Mining: Information and Pattern Discovery on the World Wide Web, *IEEE International Conference on Tools with Artificial Intelligence*, Newport Beach, pp. 558–567, 1997.

[Hal06] Halvey, M., M.T. Keane, and B. Smyth, Temporal Rules for Predicting User Navigation in the Mobile Web, *Lecture Notes on Computer Science*, Wade, Ashman, and Smyth eds, pp. 111–120, Springer, 2006.

[Hee01] Heer, J. and E.H. Chi, Identification of Web User Traffic Composition Using Multi-Modal Clustering and Information Scent, *Proceedings of the Workshop on Web Mining, SIAM Conference on Data Mining*, pp. 51–58, 2001.

[Ho97] Ho, J., Evaluating the World Wide Web: A Global Study of Commercial Web Sites, *Journal of Computer Mediated Communication*, vol. 3, no. 1, 1997.

[Hog03] Hogo, M., M. Snorek, and P. Lingras, Temporal Web Usage Mining, *Proceedings of the Web Intelligence Conference*, pp. 450–453, 2003.

[Jes02] Jespersen, T.E., J. Thorhauge, and T.B. Pedersen, A Hybrid Approach to Web Usage Mining, *Proceedings of the 4th International Conference, Data Warehousing, and Knowledge Discovery, LNCS 2454*, pp. 73–82, Germany: Springer, Verlag, 2002.

[Joa97] Joachim, T., D. Freitag, and T. Mitchell, Webwatcher—A Tour Guide for the World Wide Web, *Proceedings of the IJCAI Conference*, pp. 770–777, 1997.

[Kit02] Kitsuregawa, M., M. Toyoda, and I. Pramudiono, Web Community Mining and Web Log Mining, Commodity Cluster Page Execution, *Proceedings of the 13th Australian Conference on Database Technologies*, vol. 5, pp. 3–5, 2002.

[Lin04] Lingras, P., M. Hogo, and M. Snorek, Temporal Cluster Migration Matrices for Web Usage Mining, *Proceedings of the 2004 IEEE/WIC/ACM International Conference on Web Intelligence*, pp. 441–444, 2004.

[Lin07a] Lingras, P., Neighborhood Sets Based on Web Usage Mining, *Annual Meeting of the North American Fuzzy Information Processing Society*, pp. 659–664, 2007.

[Lin07b] Lingras, P. and R. Lingras, Adaptive Hyperlinks Using Page Access Sequences and Minimum Spanning Trees, *Proceedings of IEEE International Conference on Fuzzy Systems*, pp. 1–6, 2007.

[Liu04] Liu, J.G. and W.P. Wu, Web Usage Mining for Electronic Business Applications, *Proceedings of the 3rd International Conference on Machine Learning and Cybernetics*, vol. 2, Shanghai, pp. 1314–1318, 2004.

[Liu06a] Liu, B., *Web Data Mining*, Springer, 2006.

[Liu06b] Liu, B., R. Jones, and K.L. Klinkner, Measuring the Meaning in Time Series Clustering of Text Search Queries, *Proceedings of CIKM*, Arlington, VA, 2006.

[Mas99] Masseglia, F., P. Poncelet, and R. Cicchetti, An Efficient Algorithm for Web Usage Mining, *Networking and Information Systems Journal*, vol. 2, no. 5–6, pp. 571–603, 1999.

[Mas02] Masseglia, F., M. Teisseire, and P. Poncelet, Real Time Web Usage Mining: A Heuristic Based Distributed Miner, *Proceedings of the 12th international Workshop on Research Issues in Data Engineering: Engineering e-Commerce/ e-Business Systems (Ride'02) IEEE Computer Society*, Washington, DC, p. 169, 2002.

[Mob96] Mobasher, B. et al. Web Mining: Pattern Discovery from World Wide Transactions, *Technical Report: 96-050*, University of Minnesota, 1996.

[Mob00] Mobasher, B., R. Cooley, and J. Srivastava, Automatic Personalization Based on Web Usage Mining, *Communications of the ACM*, vol. 43, no. 8, pp. 142–151, 2000.

[Mor06] Morales, C.R. et al., Using Sequential Pattern Mining for Links Recommendation in Adaptive Hypermedia Educational Systems, *Current Developments in Technology-Assisted Education*, pp. 1016–1020, 2006.

[Mul00] Mulvenna, M.D., S.S. Anand, and A.G. Büchner, Personalization on the Net Using Web Mining, (Guest Editors), *Communications of the ACM*, vol. 38, no. 4, pp. 30–34, 2000.

[Nas08] Nasraoui, O. et al., A Web Usage Mining Framework for Mining Evolving User Profiles in Dynamic Web Sites, *IEEE Transactions on Knowledge and Data Engineering*, vol. 20, no. 2, pp. 202–215, February 2008.

[Oku09] Oku, K. et al., A Recommendation Method Considering Users' Time Series Contexts, *Proceedings of the 3rd International Conference on Ubiquitous Information Management and Communication*, pp. 467–490, 2009.

[Oya08] Oyama, S., K. Shirasuna, and K. Tanaka, Identification of Time Varying Objects on the Web, *Proceedings of Joint Conference on Digital Libraries (JCDL)*, pp. 285–294, 2008.

[Pal02] Pal, S.K., V. Talwar, and P. Mitra, Web Mining in Soft Computing Framework, Relevance, State of the Art, and Future Directions, *IEEE Transactions on Neural Networks*, vol. 13, no. 5, pp. 1163–1177, 2002.

[Per98] Perkowitz, M. and O. Etzioni, Adaptive Web Pages: Automatically Synthesizing Web Pages, *Proceedings of the 15th National Conference and Artificial Intelligence*, pp. 727–732, 1998.

[Pir96] Pirolli, P., J. Pitkow, and R. Rao, Silk from a Sow's Ear: Extracting Usable Structures from the Web, *Proceedings of Human Factors in Computing Systems*, pp. 118–125, 1996.

[Pit99] Pitkow, J. and P. Pirolli, Mining Longest Repeated Subsequences to Predict World Wide Web Surfing, *Proceedings of the 2nd USENIX Conference on the Internet*, p. 13, 1999.

[Pun01] Punin, J.R., M.S. Krishnamoorthy, and M.J. Zaki, Web Usage Mining— Languages and Algorithms, *Studies in Classification, Data Analysis, and Knowledge Organization*, Springer-Verlag, 2001.

[Qam06] Qamra, A., B. Tseng, and E.Y. Chang, Mining Blog Stories Using Community-Based and Temporal Clustering, *Proceedings of the 15th ACM International Conference on Information and Knowledge Management*, pp. 58–67, 2006.

[Spi99] Spiliopoulou, M. and L.C. Faulstich, WUM: A Web Utilization Miner, *Proceedings of EDBT Workshop on the Web and Data Bases*, pp. 109–115, Springer-Verlag, 1999.

[Spi01] Spiliopoulou, M. and C. Pohle, Data Mining for Measuring and Improving the Success of Web Sites, *Data Mining and Knowledge Discovery*, vol. 5, pp. 85–114, 2001.

[Sri00] Srivastava, J. et al., Web Usage Mining: Discovery and Applications of Usage Patterns from Web Data, *Proceedings of the SIGKDD Conference*, vol. 1, no. 2, pp. 12–23, 2000.

[Sun07] Sun, Y. et al., Causal Relation of Queries from Temporal Logs, *Proceedings of WWW*, pp. 1141–1142, 2007.

[Toy05] Toyoda, M. and M. Kitsuregawa, A System for Visualizing and Analyzing the Evolution of the Web with a Time Series of Graphs, *Proceedings of the 16th ACM Conference on Hypertext and Hypermedia*, pp. 151–160, 2005.

[Vla04] Vlachos M., C. Meek, Z. Vagena, et al. Identifying Similarities Periodicities and Bursts for Online Search Queries, *Proceedings of the ACM SIGMOD Conference*, pp. 131–142, Paris, June 2004.

[Wan08] Wang, X. and A. Kabán, A Dynamic Bibliometric Model for Identifying Online Communities, *Data Mining and Knowledge Discovery*, vol. 16, pp. 67–107, 2008.

[Wex96] Wexelblat, A., An Environment for Aiding Information-Browsing Tasks, *Proceedings of AAAI Spring Symposium on Acquisition, Learning, and Demonstration: Automatic Tasks for Users*, Birmingham, UK, AAAI Press, 1996.

[Wu98] Wu, K.L., P.S. Yu, and A. Ballman, SpeedTracer: A Web Usage Mining and Analysis Tool, *IBM Systems Journal*, vol. 37, no. 1, pp. 89–105, 1998.

[Yan07] Yang, L. et al., Link Analysis Using Time Series of Web Graphs, *Proceedings of the ACM Conference and Knowledge Management (CIKM)*, pp. 1011–1014, 2007.

[Zha06] Zhao, Q. et al., Time-Dependent Semantic Similarity Measure of Queries Using Historical Click-Through Data, *Proceedings of the 15th International Conference on WWW*, pp. 543–552, 2006.

CHAPTER 9

Spatiotemporal Data Mining

9.1 GENERAL CONCEPTS

Given the advent of telecommunications and particularly cellular phones, GPS technology and satellite imagery, robotics, Web traffic monitoring, and computer vision applications, the amount of spatiotemporal data stored every day has grown exponentially. Applications vary from tracking the movement of objects to monitoring the growth of forests.

As discussed in [Rod01], in most systems, time is considered to be *unidirectional* and *linear*. It is usually modeled as a discrete variable taking integer values that represent days, hours, and so on. On the other hand, space is considered *bidirectional* and, especially in geographic systems, *nonlinear*. Space is represented with real numbers and the granularity is usually small. For example, it can be in the scale of feet or meters in an area that represents an entire city.

Spatiotemporal data mining is a significantly more challenging task than simple *spatial* or *temporal* data mining. Spatiotemporal data is typically stored in 3-D format (2-D space information + time). As discussed in [Yao03], the following challenges are unique to spatiotemporal data:

- Spatial and temporal relationships can exist at various scales. Scale can have an effect on the existence or the strength of these relationships.

- Spatial relationships can be metric, such as distance, or nonmetric such as topology and shape. Temporal relationships can be of the form *before* and *after*. The challenging part is when these relationships are not known *a priori*, but need to be discovered.

9.2 FINDING PERIODIC PATTERNS IN SPATIOTEMPORAL DATA

Periodic pattern identification is an important problem in time series data, but also in spatiotemporal data. The reason is that many phenomena, such as the movement of tides, buses, trains, and even people, follow periodic patterns. This is a difficult problem in spatiotemporal data, because in most cases the period is not known and also it can change in *both* time and space. The reason is that moving objects do not always visit the exact same location and they do not visit it at the exact same time. One popular approach for the discovery of periodic patterns in spatiotemporal data is to treat them as time series data. For example, the changes in location of a moving object can be treated as different time-stamped values.

In [Mam04], a novel approach for discovering the periodic patterns in spatiotemporal data is discussed. In addition, another interesting problem is addressed: how to find an efficient way to index the periodic structure information for various applications, such as more efficient spatiotemporal data querying and data compression. The authors represent the periodic movement of objects as a 2-D sequence of values: (t_1, l_1), (t_2, l_2), (t_3, l_3), and so on. The locations are represented using spatial coordinates. The problem of finding periodic patterns becomes a problem of finding such a sequence that repeats every T time stamps and has a minimum support, *min_sup*. In addition, to accommodate for the fact that the locations are not certain, the location is replaced by a region. Such a discovered sequence will be of the form $R_0 R_1 R_2...R_{T-1}$, where R is a region or *, which indicates the entire spatial universe to accommodate for the fact that in some cases we are simply uncertain about the location of an object. For example, if we find a periodic pattern of the form A^*BC, it means that in the beginning of the cycle, the object is in region A, then it can be anywhere, then it is in region B, then it is in region C, and so on.

The authors first present the problem of finding frequent patterns of length 1 (*1-patterns*). Two methods are proposed: a *bottom-up* and a *top-down* approach. Originally, the sequence S is divided into T spatial data sets, one for each offset of the period T. Then the problem becomes one of clustering. A clustering algorithm, such as *DBSCAN*, can be used and clusters with less than *min_sup* are discarded. However, the authors use their own proprietary clustering algorithm which is fast and avoids R-tree construction. Then starting from the discovered *1-patterns*, an algorithm such as *Apriori* can be used to discover longer sequences.

Another contribution of the article is the construction of an efficient indexing algorithm of the periodic patterns. First, all periodic patterns are stored in an index called the *Periodic Index* (*PI*). Another index is created, called the *Location Index*, that stores the actual locations of objects that have periodic patterns in the *Periodic Index*. This can be implemented using a hash table. Finally, an *Exception Index* is created that stores the locations of objects that do not follow any periodic movements.

The algorithm for the discovery of periodic patterns was tested using a generator for object trajectories, which are periodic according to a set of parameter values.

9.3 MINING ASSOCIATION RULES IN SPATIOTEMPORAL DATA

In [Men05], spatial and temporal relationships are discovered in a set of data that characterize socioeconomic and land cover change in Denver, Colorado, from 1970 to 1990. The data were incorporated in *ArcGIS* (Environmental Systems Research Institute) and were then processed to produce a mining table. The association rule mining software *CBA* was used, which is based on the *Apriori* algorithm. *CBA* was developed at the National University of Singapore and is available for free download at http://www.comp.nus.edu.sg/~dm2/p_download.html.

CBA allows sorting of all association rules by support or confidence, and it also supports the exploration of rules by a particular combination of variables. In addition, the authors have developed a multiple level association rule mining algorithm, which supports mining rules at various levels of concept hierarchies to find the hierarchy that best supports the rule. Concept hierarchies were developed through data classification. The research also demonstrated how generated rules can be transformed into *GIS* queries.

In the first run of the algorithm, all variables and all levels of the concept hierarchies were included. As a result, an enormous number of rules were generated. Specifically, 122,970 rules were generated. The authors note that although association rule mining is promising, there are a number of issues that could be investigated further. Such an issue is the integration of diverse data sets. For example, in the case of the data investigated in the article, areal interpolation was needed and the transformation of data from one set of areal units to another.

In [Shu08], the authors discuss their research project, which is to find association rules in vegetation and climate changing data in northeastern China. Specifically, the steps of the association rule mining process are as follows:

- Raw data preprocessing, such as filtering out the noise. *Kriging* interpolation is also used for precipitation and air temperature data.

- Data conceptualization. Here, weather observation data and vegetation data are clustered separately using *fuzzy c-means clustering*.

- The data are organized in transactions and then the *Apriori* algorithm is used to extract association rules.

The extracted association rules are evaluated using such measures as support and confidence.

9.4 APPLICATIONS OF SPATIOTEMPORAL DATA MINING IN GEOGRAPHY

In [Lau05], two applications of data mining are discussed that deal with the increasing amount of spatiotemporal data in geography. These data in geography are called *geospatial lifelines* and of particular interest is the analysis of the lifeline of moving point objects (*MPOs*). Applications of *MPO* analysis can be found in homeland security for the monitoring of individuals, traffic analysis, animal behavioral science, and so on.

The particular article discusses the following two applications of data mining in geospatial lifeline data:

- *Retrieval by content*: In this application, we are interested in finding a particular motion pattern, similar to pattern template matching. Also of interest is understanding the meaningfulness of such patterns.

- *Descriptive modeling*: In this application, we are interested in describing the entire lifeline data, by dividing them into clusters.

Regarding retrieval by content, the approach discussed in the paper is based on detecting relative motion patterns (*REMO patterns*), which refer to detecting the relationships of the attributes of different object movements over space and time. Such attributes are the *speed, change of speed,* and *motion azimuth.* Therefore, the problem of finding a specific motion pattern becomes a problem of fitting to the data a motion template with specific motion attributes. Such patterns are discussed in detail in [Gud04] and include the following:

- *Flocking:* At least k objects ($k > 1$) are within a circular area of radius r ($r > 0$) and they move within the same direction.

- *Leadership:* At least k objects ($k > 1$) moving in the same direction are within a circular area of radius r ($r > 0$), and at least one of these objects has been ahead of all other objects for at least s ($s > 0$) steps.

- *Convergence*: At least k objects ($k > 1$) will pass through the same circular region of radius r ($r > 0$).

- *Encounter:* At least k ($k > 1$) objects will be during the same time inside the same circular region of radius r ($r > 0$).

These patterns can find application in animal behavior science, such as the tracking of animal flocks. They can also find application in the tracking of voting patterns in different geographical regions, where the direction of movement is either to the left or to the right.

There are three different goals in relationship to the discovery of these patterns:

1. *Detection*: Here, we just want to find whether a pattern exists.

2. *Find all*: Here, we are interested in finding all patterns that fit a certain template.

3. *Find largest*: Finally, the purpose here is to find the largest occurrence of a certain pattern.

When a pattern is reported, the following things are reported about it: the position of the pattern, the time of the occurrence of the pattern, and the complete subset of objects inside it.

Regarding *descriptive modeling* in [Lau05], its goal is to find clusters that adequately describe the lifelines. For example, in tracking an animal flock, we are interested in identifying animals that move in a similar way.

Assuming that tracking animals is of interest, one motion attribute we could track is motion azimuth. Then we could form an azimuth similarity matrix, where the elements would have the values 1, 0.5, and 0. The meanings of these values are as follows:

1. The animals move in the same direction (value = 1).

2. The animals move in perpendicular directions (value = 0.5).

3. The animals move in opposite directions (value = 0).

Laube in [Lau05] also discusses one further application of retrieval by content and descriptive modeling: their integration with *GIS* systems so that the data mining information can be merged with underlying geography information to provide better insights into the overall problem under study.

9.5 SPATIOTEMPORAL DATA MINING OF TRAFFIC DATA

In [Tre02], a method is described that generates spatiotemporal data from stationary traffic detectors. The authors describe the following applications of their method:

- Traffic state visualization

- Identification of traffic incidents

- Experimental verification of traffic models

- Traffic forecasting

The method is called *adaptive smoothing*, and it consists of a nonlinear spatiotemporal lowpass filter applied to the input detector data. The filter blocks small scale (or high frequency information in the Fourier domain) and adaptively takes into account the main direction of the information flow.

The filter takes into account the following heuristic information:

- When the traffic is moving freely, then perturbations such as velocity and flow move into the direction of traffic flow.

- When the traffic is not moving freely (i.e., in the case of congestion), then the perturbations move against the direction of traffic flow.

- Some information should be filtered out. Specifically, the user can choose to filter out all high frequency information on a time scale smaller than t and spatial changes smaller than x, where the values of t and x can be chosen by the user.

The *adaptive smoothing* method was applied to information collected from German highways. It is remarkable to note that the results of the algorithm are robust, even if 65% of the detectors are ignored. However, as the authors note, the algorithm works well if the distance between neighboring detectors does not exceed 3 km.

In [Kon07], the authors propose a conceptual multidimensional model, *CT-OLAP*, and an associated algebra that supports sequenced queries, which are conceptually evaluated at each time instant. This type of model and algebra can be useful in applications where the data exhibit continuous changes, such as traffic analysis. Previous attempts at incorporating time in *OLAP* include *TOLAP*, which is an extension of *OLAP* that extends time-related analytical queries. One of the disadvantages of *TOLAP* is that the valid time is discrete, and it cannot support queries about continuous data changes. This disadvantage is overcome by *CT-OLAP*.

9.6 SPATIOTEMPORAL DATA REDUCTION

In [Cao03], a method to compress spatiotemporal data is based on computer graphics and is known as *line simplification* is discussed. With the advent of location aware applications, the size of spatiotemporal data can be a limiting factor for efficient processing on devices such as PDAs. For example, GPS points are of the form (x, y, t) and a GPS point can be generated every second, so one can easily see the vast amount of information that is being generated.

Line simplification's main idea is that a polygonal trajectory that consists of many straight line segments can be represented sufficiently close with another trajectory that has fewer straight line segments. The authors describe the following significant advantages of the algorithm:

- Its error is deterministically bound. This is a unique advantage over other popular techniques, such as wavelets. Specifically, the error between the original trajectory and the approximated one is bounded by a parameter known as *error-tolerance*. However, the authors note that the errors of answers to queries may not be bounded and depend on the combination of the distance metric used and the type of the spatiotemporal query.

- It achieves a drastic data reduction rate. The authors applied their algorithm on real object trajectories and as the imprecision was allowed to grow to 0.1 miles, they were able to achieve 99% data reduction.

9.7 SPATIOTEMPORAL DATA QUERIES

[Cao03] discusses the most common queries on spatiotemporal data. Assuming a trajectory defined by a set of 3-D points $Q = (x_1, y_1, t_1), (x_2, y_2, t_2),...,$ one can define the following queries:

- *Where at (Q, t)?* This returns the expected location of trajectory Q at time t.

- *When at (Q, x, y)?* This returns the expected time the trajectory Q will pass through point x, y.

- *Intersects (Q, P, t_1, t_2)?* This query returns true if the trajectory Q intersects the polygon P between times t_1 and t_2.

- *Nearest Neighbor (Q, S, th)?* Given the trajectory Q, and a set of trajectories S, this query returns the trajectory whose distance from Q is less than the threshold th.

- *Spatial Join (S, th).* S is a set of trajectories and the query returns the trajectories Q_1, Q_2 whose distances are less than threshold th.

9.8 INDEXING SPATIOTEMPORAL DATA WAREHOUSES

In [Pap02], the authors describe the construction of a spatiotemporal data warehouse that supports *OLAP* operations. In most work related to spatiotemporal data, it is assumed that data are stored individually and queries address individual spatiotemporal points. The work here addresses the problem of summarization of information about spatiotemporal data. For example, animal behavior studies need the number of certain animals within an area of interest. In database terms, this constitutes an aggregate spatiotemporal query over an area of interest. The authors note two differences with traditional *OLAP* data that make the development of spatiotemporal warehouses challenging:

- There are no clear predefined hierarchies, such as product types, to facilitate aggregation.

- The spatial information requested in the query could be static, such as "Find all hotels within 5 km of the airport." However, it could also be dynamic, such as "Find the range covered by a cell phone."

Regarding static spatial information, first a data cube is constructed where one dimension is time and the other is space and the corresponding data are stored in a 2-D table. The table is stored ordered by time, and a *B-tree* index is constructed to find information about each time stamp. An alternative approach is discussed where the spatial and temporal dimensions are indexed simultaneously by a generalization of an *R-tree* to 3-D space. Each entry in the tree keeps the aggregate value of each region (such as count) and the time stamp for which it is valid. A more advanced approach is discussed where an *aggregate R-B tree* (*aRB tree*) is constructed. Here, the regions that are part of the spatial hierarchy are stored once and indexed by an *R-tree*. For each *R-tree* entry, there is a pointer to a *B-tree* in which historical aggregate data about the entry is stored.

Regarding dynamic spatial information, the *aggregate 3-D R-B tree* is proposed. The *a3DRB tree* combines *B-trees* with *3DR-trees*. A 3-D box is used to model every region, such that the projection on the temporal axis corresponds to a time interval when the spatial area is fixed.

9.9 SEMANTIC REPRESENTATION OF SPATIOTEMPORAL DATA

In [Jin07a] and [Jin07b], a semantic framework is described for spatiotemporal data. The novel features of the work are the semantic description of the characteristics of spatiotemporal data and also the extension of the *ER* database model to support spatiotemporal characteristics. Objects are identified by identifiers and their state. The state of an object is represented by spatial (*location, region*) and nonspatial attributes (*context-related*).

The authors describe six types of changes in spatiotemporal data:

1. *Continuous spatial processes*: The spatial attributes of an object, such as the spread of a forest and the movement of a car, change continuously.

2. *Discrete spatial processes*: Spatial boundaries do not change over a period of time and then suddenly change. Example is the expansion of a building.

3. *Continuous thematic processes*: The thematic (context) attributes, such as change of population of fishes in a lake, change continuously.

4. *Discrete thematic processes*: The context attributes are static, but change suddenly over time, for example, change of ownership of a building.

5. *Discrete life*: These concern sudden changes that result in the creation of new objects or the deletion of objects. Examples are the construction of a new building or the demolition of a building.

6. *Discrete topological changes*: These concern sudden changes of the topological changes of two related objects, for example, the relationship between a plane and the city it is flying over.

Another article that describes a conceptual model for spatiotemporal applications is [Try03]. The article proposes two models: a spatiotemporal entity relationship model and an extended spatiotemporal *Unified Modeling Language (UML)*. Finally, another interesting article that deals with spatiotemporal databases is [Kou03], which describes the advances in the European *CHOROCHRONOS* project.

9.10 HISTORICAL SPATIOTEMPORAL AGGREGATION

The existence of summarized information in spatiotemporal databases is a feature that is becoming increasingly desirable in many spatiotemporal applications. An example of important summarized information is the number of vehicles that pass through an area per day or the number of visitors to a tourist attraction per day. There are two ways one can obtain this summarized information:

1. In real time as the result of a query, by accessing individual tuples in the database. However, this can be very computationally expensive. Another problem is privacy concerns about storing individual user data.

2. Having the summarization information available prior to the query.

In [Tao05], the authors present methods for combining spatiotemporal indexing with preaggregation. Spatiotemporal data aggregation is more challenging than aggregation in typical data warehouses because, as the authors note, there are not predefined aggregation units. For example, the region and the temporal interval of the query can be of any size. In the proposed methods, at the finest aggregation level, there are regions associated with measures, such as the number of vehicles or the number of visitors per day. The goal of the authors is to answer the *spatiotemporal window aggregate query*, which returns an aggregate measure, such as sum, over a specified rectangle and a specified time interval. The authors propose a number of multitree indices that support the following:

- Ad hoc groupings

- Arbitrary query windows

- Historical time intervals

The authors propose two indexes: (1) a host index, which is an aggregate structure that manages region extents, and (2) measure indexes (one for each host entry), which are aggregate temporal structures, whose purpose is to store the historical values of the measures.

9.11 SPATIOTEMPORAL RULE MINING FOR LOCATION-BASED AWARE SYSTEMS

The advances of GPS technology and the plethora of wireless devices have increased the need for location-based services, where location information about users is recorded. This information has two dimensions, space and time, and is known as *Location-Based Services* (*LBS*). This way, the user can request historical or future event information about the present location.

In [Gid07], a framework is described that achieves the following: (1) it extends market basket analysis to finding rules in the spatiotemporal domain; (2) it shows the usefulness of these rules by applying them to real-world data sets, such as the "Space, Time, Man" project that records the activities of thousands of individuals on the GPS-enabled mobile phones; and (3) it addresses privacy concerns by anonymizing the location data using data cloaking and data swapping techniques. Regarding spatiotemporal rule mining, a technique called *pivoting* that utilizes a spatiotemporal data hierarchy to extend association rule mining is described. The spatiotemporal data are separated into *items* and *baskets*. The latter can be *ordinal, spatial, temporal*, or *spatiotemporal* baskets. The *pivoting* method is applied to GPS traces to obtain frequent routes. The GPS traces are processed to discover the spatiotemporal itemsets. A spatiotemporal clustering technique that utilizes heuristics to define clusters in a transportation domain problem (cab-sharing) is also described in [Gid07].

9.12 TRAJECTORY DATA MINING

In [Vla02b], the problem of finding similar trajectories of moving objects is addressed. This is an important problem in many areas, such as identification of moving objects in spatiotemporal data and matching of shapes extracted from video clips. As the authors note, a similarity measure for such kinds of data, in addition to being computationally efficient, must be able to handle the following problems:

- Different sampling rates or speeds

- Similar motions in different space regions

- Outliers and generally the presence of noise

- Data with different lengths

A similarity measure that addresses the above issues is the *Longest Common Sub-Sequence* (*LCSS*). As noted in the article, the advantages of the *LCSS* are that some pieces of the time series can remain unmatched and the *LCSS* allows a more efficient approximate computation. The authors propose a measure that is a variant of the *LCSS* that has two controlling parameters, ε and δ. The ε parameter is the matching threshold, while δ controls how far in time we can go to match a point in a trajectory. The value of ε is application dependent, and it can be set equal to the smallest standard deviation between the two trajectories. Regarding δ, the authors found out experimentally that setting to values more than 20%–30% did not yield significant improvement. The authors also proposed an indexing scheme, which is based on hierarchical clustering.

In [Vla02a], an improved similarity measure is proposed that uses a weighted matching function (a *sigmoid* function) according to the distance of points. The improved measure does not penalize points that are marginally outside the matching region, which is the case with the aforementioned *LCSS* measure. In [Vla03], an indexing scheme for trajectory data is described which works by splitting the trajectories into multidimensional *MBRs* (*Minimum Bounding Rectangles*) and storing them in an *R-tree*. The best set of *MBRs* is the one that completely contains the sequence and minimizes the volume consumption. The indexing scheme is scalable and therefore suitable for massive data sets. The index has the advantage that it supports many distance functions, such as *LCSS*, *DTW*, and *Euclidean*.

In [Kha05], the authors propose a method for the classification and clustering of trajectories using spatiotemporal functional approximations. The motion trajectories are viewed as time series and their *Fourier* coefficients are computed using the *DFT*. Trajectory clustering is performed in the *Fourier* coefficient feature space using the leading *Fourier* coefficients. Then the coefficients are used as input feature vectors to a *Self-Organizing Map* which can learn similarities between object trajectories.

In [She08], the authors propose a method for the efficient computation of similarity in trajectory historical data. Specifically, the history consists of d-dimensional time series data where d >= 1. The main idea of the method is to summarize the data at different levels of details. These summaries can be used for the indexing of the histories, but also for efficient pruning of the histories to use them as input to distance functions, such as *DTW* and *LCSS*.

In [Gom09], the authors address the problem that only a few of the frequent sequences discovered in trajectory databases are of interest to the user. To avoid post processing to discover these interesting patterns, the authors propose a language, called *Re-SPaM*, that utilizes regular expressions to allow the user to express constraints for the items to be mined. Specifically, these constraints are defined for the attributes (temporal and nontemporal) of the items.

In another article [Eln08] by Elnekave et al., trajectories of mobile objects are clustered to find groups of movement patterns using a technique based on the representation of a mobile trajectory with minimum bounding boxes and a novel similarity measure between trajectories. In a similar work, [Eln07], the authors cluster trajectories of a mobile object and use the cluster centroid as an estimate of the object's movement and as a way to predict the future locations of the object. A similarity measure is used to discover recurring trajectory patterns. The algorithm achieved an 89.6% average precision and 89.5% average recall.

9.13 THE *FLOWMINER* ALGORITHM

In [Wan04], the authors describe an algorithm, *FlowMiner*, to find flow patterns in spatiotemporal databases. Detection of flow deals with the detection of patterns in both space and time. Here is an example:

A significant increase of airline reservations to Aegean Islands destinations is noted in travel agencies in Athens during spring.

There is also an increase in unemployment in the northern city of Thessaloniki, in spring, because of the closing of many factories. As a result of the above two events, there are many tourists on the Greek islands in the summer, but few are from the city of Thessaloniki.

The authors in [Wan04] define the following:

- R is a neighborhood region. In the example above, R is Greece.

- (loc,t) is a location-based event that happens at location loc at time t. In the example above, there are the following location-based events:

  Event 1: Increase of reservations at travel agencies: (*Athens, Spring*)

  Event 2: Increase of tourists: (*Greek Islands, Summer*).

  Event 3: Significant increase of unemployment (*Thessaloniki, Spring*)

  Event 4: Reduced numbers of Thessalonians (*Greek Islands, Summer*)

- Two events (loc_1, t_1), (loc_2, t_2), where $t_1 <= t_2$, are related if and only if both loc_1 and loc_2 belong to R and t_1 is near t_2.

- A set of events that happen at the same time make up an *eventset*. An eventset EA at time t_1 is said to *flow* to eventset EB if every event in EA results into an event in EB. In the example above we have two eventsets: EA = (*Event 1, Event 3*), EB = (*Event 2, Event 4*). Because *Event 1* results into *Event 2* and *Event 3* results into *Event 4*, we say that *EA flows to EB*.

The *FlowMiner* algorithm consists of the two typical steps of candidate generation and support counting. First, all frequent events are discovered and events are sorted in descending order. Then, all frequent *length 2-sequences* are found. Finally, longer flows are discovered.

Several techniques are used to optimize *FlowMiner*: (1) Infrequent candidates are pruned, (2) the support counting process is optimized by not hashing nonpromising events, and (3) the number of database scans is minimized. Experimental results on synthetic and real data sets show that *FlowMiner* is a linearly scalable algorithm and it can find all flow patterns in an efficient manner.

9.14 THE *TOPOLOGYMINER* ALGORITHM

In [Wan05], the authors address the problem of mining spatial patterns in spatiotemporal databases, taking into account the temporal dimension. For example, a spatial pattern with a temporal dimension is the following:

> An increase of British tourists on the island of Rhodes is usually followed by a simultaneous decrease of British tourists on the island of Kos during the months of July and August. It looks like these two islands are in competition to attract British tourists during the peak tourism season in Greece.

The two main contributions of the work presented in [Wan05] are as follows:

1. A summary structure that keeps track of the count of the instances of a feature within a spatial region and a time window. This is a concise way to represent the information in the database.

2. An algorithm to mine topological patterns, called *TopologyMiner*. The algorithm does not follow an *Apriori* approach (candidate generation, testing). Instead it operates in a *depth-first* approach, and it grows patterns similar to the *PrefixSpan* algorithm we discussed in Chapter 5. Specifically, the algorithm first divides the space time in a set of disjoint cubes. Then the summary structure is created using only one database scan and hashing the feature instances in the corresponding cubes. In the mining of topological structures, the authors use the concept of a projected database of a topological database. This is defined as the set of cubes, which contain the instances of the features of the pattern.

Experimental results show that *TopologyMiner* is an efficient algorithm and outperforms *Apriori-like* algorithms by orders of magnitude. In addition it requires much less memory than *Apriori-like* algorithms, which store the candidates in main memory and eventually run out of memory, when the size of the database increases significantly.

9.15 APPLICATIONS OF TEMPORAL DATA MINING IN THE ENVIRONMENTAL SCIENCES

[Ste02] describes a clustering technique to discover Ocean Climate Indices (*OCIs*), which are time series that describe the behavior of the ocean climate in selected areas. Earth science data consist of the measurement of such variables as sea surface temperature and precipitation. Typically, principal component analysis (*PCA*) is used to discover *OCIs*. However, as the authors in [Ste02] note, the problem with *PCA* is that it discovers only the strongest signals, which in addition have to be orthogonal to each other. The proposed clustering technique groups ocean areas with relatively homogeneous behavior into clusters, where the centroids are time series that summarize the behavior of the ocean areas. The cluster centroids are divided into the following categories:

- Centroids that correspond to known *OCIs*

- Centroids that correspond to variants of known *OCIs*

- Centroids that correspond to potentially new *OCIs*

The clustering algorithm can be described in the following steps:

- *Cluster the ocean areas that have homogeneous behavior.* The *OCI* that corresponds to each cluster is represented by the centroid of each cluster. The centroid is computed, by averaging the time series that belong to the cluster.

- *Compute the influence that potential OCIs can have on well-defined land points.* The only *OCIs* of interest are those that have a strong correlation with specific land points.

- *Compute the influence of candidate OCIs on known OCIs.* In terms of correlation with known *OCIs*, the candidate *OCIs* are divided into the following categories: *very high, high, medium, low.* The *OCIs* of most interest are the ones that have the lowest correlation with known *OCIs* because these *OCIs* might represent *undiscovered OCIs*. The *OCIs* that have very high correlation with known *OCIs* are of little interest because they represent already known *OCIs*. Finally, the *OCIs* with high or medium correlation represent alternatives or better representations of current *OCIs*.

The *Pearson* correlation coefficient is used as a measure of similarity. The data are preprocessed to remove seasonality. In particular, the authors use *z*-score normalization for each month. In other words, for the values of each month, the mean and standard deviation are computed and then used to normalize the data.

In [Hof03] and [Hof08], the authors describe a *multivariate k-means-based* clustering technique applied to a variety of environmental applications, such as the following:

- Comparison of climate model and climate measurements and comparison of different climate models.

- Analysis of remote sensing data obtained from satellites or airborne sensors or ground sensors.

- Ecological regionalization. An *ecoregion* is a simplified generalization of complex combinations of climatic and geological features that are salient to the growth and reproduction of animal and plant species.

The authors modified the original *k-means* algorithm, such that it has a highly scalable and parallelizable implementation. Originally, it was applied to spatial (geographical) data only, but then it was extended to spatial and temporal data and was given the name *Multivariate Spatiotemporal Clustering (MSTC)*. The algorithm was parallelized by parallelizing the seed-finding procedure.

In [Ver04], the authors address the problem of developing effective vegetation moisture monitoring tools using time series analysis. This is an important problem, because of the increasing number of severe wild fires. Satellite vegetation data and rainfall/temperature data are related to indicators of vegetation moisture dynamics. The authors point out that, because time series data are autocorrelated, cross correlation of time series can be problematic and unable to derive meaningful results. For this reason, they focus on optimizing methods to derive independent measures from time series. Then these measures can be used to extract valid statistical relationships between remote sensing and meteorological time series.

The authors of [Li03] address a similar problem: that of drought risk management. For this purpose, they incorporate time series data mining techniques in a geospatial decision support system. Specifically, two association rule mining techniques are utilized (*REAR* and *MOWCATL*) to predict local drought conditions based on global climatic conditions.

In [Bou06], time and space are integrated in geo-statistics to map land cover changes. The authors propose a novel spatiotemporal analysis technique to take advantage of the 30+ years Earth monitoring satellite data.

In their work, there are three sources of information for the mapping of land cover changes:

- The remotely sensed data

- The spatial pattern through which land cover classes are related

- The temporal pattern of classes

Therefore, at any location there are two types of information considered: (1) nontemporal information (at a specific instant), which consists of the satellite response and the neighboring land cover indicators, and (2) temporal information, which is time series information that consists of transition probabilities that connect the time series indicators through time.

[Fer04] addresses the problem of analyzing ice sheet elevation changes in time series data constructed from satellite radar or laser altimeter. It proposes an auto-regression method that can characterize seasonal and interannual changes, since autoregression can be used to model cyclical variations. To account for long-term linear variations, they combine the autoregression model with weighted squares linear regression. They specifically utilized an *autoregressive model* of order 2. In the experimental results, the autoregressive model yielded statistically accurate fits to actual ice sheet elevation change data, depicting a variety of conditions. The authors also conducted *Monte Carlo* simulations that closely resemble the 5-year elevation data from Antarctica. It was shown that the autoregressive method gave a linear trend that was less biased than two other techniques that were used for the analysis for these data.

9.16 ADDITIONAL BIBLIOGRAPHY

Additional references on spatiotemporal data mining can be found in the *2000 Proceedings of the International Workshop on Temporal, Spatial, and Spatiotemporal Data mining* [Hor00] . For example, [Bit00] describes the use of rough sets in spatiotemporal data mining and [Rod00] describes an updated bibliography of temporal, spatial, and spatiotemporal data mining research. A review of spatiotemporal literature can be found in [Hsu07]. In [Tan08], recent advances in spatiotemporal analysis are presented, while in [And05], exploratory analysis of spatial and temporal data is discussed.

9.16.1 Modeling of Spatiotemporal Data and Query Languages

In [Gue07], a method to extract relevant information from satellite image time series, using ratio-distortion analysis with *Gauss-Markov* random fields and *auto-binomial* random fields, is described. In [Con06], a spatiotemporal *Bayesian* model is used to forecast health indicators. The model allows both spatial and temporal dependencies. In [Kau08], an ontology is used to remove ambiguities from spatiotemporal locations. In [Viq07], an *SQL* extension to handle spatiotemporal data is proposed. Data types are defined for time (*instant, period*) and spatial quanta. There are three types of space (*point, pure line, pure surface*), two types of line (*pure line* or *point*) and two types of surface (*pure surface* or *line*). In [Cam03], Camossi et al. discuss the application of multigranularity concepts in spatiotemporal data. Specifically, they discuss a multigranular data model and a multigranular query language.

In [Yan05], the authors present a framework to discover spatial associations and spatiotemporal patterns in scientific data sets. A novel contribution of the work is that features are modeled as geometric objects instead of points. In addition, the authors define multiple distance metrics that take into account the shape and extent of objects and four different types of spatial object association patterns.

9.16.2 Moving Object Databases

In [Lee04], the authors describe a technique that views users of location-based services as moving objects and then employs an algorithm for the computation of temporal patterns from a series of locations of objects. In [Pra07], the authors discuss a modeling technique for movements of spatiotemporal objects in moving object databases. Their model is called the *Balloon* model, and it is a way to model both historical and future movements of objects. In [Sin09], the temporal behavior analysis of mobile ad hoc networks is analyzed using three different mobility patterns: (1) *Gaussian* mobility model, (2) *Random Walk*, and (3) *Random Way Point*. For any of these mobility patterns, the number of neighbor nodes of a node behaves as a random variable, whose variation can be modeled using an autoregressive model.

In [Ver08] the authors address the problem of mining spatiotemporal patterns in object-mobility databases. They define *STARS* (*Spatio-Temporal Association Rules*), which allow the description of object movement between regions over time. Then, they define an algorithm, *STARMiner*, to harvest *STARS* information from data. Finally, in [Che00], the authors propose a time series forecasting technique for temperature prediction. The technique is based on a fuzzy time series model that utilizes two factors (daily average temperature and daily cloud density).

REFERENCES

[And05] Andrienko, N. and G. Andrienko, *Exploratory Analysis of Spatial and Temporal Data*, Springer, 2005.

[Bit00] Bittner, T., Rough Sets in Spatiotemporal Data Mining, *Proceedings of the First International TSDM Workshop*, Lyon, France, pp. 89–104, 2000.

[Bou06] Boucher, A., K.C. Seto, and A.G. Journel, A Novel Method for Mapping Land Cover Changes: Incorporating Time and Space with Geostatistics, *IEEE Transactions on Geoscience and Remote Sensing*, vol. 44, no. 11, pp. 3427–3435, Nov. 2006.

[Cao03] Cao, H., O. Wolfson, and G. Trajcevski, Spatio-temporal Data Reduction with Deterministic Error Bounds, *International Conference on Mobile Computing and Networking*, San Diego, CA, pp. 33–42, 2003.

[Cam03] Camossi, E., M. Bertolotto, and E. Bertino, Multigranular Spatiotemporal Models: Implementation Challenges, *Proceedings of the 11th ACM International Symposium on Advances in Geographic Information Systems*, pp. 94–101, 2003.

[Che00] Chen, S.M. and J.R. Hwang, Temperature Prediction Using Fuzzy Time Series, *IEEE Transactions on Systems, Man, and Cybernetics*, vol. 30, no. 2, pp. 263–275, April 2000.

[Con06] Congdon, P., A Spatio-temporal Forecasting Approach to Health Indicators, *Journal of Data Science*, vol. 4, pp. 399–412, 2006.

[Eln07] Elnekave, S., M. Last, and O. Maimon, Predicting Future Locations Using Clusters' Centroids, *Proceedings of the 15th International Symposium on Advances in Geographic Information Systems*, 2007.

[Eln08] Elnekave, S., M. Last, and O. Maimon, Measuring Similarity between Trajectories of Mobile Objects, *Studies in Computational Intelligence*, vol. 91, pp. 101–128, 2008.

[Fer04] Ferguson, A.C., C.H. Davis, and J.E. Cavanaugh, An Autoregressive Model for Analysis of Ice Sheet Elevation Change Time Series, *IEEE Transactions on Geoscience and Remote Sensing*, vol. 42, no. 11, pp. 2426–2436, November 2004.

[Gid07] Gidofalvi, G., Spatio-temporal Rule Mining for Location-Based Aware Services, Ph.D. thesis, Aalborg University, 2007.

[Gom09] Gomez, L.I. and A.A. Vaisman, Efficient Constraint Evaluation in Categorical Sequential Patterns Mining for Trajectory Databases, *Proceedings of EDBT*, pp. 541–552, 2009.

[Gue07] Gueguen, L. and Datcu, M., Image Time-Series Data Mining Based on the Information-Bottleneck Principle, *IEEE Transactions on Geoscience and Remote Sensing*, vol. 45, pp. 827–838, 2007.

[Gud04] Gudmundsson, J., M. van Kreveld, and B. Speckmann, Efficient Detection of Motion Patterns in Spatio-temporal Data Sets, *Proceedings of the ACM International Workshop on Geographic Information Systems*, pp. 250–257, 2004.

[Hor00] Hornsby, K. and J.F. Roddick, *Temporal, Spatial, and Spatio-temporal Data Mining: First International TSDM Workshop,* Springer, 2000.

[Hof03] Hoffman, F.M. and W.W. Hargrove, Multivariate Spatiotemporal Clustering of Time Series Data: An Approach for Diagnosing Cloud Properties and Understanding ARM Site Representativeness, *13th ARM Science Team Meeting Proceedings,* 2003.

[Hof08] Hoffman, F.M. et al., Multivariate Spatiotemporal Clustering as a Data Mining Tool for Environmental Applications, *International Congress on Environmental Modeling,* 2008.

[Hsu07] Hsu, W., M.L. Lee, and J. Wang, *Temporal and Spatio-temporal Data Mining,* IGI Publishing, 2007.

[Jin07a] Jin, P., S. Wan, and L. Yue, A Semantic Framework for Spatiotemporal Data Representation, *3rd International IEEE Conference on Signal-Image Technologies and Internet Based Systems,* pp. 10–17, 2007.

[Jin07b] Jin, P., S. Wan, and L. Yue, Ontology-Driven Conceptual Modeling for Spatiotemporal Database Applications, *First International Symposium on Data, Privacy, and e-Commerce,* pp. 350–352, 2007.

[Kau08] Kauppinen, T., Ontology-Based Disambiguation of Spatiotemporal Locations, *Proceedings of the 1st international workshop on Identity and Reference on the Semantic Web (IRSW2008), 5th European Semantic Web Conference 2008 (ESWC 2008),* Tenerife, Spain, June 1–5, 2008.

[Kha05] Khalid, S. and A. Naftel, Classifying Spatiotemporal Object Trajectories Using Unsupervised Learning of Basis Functions Coefficients, *Proceedings of the 3rd ACM International Workshop on Video Surveillance and Sensor Networks,* pp. 45–52, 2005.

[Kon07] Kondratas, E. and I. Timko, CT-OLAP: Temporal Multidimensional Model and Algebra for Moving Objects, *Proceedings of ACM Workshop on Data Warehousing and OLAP (DOLAP),* Lisboa, Portugal, pp. 81–88, 2007.

[Kou03] Koubarakis, M., Y. Theodoridis, and T. Sellis, Spatiotemporal Databases in the Years Ahead, *Lecture Notes on Computer Science,* pp. 345–347, Springer, 2003.

[Lau05] Laube, P., Spatio-temporal Data Mining-Coping with the Increasing Availability of Motion Data in Geography, *SIRC, the 17th Annual Colloquium of the Spatial Information Research Centre,* University of Otago, Dunedin, New Zealand, 2005.

[Lee04] Lee, J.W., O. H. Paek, and K.H. Ryu, Temporal Moving Pattern Mining for Location-Based Service, *Journal of Systems and Software,* vol. 73, no. 3, pp. 481–490, 2004.

[Li03] Li, D. et al., Time Series Data Mining in a Geospatial Decision Support System, *Proceedings of the 2003 Annual National Conference on Digital Government Research,* pp. 1–4, 2003.

[Mam04] Mamoulis, N. et al., Mining, Indexing, and Querying Historical Spatiotemporal Data, *Proceedings of the Knowledge Discovery and Data Mining Conference (KDD),* August 22–25, pp. 236–245, Seattle, Washington, 2004.

[Men05] Mennis, J. and J. W. Liu, Mining Association Rules in Spatiotemporal Data, *Proceedings of the 17th International Conference on GeoComputation,* 2005.

[Pap02] Papadias, D. et al., Indexing Spatiotemporal Data Warehouses, *Proceedings of ICDE*, pp. 166–175, 2002.

[Pra07] Praing, R. and M. Schneider, Modeling Historical and Future Movements of Spatio-temporal Objects in Moving Object Databases, *Proceedings of the ACM Conference on Information and Knowledge Management (CIKM)*, pp. 183–192, Lisboa, Portugal, 2007.

[Rod00] Roddick, J.F., K. Hornsby, and M. Spyliopoulou, *Proceedings of the First International TSDM Workshop*, Lyon, France, 2000.

[Rod01] Roddick, J.F. and B. G. Lees, Paradigms for Spatial and Spatiotemporal Data Mining, *Geographic Data Mining and Knowledge Discovery*, H. Miller and J. Han, eds., Taylor & Francis 2001.

[She08] Sherkat, R. and D. Rafiei, On Efficiently Searching Trajectories and Archival Data for Historical Similarities, *Proceedings of the VLDB Conference*, pp. 896–908, 2008.

[Shu08] Shu, H., X. Zhu, and S. Dai, Mining Association Rules in Geographical Spatial Data, www.isprs.org/congresses/beijing2008/proceedings/2_pdf/2_WG- II-2/10.pdf, 2008.

[Sin09] Singh, J.P. and P. Dutta, Temporal Behavior Analysis of Mobile Ad Hoc Network with Different Mobility Patterns, *International Conference on Advances in Computing, Communication, and Control*, pp. 696–702, 2009.

[Ste02] Steinbach, M., P.N. Tan, and V. Kumar, Temporal Data Mining for the Discovery and Analysis of Ocean Climate Indices, *Proceedings of the KDD Temporal Data Mining Workshop*, pp. 2–3, 2002.

[Tan08] Tang, X., Liu, Y., and W. Kainz, *Advances in Spatiotemporal Analysis*, Taylor & Francis, 2008.

[Tao05] Tao,Y. and D. Papadias, Historical Spatio-temporal Aggregation, *ACM Transactions on Information Systems*, vol. 23, no. 1, pp. 61–102, January 2005.

[Tre02] Treiber, M. and D. Helbing, Reconstructing the Spatio-Temporal Traffic Dynamics from Stationary Detector Data, *Cooperative Transportation Dynamics*, vol. 1, pp. 3.1–3.24, 2002.

[Try03] Tryfona, N., R. Price, and C.S. Jensen, Conceptual Model for Spatio-temporal Applications, *Lecture Notes on Computer Science 2520, Spatiotemporal Databases*, T. Sellis et al., eds. pp. 79–116, Springer, 2003.

[Ver04] Verbesselt, J. et al., Biophysical Drought Metrics Extraction by Time Series Analysis of SPOT Vegetation Data, *IEEE International Geoscience and Remote Sensing Symposium*, vol. 3, pp. 2026–2065, 2004.

[Ver08] Verhein, F. and S. Chawla, Mining Spatiotemporal Patterns in Object Mobility Databases, *Data Mining and Knowledge Discovery Journal*, vol. 16, no.1, pp. 5–38, 2008.

[Viq07] Viqueira, J. R. and N.A. Lorentzos, SQL Extension for Spatiotemporal Data, *The VLDB Journal*, vol. 16, no. 2, pp. 179–200, 2007.

[Vla02a] Vlachos, M., D. Gunopulos, and G. Kollios, Robust Similarity Measures for Mobile Object Trajectories, *DEXA Workshops*, pp. 721–728, 2002.

[Vla02b] Vlachos, M., G. Kollios, and D. Gunopulos, Discovering Similar Multidimensional Trajectories, *Proceedings of the International Conference on Data Engineering (ICDE)*, pp. 673–684, 2002.

[Vla03] Vlachos M. et al., Indexing Multi-dimensional Time Series with Support for Multiple Distance Measures, *Proceedings of the ACM SIGKDD Conference*, Washington DC, pp. 216–225, August 2003.

[Wan04] Wang, J. et al., FlowMiner: Finding Flow Patterns in Spatio-temporal Databases, *Proceedings of ICTAI*, pp. 14–21, 2004.

[Wan05] Wang, J., W. Hsu, and M.L. Lee, A Framework for Mining Topological Patterns in Spatio-temporal Databases, *Proceedings of CIKM*, pp. 429–436, 2005.

[Yan05] Yang, H., S. Parthasarathy, and S. Mehta, A Generalized Framework for Mining Spatio-temporal Patterns in Scientific Data, *Proceedings of the KDD Conference*, Chicago, IL, pp. 716–721, 2005.

[Yao03] Yao, X. Research Issues in Spatio-Temporal Mining, http://www.ucgis.org/visualization/whitepapers/yao-KDVIS2003.pdf (white paper submitted to the UCGIS workshop 2003).

Appendix A

In A.1, we discuss interpretation of data mining results from two different aspects: (1) how the derived results represent the entire population and (2) how data mining fits in the overall goals of an organization. The first is important because data mining is based on a sample of data from a certain population (customers, patients); however, in most cases we are interested in deriving conclusions about the entire population.

In A.2, Internet sites that contain time series data sets are referenced, for readers interested in performing temporal data mining research.

A.1 INTERPRETATION OF DATA MINING RESULTS

Having examined different temporal data representation schemes, let us now consider what the *data mining results themselves represent* in the larger scheme of an organization's plan. We will examine two aspects of the data mining results interpretation.

A.1.1 How Representative Are the Mined Results of the Actual Targeted Population?

Data mining is performed on database data that usually represent samples from a population (e.g., a population of patients, a population of customers). Whatever the data mining operation might be (clustering, classification, etc.), it is based on these samples and not on the entire population. For example, an Internet shopping site has collected data on the average age of its shoppers since its launching day two months ago. It found that the average age is 29 years. A question that arises often is how different the sample mean is from the population mean. The error in estimating the population mean through the sample mean is known as the standard error (SE) and given by [Alb06], [Gla05]:

$$SE = \frac{s}{\sqrt{n}}$$

where s is the standard deviation of the sample and n is the size of the sample. Another frequent question is related to the confidence interval for the population mean, which is given by

$$Sample\ Mean \pm multiple \times SE$$

Specifically, the 95% confidence interval for the mean is

$$Sample\ Mean \pm \frac{2s}{\sqrt{n}}$$

Another issue that often comes into focus, when one uses samples to make estimates about an underlying population, is the kind of assumptions one makes about the population. A common assumption is that of normality, i.e., that the underlying data distribution is normal. Consider the following example: A Web site owner has devised a classification scheme, according to which users of the site are placed in two classes: (1) Class A, the user is likely to buy upgrade membership for the site; (2) Class B, the user is not likely to buy the upgrade. Each user in each class is represented with a feature vector. Then the owner of the Web site wants to know whether the mean feature vectors of the two classes are different in a *statistically significant* way. For this purpose, he or she performs *analysis of variance* [Gla05], for which it is assumed that the underlying populations are normally distributed. There are several ways one can test for normality:

- Chi-square test for normality (histogram-based) [Alb06], Kolmogorov–Smirnov test (cumulative distribution function-based) [Pre02], Shapiro–Wilk test [Fie09]. The main idea of these tests is to compare the sample data against data of a normal distribution with the same mean and standard deviation.

- Visual inspection of the data and computation of measures, such as skewness and kurtosis (discussed in Chapter 2), which show deviation in the distribution's shape from a normal distribution [Fie09].

Finally, if one wants to generallize the results of regression to an entire population, a number of assumptions must be true. Besides the assumptions discussed in Chapter 4, such as normally distributed errors, other assumptions including uncorrelated independent variables with external variables and linearity must be checked. For further discussion on this topic, see [Fie09].

A.1.2 What Is the Goal of Temporal Data Mining?

The goal of temporal data mining is knowledge discovery. Two common reasons for beginning a knowledge discovery process are (1) prediction and (2) hypothesis testing [Leh08], [Fie09], [Gla05], [Alb06]. Prediction refers to the ability to forecast the behavior of an entity, such as a company stock, and it is examined in detail in Chapter 4. Let us now focus on hypothesis testing and see some examples:

1. An Internet company just upgraded its Web site with the goal to attract more customers. The company believes that if the site averages 500 hits/day, then the upgrade was successful.

2. A drug company wants to confirm the hypothesis that patients on a new drug that treats heart arrhythmia have a normal ECG on average after being on the drug for two months.

Let us look at the first example and assume that the data collected during the first 15 days of the new site are as follows:

Site visits/day: 490, 550, 400, 600, 632, 400, 765, 578, 467, 623, 534, 577, 645, 456, 589.

The hypothesis that the company wants to confirm is that the mean of the site visits/day is greater than 500. We will call this the *alternative hypothesis*. The *null hypothesis* (or status quo) is that the mean of the site visits/day $\leq$ 500. The mean and standard deviation of the site visits/day above are as follows:

$$Mean = 553.7, \text{ s.d. } = 99$$

The mean is indeed greater than 500. However, we do not know whether this result is statistically significant. In other words, the company wants to know whether this mean is an accurate estimate of the mean of site visits/day for the entire time (until the next site upgrade). To express this in statistical terms, the company wants to have a 95% confidence in the confirmation of the alternative hypothesis. In other words, the significance level, α, is 0.05 ($\alpha = 1-0.95$). Having defined the confidence level, we can run a statistical test known as the Student's t-test [Kac86], [Alb06], which gives us the following results:

$$t\text{-}test \text{ value } = 2.1 \quad p\text{-value } = 0.0271$$

The *t*-test value indicates how many standard errors the sample mean is from the population mean. The *p*-value indicates the probability that the *t*-test statistic gets its value by chance. The smaller this probability, the more unlikely the null hypothesis and the more likely the alternative hypothesis (the one we want to prove). Specifically, if the *p*-value is less than the significance level then the alternative hypothesis is true. In our case the *p*-value is indeed less than the significance level (0.05), which means that the company is 95% confident that the site visits/day is > 500. However, if we had said that we wanted to be 99% confident, then the significance level would have been 0.01 and *p*-value would have been greater than the significance level.

Let us now look at the second example. Let us assume that each ECG is represented by the following features: *wavelet coefficients* and *fractal dimension*. Here the hypothesis that the drug company wants to confirm is that the ECG of a patient on the new drug has all the characteristics of a normal ECG. To do this, the company performs an initial clinical trial with 19 patients for whom it collects ECG data and measures the wavelet coefficients and fractal dimension. Then the features are normalized using the min-max normalization method. Finally, the *Euclidean distance* between each ECG's features and the corresponding features of a guideline ECG is computed. The doctor in charge of the trial has decided that if the Euclidean distance between the patient ECG's features and the guideline ECG's features is less than 0.5 then the patient ECG can be considered normal. Therefore, the alternative hypothesis (the one he wants to confirm) is that the Euclidean distance between the patients' ECG and the guideline ECG is less than 0.5. Below are the Euclidean distance data for the 19 patients:

Euclidean distance data: 0.2, 0.3, 0.4, 0.5, 0.6, 0.3, 0.5, 0.3, 0.3, 0.4, 0.2, 0.5, 0.4, 0.6, 0.3, 0.4, 0.5, 0.2, 0.3, 0.4

The mean of these data is 0.38 and the standard deviation is 0.12. The mean is indeed less 0.5. However, to confirm that this result is representative of the entire patient population, we must perform a *t*-test. He chooses a significance level of 0.01. The results of this test are as follows:

$$t\text{-}test \text{ value} = -4.328 \qquad p\text{-value} = 0.0002$$

Because the *p*-value is less than 0.01, this means we are 99% confident that the patient population using this new drug will get a normal-looking ECG after two months on the drug.

A.2 INTERNET SITES WITH TIME SERIES DATA

A.2.1 Time Series Data for Classification/Clustering

http://www.cs.ucr.edu/~eamonn/time_series_data/

This site contains a diverse set of time series data appropriate for classification/clustering purposes. The number of classes in each time series is given.

A.2.2 Diverse Time Series Data

http://kdd.ics.uci.edu/summary.data.type.html

This site contains eight data sets of diverse nature.

A.2.3 Physiological Data

http://www.physionet.org/physiobank/database/

This site contains a variety of physiological signals, such as ECG signals and gait signals.

A.2.4 List of Data Set Sites

http://www.kdnuggets.com/datasets/

This site contains references to many sites that contain data sets for data mining, including temporal data mining.

REFERENCES

[Alb06] Albright, S.C., W. L. Winston, and C. Zappe, *Data Analysis & Decision Making*, Thomson Higher Education, 2006.

[Fie09] Field, A., *Discovering Statistics Using SPSS*, 3rd edition, Sage Publishing, 2009.

[Gla05] Glantz, S., *Primer of Biostatistics*, 6th edition, McGraw-Hill Medical, 2005.

[Kac86] Kachigan, S.K., *Statistical Analysis: An Interdisciplinary Introduction to Univariate and Multivariate Methods*, Radius Press, 1986.

[Leh08] Lehman, E.L. and J.P. Romano, *Testing Statistical Hypotheses*, Springer, 2008.

[Pre02] Press, W.H., S.A. Teukolsky, W.T. Vetterling, B.P. Flannery, *Numerical Recipes in C*, 2nd edition, Cambridge University Press, 2002.

Appendix B

To the best of the author's knowledge the programs work (after the appropriate database driver information is entered). However, runtime or compilation time errors can not be excluded.

CHAPTER 1 PROGRAMS

Program 1. Program for the implementation of the *before* temporal relationship. *Note*: It uses an Oracle driver.

```java
import java.sql.*;
import java.io.*;
import java.text.*;
import java.net.*;
// import your driver here
//The program checks whether patient Jones was released
//before Smith.
import oracle.jdbc.driver.*;
//Author: Theophano Mitsa
public class DateComp {
  static String url = "Your database's url here";
  Connection connection;
  Statement statement;
  DateComp() {
  connection = null;
  statement = null;
  }
  public void initialize() {
  try {
  DriverManager.registerDriver (new oracle.jdbc.driver.
OracleDriver());
  connection = DriverManager.getConnection(url);
  }catch(SQLException e) {}
  }
```

```
public void query() {
try {
Statement statement = connection.createStatement();
String sqlString = "SELECT RELEASE_DATE FROM PATIENTS
WHERE
FIRSTNAME='Ed' AND LASTNAME = 'Jones' ";
ResultSet rs = statement.executeQuery(sqlString);
Date date1 = rs.getDate("RELEASE_DATE");
String sqlString2 = "SELECT RELEASE_DATE FROM PATIENTS
WHERE
FIRSTNAME = 'John' AND LASTNAME= 'Smith' ";
ResultSet rs2 = statement.executeQuery(sqlString2);
Date date2 = rs2.getDate("RELEASE_DATE");
if(date1.before(date2)) {
System.out.println("Ed Jones was released before John
Smith");
 }
}catch(SQLException e) {}
 }
public void close(){
try {
connection.close();
} catch (SQLException e) {}
 }
public static void main(String arg[]) {
DateComp t1 = new DateComp();
t1.initialize();
t1.query();
t1.close();
 }
}
```

Program 2. Program for the implementation of a conversion of anchored data to an interval. *Note*: It uses an Oracle driver.

```
import java.sql.*;
import java.io.*;
import java.text.*;
import java.net.*;
import java.util.*;
// import your driver here
```

```java
import oracle.jdbc.driver.*;
//Author: Theophano Mitsa
//The program checks whether two patients stayed in the
// hospital the same number of days. It assumes that
// the patients were hospitalized the same
// year.
public class TempConv {
 static String url = "Your database's url here";
 Connection connection;
 Statement statement;
 TempConv() {
 connection = null;
 statement = null;
 }
 public void initialize() {
 try {
 DriverManager.registerDriver (new oracle.jdbc.driver.
OracleDriver());
 connection = DriverManager.getConnection(url);
 }catch(SQLException e) {}
 }
 public void query() {
 try {
 Statement statement = connection.createStatement();
 String sqlString1 = "SELECT RELEASE_DATE FROM PATIENTS
WHERE
 FIRSTNAME='Ed' AND LASTNAME = 'Jones' ";
 ResultSet rs1 = statement.executeQuery(sqlString1);
 java.sql.Date releaseDate1 = rs1.getDate("RELEASE_DATE");
 String sqlString2 = "SELECT ADMISSION_DATE FROM
PATIENTS WHERE
 FIRSTNAME='Ed' AND LASTNAME = 'Jones' ";
 ResultSet rs2 = statement.executeQuery(sqlString2);
 java.sql.Date admissionDate1 = rs2.getDate("ADMISSION_
DATE");
 String sqlString3 = "SELECT RELEASE_DATE FROM PATIENTS
WHERE
 FIRSTNAME = 'John' AND LASTNAME= 'Smith' ";
 ResultSet rs3 = statement.executeQuery(sqlString3);
 java.sql.Date releaseDate2 = rs3.getDate("RELEASE_DATE");
 String sqlString4 = "SELECT ADMISSION_DATE FROM
PATIENTS WHERE
```

```
FIRSTNAME = 'John' AND LASTNAME= 'Smith' ";
ResultSet rs4 = statement.executeQuery(sqlString4);
java.sql.Date admissionDate2 = rs4.getDate("ADMISSION_
DATE");
//Convert to unanchored data
Calendar c1 = Calendar.getInstance();
c1.setTime(releaseDate1);
Calendar c2 = Calendar.getInstance();
c2.setTime(admissionDate1);
Calendar c3 = Calendar.getInstance();
c3.setTime(releaseDate2);
Calendar c4 = Calendar.getInstance();
c4.setTime(admissionDate2);
int noOfDay1 = Math.abs(c1.get(Calendar.DAY_OF_YEAR)-
c2.get(Calendar.DAY_OF_YEAR));
int noOfDay2 = Math.abs(c3.get(Calendar.DAY_OF_YEAR)-
c4.get(Calendar.DAY_OF_YEAR));
if(noOfDay1 == noOfDay2) {
System.out.println("The patients stayed in the hospital
an equal number of days");
}
else {
System.out.println("The patients stayed in the hospital
an unequal number of days");
}
}catch(SQLException e) {}
}
public void close(){
try {
connection.close();
} catch (SQLException e) {}
}
public static void main(String arg[]) {
TempConv t1 = new TempConv();
t1.initialize();
t1.query();
t1.close();
}
}
```

XML file that contains the ontological description of the geologic eras.

```xml
<?xml version="1.0" encoding="UTF-8"?>
<Genealogy>
 <Era>
 <Name> Cenozoic </Name>
 <Period parent="Cenozoic">
 <Name>Quarternary</Name>
 <BeginDate>1.8</BeginDate>
 <EndDate>0.0</EndDate>
 </Period>
 <Period parent="Cenozoic">
 <Name>Neogene</Name>
 <BeginDate>24.0</BeginDate>
 <EndDate>1.8</EndDate>
 </Period>
 <Period parent="Cenozoic">
 <Name>Paleogene</Name>
 <BeginDate>65.0</BeginDate>
 <EndDate>24.0</EndDate>
 </Period>
 </Era>
 <Era>
 <Name> Mesozoic </Name>
 <Period parent="Mesozoic">
 <Name>Cretaceous</Name>
 <BeginDate>146.0</BeginDate>
 <EndDate>65.0</EndDate>
 </Period>
 <Period parent="Mesozoic">
 <Name>Jurassic</Name>
 <BeginDate>208.0</BeginDate>
 <EndDate>146.0</EndDate>
 </Period>
 <Period parent="Mesozoic">
 <Name>Triassic</Name>
 <BeginDate>245.0</BeginDate>
 <EndDate>208.0</EndDate>
 </Period>
 </Era>
</Genealogy>
```

Program 3. Program for the parsing of XML ontology file and extraction of temporal information. The program prints out the begin and end dates for the Mesozoic and Jurassic periods.

```
//Author: Theophano Mitsa
import javax.xml.parsers.DocumentBuilder;
import javax.xml.parsers.DocumentBuilderFactory;
import javax.xml.parsers.ParserConfigurationException;
import org.w3c.dom.Document;
import org.w3c.dom.Element;
import org.w3c.dom.NodeList;
public class Parser {
      Document dom;
      public Parser(){
      }
      public void doParsing() {
        //parse the xml file and get the dom object
            parseXmlDoc();
        //get the elements out of the dom object
            obtainElements();
      }

      private void parseXmlDoc(){
        DocumentBuilderFactory dbf =
DocumentBuilderFactory.newInstance();
        try {
              DocumentBuilder db = dbf.
newDocumentBuilder();
              // get DOM representation of the XML file
              dom = db.parse("ontology.xml");
              }catch(Exception e) {
                    e.printStackTrace(); }
      }
      private void obtainElements(){
        //get the root elememt
        Element docEle = dom.getDocumentElement();
        float maxBeginDate=0.0f, minEndDate=100.0f, Ju_
BeginDate=0, Ju_EndDate=0;
  //get the Period elements
        NodeList nl =
docEle.getElementsByTagName("Period");
```

```
if(nl != null && nl.getLength() > 0) {
          for(int i = 0 ; i < nl.getLength();i++) {

               Element el = (Element)nl.item(i);
//Find the date range for Jurassic period
          String name = getText(el,"Name");
if(name.equals("Jurassic")) {
Ju_BeginDate = getFloatValue(el,"BeginDate");
Ju_EndDate = getFloatValue(el,"EndDate");
}

          //Find the date range for the Mesozoic era
               String type = el.getAttribute("parent");
if(type.equals("Mesozoic")) {
if(maxBeginDate < getFloatValue(el,"BeginDate")) {
maxBeginDate = getFloatValue(el, "BeginDate");
}
if( minEndDate > getFloatValue(el,"EndDate")) {
minEndDate = getFloatValue(el,"EndDate");
}
}

          }
System.out.println("For the Jurassic period, the
BeginDate is" + Ju_BeginDate+
 "and the EndDate is" + Ju_EndDate);
 System.out.println(" For the Mesozoic period BeginDate
is:" + maxBeginDate +
 "EndDate is:" + minEndDate);
 }
      }
     private String getText(Element ele, String
tagName) {
          String text = null;
          NodeList nl = ele.
getElementsByTagName(tagName);
          if(nl != null && nl.getLength() > 0) {
               Element el = (Element)nl.item(0);
               text = el.getFirstChild().
getNodeValue();
          }
```

```java
                return text;
        }
        private float getFloatValue(Element ele, String
tagName) {
                return Float.parseFloat(getText(ele,tagName)
);
        }

        public static void main(String[] args){
                Parser p = new Parser();
                p.doParsing();
        }
}
```

Index